Ceramics for Nuclear Applications

Elsevier Series in Advanced Ceramic Materials

Ceramics for Nuclear Applications

Edited by

Peng Xu

Idaho National Laboratory, Idaho Falls, United States

Christian P. Deck

Nuclear Technologies and Materials Division, General Atomics, Electromagnetic Systems, San Diego, United States

William E. Lee

Nuclear Futures Institute, Bangor University, Bangor, United Kingdom

Michael J.D. Rushton

Nuclear Futures Institute, Bangor University, Bangor, United Kingdom

ELSEVIER

Elsevier
Radarweg 29, PO Box 211, 1000 AE Amsterdam, Netherlands
125 London Wall, London EC2Y 5AS, United Kingdom
50 Hampshire Street, 5th Floor, Cambridge, MA 02139, United States

ISBN: 978-0-12-823865-3

For information on all Elsevier publications visit our website at
https://www.elsevier.com/books-and-journals

Publisher: Matthew Deans
Acquisitions Editor: Gwen Jones
Editorial Project Manager: Debarati Roy
Production Project Manager: Sujatha Thirugnana Sambandam
Cover Designer: Bakyalakshmi S

Typeset by TNQ Tech

Contents

Contributors

Tyler Abrams General Atomics, San Diego, CA, United States

Thomas P. Davis Oxford Sigma, Oxford, United Kingdom; Nuclear Futures Institute, Bangor University, Bangor, United Kingdom

Christian Deck General Atomics, Electromagnetic Systems, San Diego, CA, United States

Lee J. Evitts United Kingdom Atomic Energy Authority (UKAEA), Abingdon, United Kingdom

Samuel Humphry-Baker Department of Materials, Imperial College London, London, United Kingdom

E. Kardoulaki Materials Science and Technology Division, Los Alamos National Laboratory, Los Alamos, NM, United States

Yutai Katoh Materials Science and Technology Division, Oak Ridge National Laboratory, Oak Ridge, TN, United States

Takaaki Koyanagi Materials Science and Technology Division, Oak Ridge National Laboratory, Oak Ridge, TN, United States

W.E. Lee Nuclear Futures Institute, Bangor University, Bangor, United Kingdom

Dong Liu Department of Engineering Science, University of Oxford, Oxford, United Kingdom

S. Middleburgh Nuclear Futures Institute, Bangor University, Bangor, United Kingdom

T. Ming Nuclear Futures Institute, Bangor University, Bangor, United Kingdom; Department of Materials, Clemson University, Clemson, SC, United States

Samuel T. Murphy School of Engineering, Lancaster University, Lancaster, United Kingdom

Fiona Pearce Nottingham University Hospitals NHS Trust, Nottingham, United Kingdom

Michael J.D. Rushton Nuclear Futures Institute, Bangor University, Bangor, United Kingdom

Peng Xu Idaho National Laboratory, Idaho Falls, ID, United States

About the editors

Dr. Peng Xu is the technical lead for the SiC program and Manager of In Core Experiments at Idaho National Laboratory (INL) in Idaho Falls, USA. He earned his PhD in Materials Science and Engineering from the University of Florida in 2009, was a postdoctoral researcher at the University of Wisconsin-Madison from 2009 to 2011, and served as fellow engineer at Westinghouse Electric Company until 2020. With over 15 years of experience in nuclear materials and fuel R&D, he has held leadership roles at INL and Westinghouse in accident-tolerant fuel development. Dr. Xu has authored more than 100 publications and filed over 30 patents.

Dr. Christian P. Deck manages the Advanced Materials group within the Nuclear Technologies and Materials division at General Atomics Electromagnetic Systems. He attained his PhD in Materials Science and Engineering from the University of California, San Diego in 2009, and then directly began work at General Atomics, where he has established and grown the ceramic matrix composite fabrication facilities. His work focuses on maturation and scale-up of materials for extreme environments, including engineered silicon carbide composites for accident tolerant fuel cladding and advanced nuclear fission and fusion reactor designs, and CMCs for space nuclear and aerospace applications.

Professor William E. Lee is Founding Director of the Nuclear Futures Institute at Bangor University and Emeritus Professor in the Department of Materials at Imperial College London. He has held a number of UK Government advisory roles including with the Committee on Radioactive Waste Management, the Nuclear Innovation and Research Advisory Board and as Chief Scientific Advisor (Nuclear) to the Ministry of Defence.

Prof. Michael J.D. Rushton is Professor in the Nuclear Futures Institute at Bangor University UK, having previously been employed at Imperial College London in the Centre for Nuclear Engineering. His research interests are in applying atomic-scale simulation methods to energy materials with a particular emphasis on nuclear materials such as fuel and waste forms. Of particular relevance is his work to derive new potential models providing high-fidelity atomic descriptions of actinide oxides over wide temperature ranges. In addition, by using molecular dynamics and static lattice methods, he is considering the interaction between small fission gas bubbles and the lattice of nuclear fuel to improve our understanding of fuel performance under normal and extreme conditions. His work considers both crystalline and amorphous materials, and Michael uses a variety of methods (e.g., MD, MC, RMC) to characterize the interfaces between crystals and glasses. In addition to his work on nuclear materials, Michael has published extensively on battery and fuel cell materials and continues to look at using engineered strains to improve their performance.

Preface

Ceramics have been used in nuclear technology since its inception due to their ability to function in severe, high-temperature, corrosive, and radiation environments. They find use in fission, fusion, medicine, civil, and defense applications and will do so increasingly in the future as mankind realizes the benefits of the most intense energy source available. This is particularly so if we are to reduce our reliance on fossil fuels for civil power production. Current uses in fission-based nuclear power include oxide nuclear fuels based on UO_2 and PuO_2/UO_2 mixed oxides, as glass and glass composite high-level waste forms, and in composite cements to encapsulate intermediate-level wastes (ILW). They have also been used as porous filters with the ability to absorb radionuclides (RN) from air and liquids and are playing a key role in the cleanup at Fukushima. Non-oxides also find current fission applications, including in graphite moderators and B_4C control rods. Ceramics will continue to be significant in the near-term expansion of nuclear power via next-step developments of fuels that are accident tolerant and use HALEU (High Assay Low Enriched Uranium) for application in smaller reactors (small modular reactors and earthbound and space-based microreactors) and in waste forms using alternative composite cements or single or multiphase ceramics that can host Pu and other difficult RN. Longer-term advances for Generation IV reactors, which will operate at higher temperatures and with higher fuel burn-up, require innovative fuel developments potentially via carbides and nitrides or composite fuel systems. A rich future exists for nuclear ceramics in medicine as new uses for radionuclides are found in cancer diagnosis and treatment and in space technology for propulsion and planetary bases due to the intensity of heat and power they can produce in a small volume and mass. Another area where nuclear ceramics will play an important role is in marine transportation, where nuclear-powered bulk carriers and tankers will be developed to replace the current diesel technology producing a significant proportion of global emissions.

Opportunities also exist for ceramics in structural applications in Generation IV and Small Modular Reactors, such as composite SiC/SiC and C/C for fuel cladding and control rods, and MAX phases and ultrahigh-temperature ceramics (UHTCs) may find near-core fuel coating and cladding applications. Uses of ceramics in fusion reactor systems will be both functional (ceramic superconductors in magnet systems for plasma control and in Li silicate breeder blankets in tokamaks) and structural, including as sapphire diagnostic windows, graphite divertors, and plasma-facing C and UHTCs. In all these cases, performance is limited by poorly understood radiation damage and interface-controlled processes, which demand a combined modeling/

experimental approach. With increasing dependence on digital technology and artificial intelligence, this need to marry modeling and experiments will be increasingly important. In this book, in chapters written by leading experts from across the globe, we provide a snapshot of some of the latest developments in ceramic materials in nuclear applications.

Overview of current nuclear ceramics and outlooks

1

Samuel T. Murphy[1] *and Peng Xu*[2]
[1]School of Engineering, Lancaster University, Lancaster, United Kingdom; [2]Idaho National Laboratory, Idaho Falls, ID, United States

1. Introduction to the book

Nuclear ceramics have evolved as a critical component in the advancement of nuclear technology, providing essential materials for fuel, structural applications, and waste management. These ceramics are distinguished by their ability to withstand extreme environments, including high temperatures, radiation, and corrosive conditions typical of nuclear reactors. This book will delve into the current applications of nuclear ceramics, specifically focusing on their performance, fabrication techniques, and the challenges they face. Additionally, the potential for future developments and innovations in the medical field, detection, and fusion will be discussed. A quick summary of each book chapter is provided below as a guide for the readers.

This first chapter provides an overview of current nuclear ceramics and outlooks, and sets the stage by providing an overview of the current state of nuclear ceramics and their future prospects. It highlights the latest developments in ceramic materials designed for nuclear applications and discusses the potential advancements that could shape the future of the industry.

Chapter 2 highlights that advanced ceramic and composite fuels, a class of ceramics with complex structures, are essential for the operation of advanced nuclear reactors. This chapter delves into different types of nuclear fuels, including uranium nitride, uranium carbide, and uranium boride, as well as composite fuels and coated particle fuels. It covers their history, evolution, synthesis, and the challenges associated with their processing and performance.

Chapter 3 focuses on fusion energy. As fusion energy moves toward its engineering and implementation phase, the demand for advanced materials capable of withstanding extreme conditions increases. This chapter addresses the emerging trends in high-temperature structural ceramics, radiation shielding materials, and new first-wall materials, emphasizing their structure-property relationships and irradiation damage characteristics.

Chapter 4 introduces one of the most significant and studied nuclear materials, silicon carbide (SiC) fiber-reinforced SiC matrix composites. The SiC composites are gaining attention for their high-temperature resistance, chemical inertness, and low neutron absorption. This chapter focuses on designing radiation-resistant SiC-based

Ceramics for Nuclear Applications. https://doi.org/10.1016/B978-0-12-823865-3.00003-1

composites, exploring the effects of irradiation on their dimensional stability and mechanical properties.

Chapter 5 describes nuclear graphite, a fascinating material due to its complex nano- and microscale structures and its ability to withstand extreme conditions for nuclear applications. This chapter discusses the use of nuclear graphite in various reactor designs, its irradiation-induced dimensional and property changes, and the challenges associated with its performance in service.

Chapter 6 discusses a critical issue for nuclear energy, the waste disposition and roles of advanced ceramics as waste forms. This chapter describes the mechanisms by which radioactive wastes are immobilized in glass and ceramic hosts to create waste forms. It covers the various processing methods and recent advances in modeling waste form behavior.

Chapter 7 introduces an emerging field for ceramics in medical applications. This chapter explores the use of ceramics in the diagnosis and treatment of diseases, including the detection of X-rays and γ-rays, the delivery of isotopes for radiotherapy, and the historical use of nuclear-powered heart pacemakers.

Finally, Chapter 8 summarises novel ceramic processes for nuclear applications. The final chapter presents recent progress in novel processing methods for nuclear ceramics. It focuses on state-of-the-art powder sintering technologies and additive manufacturing, highlighting their potential to create unique ceramic microstructures and properties that traditional methods cannot achieve.

As evident in this book, the field of nuclear applications has seen significant advancements over the years, with ceramics playing a crucial role in enhancing the efficiency, safety, and functionality of nuclear systems.

2. Overview of *nuclear fuels*

The pivotal role of ceramic fuels in the modern nuclear industry cannot be overstated. These fuels, with their intricate structures and varied compositions, serve as the backbone of advanced nuclear reactors, meeting the rigorous demands of stability and performance under extreme conditions. This section delves into the diverse realm of ceramic fuels, beginning with uranium dioxide, the predominant choice for nuclear power plant fuel across the globe.

2.1 Uranium dioxide

Uranium dioxide, UO_2, currently fuels more than 90% of the world's nuclear power plants. The reasons for UO_2's dominance as a fuel are a result of a number of important properties, including excellent stability, tolerance to irradiation and corrosion resistance to water and steam. However, UO_2 has a low thermal conductivity that can lead to very high temperatures at the centers of the fuel pellets.

UO_2 adopts the fluorite structure within the FM3m space group (number 225), where the uranium ions occupy the *4a* Wyckoff sites forming a face centered cubic (FCC) arrangement, and the tetrahedrally coordinated oxygen ions occupy the *8c* sites [1].

This fluorite-structured phase is stable up to the melting temperature of 2865°C and can accommodate significant nonstoichiometry, with the U/O ratio ranging from 1.7 to 2.2 at high temperatures. On the hypostoichiometric (UO_{2-x}) side, the oxygen deficiency is accommodated by oxygen vacancy defects, while hyperstoichiometry (UO_{2+x}) side, the excess is facilitated by oxygen interstitials [2]. These interstitials aggregate, resulting in the so-called Willis clusters [3]. Further increases in the oxygen content result in the formation of other oxygen-rich phases, including UO_3, U_4O_9 and U_3O_8.

Given that reactors using UO_2 have been in operation for nearly 70 years, the fabrication techniques are long-established and a summary can be found in Abe and Asakura [4]. Naturally occurring uranium contains only 0.72% of the fissile ^{235}U, while the remainder is predominantly the fertile ^{238}U. While some reactors can operate using natural uranium (notably heavy water-cooled reactors, such as CANDU) most light water reactors require enrichment of ^{235}U to between 3% and 5%. Enrichment is predominantly achieved by converting the ore into uranium hexafluoride, which is then separated in a centrifuge.

The enriched uranium is converted to UO_2 in the form of a powder that is then granulated and cold-compacted before sintering at high temperature (1750°C) in a hydrogen atmosphere for approximately 5 h to ensure pellets are close to the stoichiometric composition. The pellets are then ground and inspected before being loaded into a cladding tube (typically zircalloy) and sealed [5]. These pins are then grouped into an assembly for introduction into the reactor [6].

During in-pile operation, the chemical composition and microstructure of the fuel evolve in response to high temperatures, neutron irradiation and the incorporation of fission products [7]. Perhaps the most apparent change in the fuel pellets is the change in their physical appearance. The combination of swelling due to the appearance of fission products and the thermal stresses induced by the low thermal conductivity of UO_2 results in pellet cracking as shown in Fig. 1.1 [8]. Further, the switch from plane-strain conditions near the pellet midplane to plane-stress at the top and bottom results in an "hourglass" effect that deforms the cladding, giving it the appearance of a bamboo stalk. This pellet-cladding interaction (PCI) can lead to failure of cladding and release of fission products to the coolant.

The fission products display a wide variety of different chemistries but can generally fall into four categories as illustrated in Table 1.1 [9].

Which class a fission product finds itself in depends on thermodynamics; therefore, it should not be a surprise that some fission products can switch between classes based on conditions, such as temperature and oxygen activity. Ba, for example, has a higher solubility in the fuel at high temperature but may exsolve at lower temperatures to form precipitates such as the gray phase ($BaZrO_3$ perovskite).

The presence of these fission products modifies the properties of the fuel and can have serious safety implications. One area of particular interest is the behavior of the fission gas atoms. If released from the fuel into the gap between the fuel pellet and the clad, this will lead to swelling and inhibit heat transfer; however, if fission gases remain in the pellet, this will lead to fuel swelling and increase the mechanical interaction between the fuel and the clad [10]. The fission gas release process is complex and encompasses a range of mechanisms, from the diffusion of individual Xe and Kr

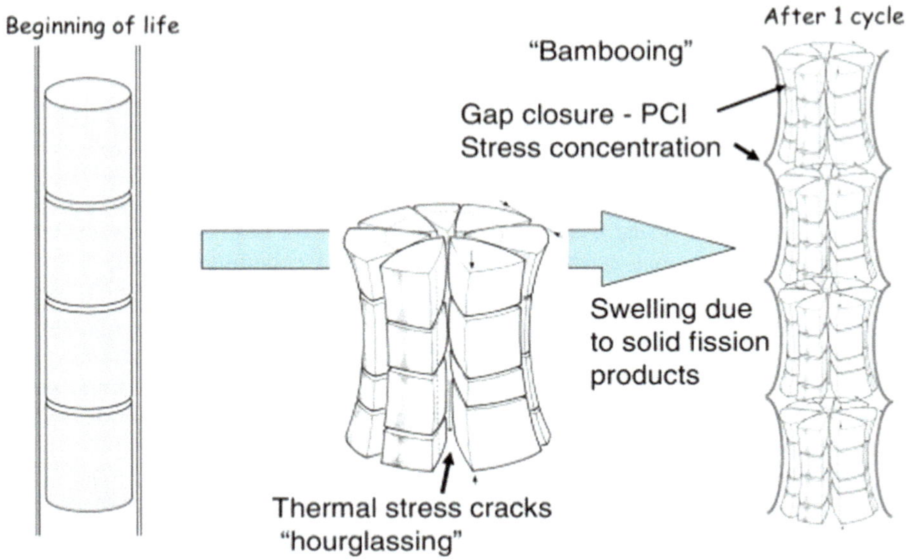

Figure 1.1 The effect of fuel swelling and thermal stress.
Image taken from Olander [8].

Table 1.1 List of fission products grouped based on their behavior in the UO_2 fuel.

Class	Elemental examples
Dissolved	Ba, Zr, Y, and lanthanides
Precipitated	Ba, Sr, Sn, Zr, Mo(Tc), Cs, and Rb
Metal particulates	Pd, Ru, Rh, Pt, and Mo(Tc)
Volatiles/gaseous	He, Kr. Xe, te, I, Br, and Cs

atoms, to the formation of inter and intragranular bubbles and radiation-enhanced resolution. The process is discussed in detail in the review of Rest et al. [11].

Release of fission gas is of particular importance in the rim structure that occurs toward the pellet exterior. Here, the absorption of epithermal neutrons by ^{238}U results in the production of ^{239}Pu, which subsequently undergoes fission and increases the level of burn-up at the periphery. This increased level of burn-up results in grain subdivision and redistribution of fission gases and extended defects as illustrated in Fig. 1.2 [12].

In order to improve the properties of UO_2 fuels, there have been several efforts to incorporate dopant species, such as Westinghouse's Advanced Doped Pellet Technology (ADOPT) fuel [13] and AREVA's chromia and gadolinium doped fuels [14]. The inclusions of small amounts of Cr, Gd and Al lead to a significant increase in grain size and densities about 0.5% greater than standard UO_2 pellets. This results in a

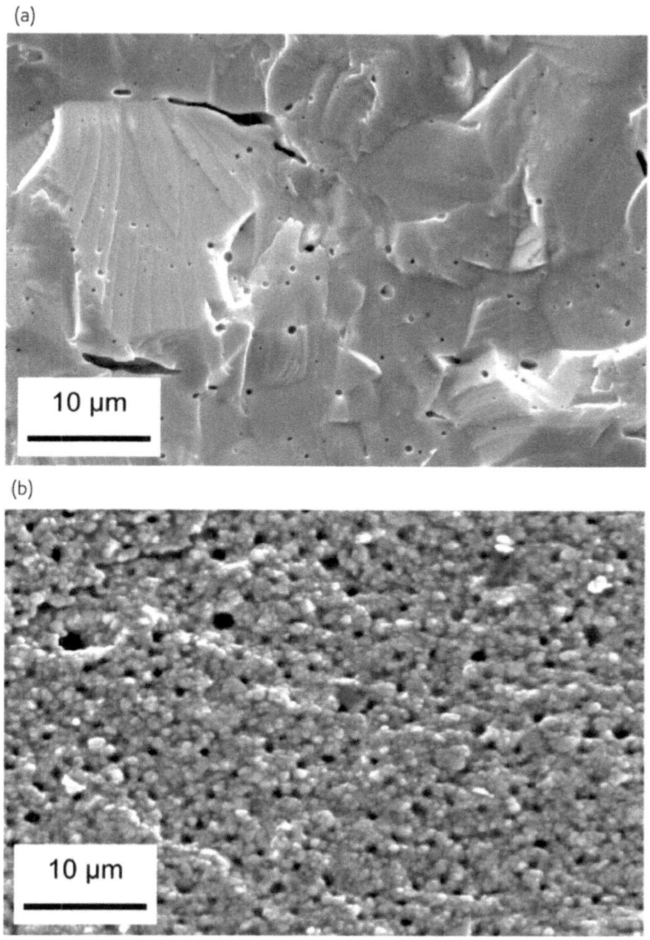

Figure 1.2 Scanning Electron Microscope micrographs showing the surfaces: (a) unirradiated UO_2 fuel and (b) UO_2 with a local burn-up of ~75 GWd/tHM showing the typical high burn-up structure.
Image taken from Ref. [12].

reduction in fission gas release as well as increased corrosion resistance and improved PCI performance.

2.2 Mixed oxide fuel

Mixed oxide fuels (MOX) used in current light water reactors correspond to a mixture of uranium and plutonium oxides [6]. It should be noted that MOX fuels can also utilize thorium, and this will be discussed in more detail later on. One of the principal advantages of adding small amounts (~3–5%) of Pu to UO_2 is that it increases the fissile concentration of the fuel, without the expensive enrichment with ^{235}U [15]. Further, it provides an opportunity to use Pu recovered from used reactor fuel or weapons programmes. For example, in the UK, the Government's preferred strategy for dealing with its significant (~100 t) of separated Pu [16] is reuse as MOX; however, the shutdown of the majority of fast reactor programmes leaves use in light water reactors as the most viable option. MOX fuels have been widely deployed in light water reactors in Europe and Japan, and currently represent about 2% of new nuclear fuel used today [17].

In general, the fabrication of MOX fuels is similar to the process employed for UO_2, with powders of UO_2 and PuO_2 blended together before being pressed and then sintered at ~1700°C to form pellets [18,19]. These pellets are then grouped together to form MOX assemblies, which then make up typically 1/3 of the assemblies in a reactor. The uranium used in MOX can be either natural or depleted. Ideally, the reprocessed plutonium will not have been in storage for more than 5 years to avoid the build-up of ^{241}Am from decay of ^{241}Pu. ^{241}Am is a strong γ-emitter and would require costly repurification or introduction of strong shielding measures during manufacturing. Further limits are placed on the quantity of ^{238}Pu present, as it has an activity ~275 times that of ^{239}Pu and poses a risk as a source of both neutrons and heat [20].

There have been a series of processes developed for MOX fabrication. The Optimized CO-Milling (OCOM) and Ammonium Uranyl-Plutonyl Carbonate (AU/PuC) process were developed in the early 1980s by Alkem [21]. These two methods adopt different approaches to achieve homogeneity. In the OCOM process, the powders of UO_2 and PuO_2 are subjected to intensive milling, while the AU/PuC process employs precipitation followed by adding ammonia and carbon dioxide to a solution of uranium nitrate and plutonium nitrate, which is then calcined. While these approaches provided pellets that appeared homogenous on the mm length scale [22], subsequent Electron Probe Micro Analysis (EPMA) suggests Pu-rich areas of the order of 50–100 μm [23]. The MIcronized MASter (MIMAS) process was developed by Belgonucleaire and involves the microinzation of PuO_2 (i.e., mechanical blending PuO_2 and UO_2 powders where the crystallites are well mixed), which is then mixed with Ammonium Di-Uranyl Carbonate or Ammonium Di-Uranate [24]. As for the OCOM and AU/PuC process, MIMAS pebbles are homogeneous on the mm scale; however, Pu-rich particles average 30 μm [25]. Finally, the Short Binderless Route (SBR), developed by British Nuclear Fuel Ltd. (BNFL), uses a high-energy attritor mill and a spherodizer to condition the powders before pressing [26]. Pellet density is controlled during milling using a lubricant and Conpore pore former, ensuring a grain size of 7.4 μm. SBR MOX pellets exhibit an excellent homogeneity with the largest Pu agglomerations observed as between 30 and 50 μm [27,28].

From an operational perspective, the neutron capture cross sections for the various plutonium isotopes are different from those for UO_2, resulting in different neutronic behavior that should be considered in the design of the reactor. The neutron spectrum is generally harder than in UO_2, and consequently, the flux resonance region for MOX fuel is higher. Importantly, the fraction of delayed neutrons is decreased, and the lifetime of the prompt neutrons is also decreased. These factors lead to a decrease in the control rod efficiency that must be managed.

In general, the thermophysical properties, including thermal expansion and specific heat, are similar to those for UO_2 [21,30]. However, the introduction of Pu into the UO_2 lattice leads to a reduction in thermal conductivity of the fuel [29]. Overall, the evolution of MOX fuel during operation is similar to that observed for UO_2, with the exception of increased levels of He generation due to the increased concentration of minor actinides [30].

One of the major attractions of MOX fuel is the ability to utilize reprocessed Pu. A PWR operating with standard UO_2 fuel will produce Pu at a rate of 25 kg/TWhe (terawatts of heat energy generated). For a similar PWR fueled with 30% MOX, the net consumption rate of Pu production is roughly equal to consumption, resulting in no net decrease in Pu [31]. Therefore, utilization of MOX fuel as a pathway for reduction in Pu stockpiles requires the development of reactors capable of 100% MOX loading, such as the European Pressurized Water Reactor (EPR) [32].

2.3 Thorium fuel

In the early days of the nuclear industry, there was concern about the availability of sufficient uranium to fuel a global fleet of reactors; therefore, considerable effort was expended on developing a fuel cycle around the only other, naturally occurring candidate, thorium. Naturally occurring thorium is predominantly ^{232}Th (99.98%), which has a half-life on the order of 14 billion years. ^{232}Th is not a fissile nuclide; however, it is fertile and neutron absorption and subsequent double β-decay (see reaction 1.1), leads to the formation of the fissile ^{233}U.

$$^{232}_{90}Th \xrightarrow{(n,\gamma)} {}^{233}_{90}Th \xrightarrow{\beta^-} {}^{233}_{91}Pa \xrightarrow{\beta^-} {}^{233}_{92}U \qquad (1.1)$$

Among a number of advantageous neutron characteristics, ^{232}Th's high thermal neutron captures crosssection leads to greater production of fissile ^{233}U than would be expected from conversion to ^{239}Pu from ^{238}U. Further, ^{233}U offers a higher neutron yield per fission than ^{238}U, thereby requiring lower levels of overall fuel enrichment [33].

As a consequence of only being fertile, Th fuels must include an external startup fissile material. Therefore, thorium is typically envisaged in the form of a mixed oxide fuel, i.e., $(Th,U)O_2$ or $(Th,Pu)O_2$. Once generated, fissile ^{233}U can be either burned in situ or recovered from the reprocessing of ThO_2 fuel. Build-up during reactor cycles gradually reduces the need for external fissile ^{235}U and ^{239}Pu, as discharged fuel contains sufficient ^{233}U for subsequent cycles, leading to a self-sustaining thorium fuel cycle [34,35].

The thorium fuel cycle has a number of advantages over the U and U-Pu MOX fuels, including greater natural abundance, high actinide burn-up, greater stability due to low fuel temperatures, reduced generation of long-lived radioactive minor actinides, low fuel failure rates and inherent proliferation resistance [35]. These are balanced by particular disadvantages, which include the presence of ^{232}U in reprocessed fuel whose decay products are hard gamma emitters, requiring extensive shielding. Further, the relatively long half-life of ^{233}Pa (27 days) will lead to an increase in reactivity after the reactor has been shut down. Finally, the presence of ^{228}Th, with its

1.91 year half-life, will result in the accumulation of the hard gamma emitters ^{212}Bi and ^{208}Tl, requiring cooling over a roughly 10-year period [36].

Fabrication of thoria and thoria-based MOX fuel pellets has been developed using a range of methods, including a powder pellet route, sol-gel microsphere pelletization, coated agglomeration pelletization and an impregnation technique [37]. Of these, the powder route is the most common (i.e., similar to the MIMAS method [38]) and research from Bhabha Atomic Research Center (BARC) has shown that high-density ThO_2-UO_2 pellets can be obtained from ThO_2–U_3O_8 green compacts [39]. It should be noted that reprocessed ^{233}U will contain as much as 1000 ppm ^{232}U (assuming 24,000 MWD/t) and thus handling of fine powders will generate radiotoxic aerosols and powder on surfaces that pose safety risks to personnel [37]. Avoiding this issue has provided an impetus to explore other fabrication routes leading to processes, such as BARC's Coated Agglomerate Pelletization (CAP) Technique, which is illustrated in Fig. 1.3 [40].

The CAP method minimizes both the role of powders and the number of steps that must be performed inside a shielded cell or glove box. The resulting pellets exhibit an interesting "rock in sand" structure [41], with large grains surrounded by colonies of fine grains. As might be expected, this structure results in an inhomogeneity in the distribution of uranium, leading to the development of an Advanced CAP process [42].

The physical properties of thoria-based fuels provide a number of advantages, including high melting point [43], low solubility in water (resulting from the stable 4+ oxidation state) [44] and high thermal conductivity [45]. It should be noted that the introduction of U and Pu cations will impact the thermal conductivity due to phonon scattering [46]. The higher thermal conductivity allows the reactor to operate at a lower temperature for the same linear power. This has the impact of reducing fission gas release as it is a thermally activated process [47].

2.4 Accident tolerant fuels

Lastly, it is worth noting the recent advancement made in accident-tolerant fuel development. Zr-clad, UO_2 fuel systems have undergone 60 years of continual development, and so, while further improvements are possible, dramatic enhancements in accident tolerance are unlikely. Consequently, there is considerable international interest in the development of both new fuel and cladding materials under the umbrella of Accident Tolerant Fuels (ATFs). ATFs are being designed to reduce overheating of the fuel and cladding during a LOCA, while ensuring the structural integrity of the cladding for longer, allowing operators more time to restore cooling. Ultimately, this should prevent the large-scale release of radioactive material witnessed at Fukushima, while reducing damage to the reactor itself, increasing the probability that the plant could be returned to operation.

Ideally, improvements in the accident tolerance should be accompanied by improvements in performance under normal operating conditions. The desirable properties of a new, ATF can be inferred by examining the shortcomings of UO_2. The low volumetric specific heat combined with the low thermal conductivity results in very high temperatures (1000°C) at the center of the fuel pellets, despite the temperatures

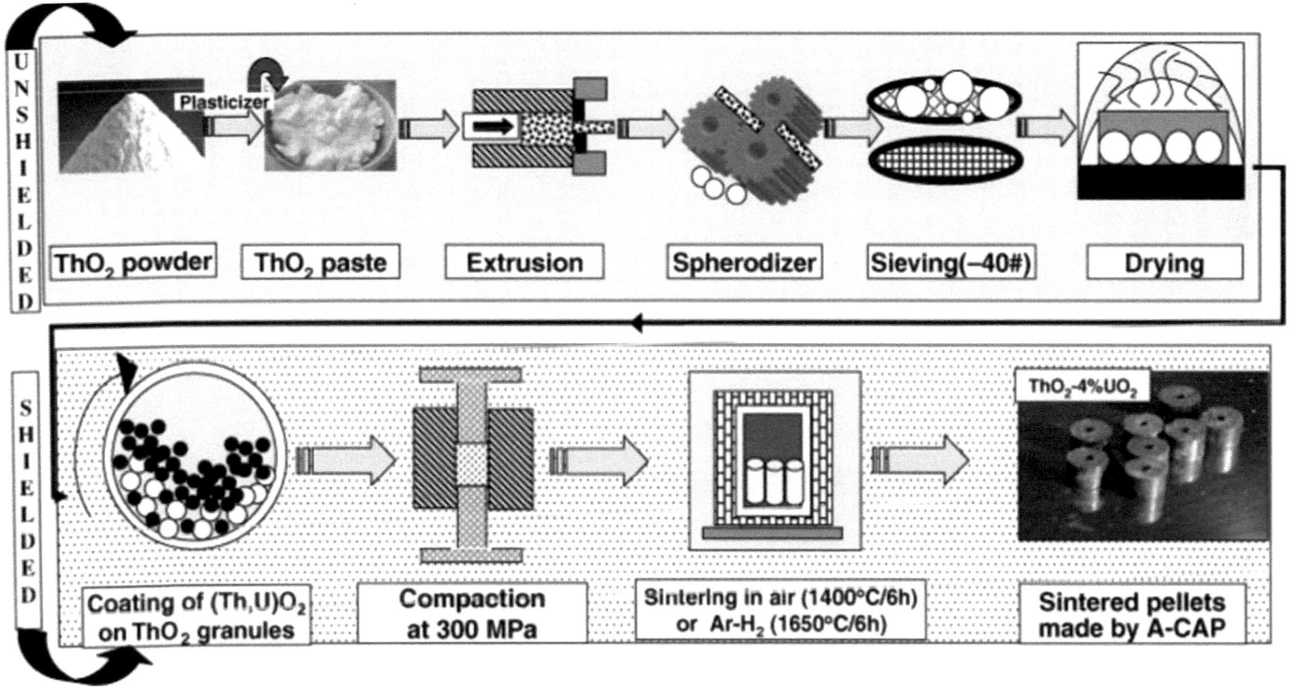

Figure 1.3 Schematic showing the Coated Agglomerate Pelletization method for fabrication of $(Th,U)O_2$. Figure taken from Kutty et al. [41].

at the pellet surface of only 400°C. This high temperature gradient induces large thermal stresses, leading to cracking and relocation, allowing the release of fission gases and other volatile fission products, notably Cs and I. The high thermal expansion of the fuel leads to swelling that exerts pressure on the cladding even under normal conditions. Finally, as UO_2 has a low U density, some enrichment is necessary for use in LWRs. Enrichment is expensive and presents proliferation issues.

The two leading ATF candidates are silicides and nitrides. Uranium silicide fuels were developed as Low Enrichment Uranium (LEU) fuels to replace the very highly enriched UO_2 used in research reactors. Due to its very high uranium density, U_3Si 'meat' was dispersed in Al; however, the dramatic swelling due to fission gas bubbles rendered U_3Si inappropriate for use in LWRs. As a consequence, attention has focused on U_3Si_2, which does not exhibit such dramatic swelling, despite becoming amorphous when irradiated. U_3Si_2 has a theoretical metal density 17% greater than UO_2 and a thermal conductivity of 17.3 $Wm^{-1}\ K^{-1}$ at 1100°C, compared to 2.8 $Wm^{-1}\ K^{-1}$ for the oxide. Its melting point of 2000 K is lower than that of UO_2, and it has a higher volumetric specific heat capacity, ensuring that the temperature under normal operating conditions is well below the melting point. Of particular concern with U_3Si_2 is its high thermal expansion coefficient, which is exacerbated by the swelling that accompanies amorphization.

More novel fuel forms will be discussed in detail in Chapter 2.

3. Nuclear fusion

The materials challenges associated with the construction of a device capable of sustaining conditions anticipated in a fusion reactor are formidable. In the fusion process, two light nuclei are fused together to give a single, heavier one, producing a large amount of energy that can be harnessed for electricity generation. However, as the nuclei are both positively charged, it is necessary to overcome the electrostatic repulsion between them, requiring temperatures in the range 100–200 million degrees for the reaction between deuterium and tritium.

$$^{2}_{1}D + ^{3}_{1}T \rightarrow ^{4}_{2}He + ^{1}_{0}n \tag{1.2}$$

The majority of the energy released (i.e., 14.1 out of 17.6 MeV) during the reaction is in the form of the kinetic energy of the neutron. Therefore, the materials operating in a fusion reactor will be subjected to high heat loads and intense radiation damage by these high-energy neutrons. Further, the choice of materials for use in a fusion reactor is constrained by the imperative to minimize the creation of long-lived radioactive waste.

Within a Magnetic Confinement Fusion (MCF) reactor, the fusion plasma is suspended in a magnetic field in either a tokamak or stellarator. Most current MCF experimental reactors, including the Joint European Torus (JET), have employed copper magnets, while the ITER reactor currently under construction will use Nb_3Sn and

NbTi superconducting materials. These materials have critical temperatures lower than 30 K and are classified as low-temperature superconductors. Based on the critical fields offered by the low temperature superconductors (LTS) it was expected that large plasma volumes (and hence reactors) would be required to achieve ignition. Hence ITER's major radius of 6.2 m results in a plasma volume of 840 m^3 relative to JET's 2.96 m and 100 m^3. In 1986, Bednorz and Müller demonstrated that the ceramic lanthanum barium copper oxide ($LaBa_2Cu_3O_7$) had a $T_c = 35$ K. This initial discovery led to a family of rare earth barium cuprates (REBCOs) with critical temperatures above 90 K. These discoveries suggest that it is possible to construct fusion reactors that are much smaller and hence cheaper than those using LTSs. An example of a tokamak utilizing HTS is the ST25-HTS device pictured in Fig. 1.4.

The brittle nature of these ceramic materials makes them difficult to form into tapes that could be used in the magnets of a fusion reactor. The first generation of commercial tapes used bismuth-strontium-calcium-copper-oxide superconductors (typically $Bi_2Sr_2Ca_2Cu_3O_{10+x}$) with typical cross sections of 0.2 mm by 4 mm and can have lengths greater than 1 km. Second-generation tapes use REBCO HTS materials in complex layered films that can be as thin as 55 μm and 4 mm wide [48]. A schematic of a SuperPower $GdBa_2Cu_3O_{7-\delta}$ is presented in Fig. 1.5.

The superconducting materials in both first and second generation tapes have anisotropic crystal structures based around CuO_2-layers in the *xy*-plane. Given the grain orientations applied during manufacture, this results in highly anisotropic fields, particularly for the REBCO-based tapes. As indicated in the formulae, the ceramics can display substantial deviations in oxygen stoichiometry that are closely linked to their superconducting properties. In addition to these oxygen vacancy defects that accommodate this nonstoichiometry there are large concentrations of antisite defects and stacking faults. Further manipulation of the HTS microstructure can be achieved through irradiation with swift heavy ions (SHI), leading to the formation of ion tracks that act as flux pinning centers thereby enhances the critical current and critical field.

Figure 1.4 The ST25-HTS spherical tokamak.

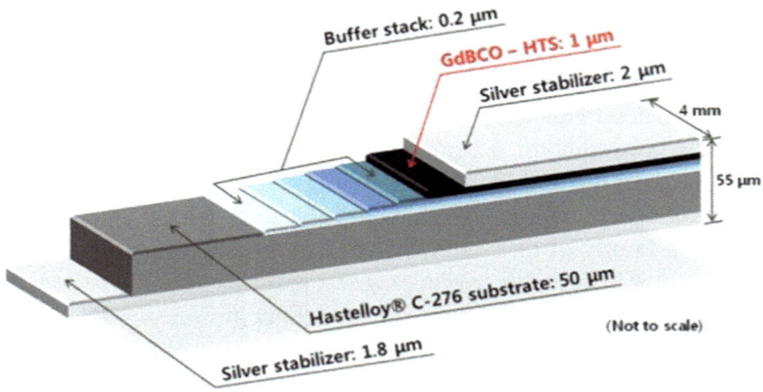

Figure 1.5 A schematic showing the configuration of the GdBCO-coated conductors (CCs) SuperPower.
Taken from Ref. [48].

A key consideration for the application of HTS in a fusion reactor is their response to irradiation. A series of studies of REBCO materials in the TRIGA reactor in Vienna suggested that the critical temperature reduces almost linearly with fast neutron fluence to high fluences. The decrease in critical temperature is accompanied by an increase in the critical current density, thereby decreasing the initially anisotropic resulting from the crystal structure. This increase in the critical current densities is thought to arise from the introduction of nanoscale defect regions that act as flux pinning centers. As the neutron fluence increases, there is a sharp decrease in the superconducting properties of YBCO due to the loss of the underlying crystal structure. It has been noted that HTS tapes that have been manipulated to include high concentrations of flux pinning defects prior to irradiation see a decrease in superconducting properties at lower fluences than those that are unmodified, as they are initially closer to the saturation point. The very large neutron-capture cross sections for ^{155}Gd and ^{157}Gd tapes in the thermal neutron regime lead to a rapid decrease in superconducting properties in $GdBa_2Cu_3O_7$-based tapes. This damage arises due to the decay of the activated Gd, transferring sufficient energy to the nucleus to displace it from its lattice site, creating point defects.

Ceramics are also expected to play a significant role in maintaining the tritium economy in future fusion reactors. While deuterium is abundant and can be easily extracted from seawater, the supply of tritium is constrained due to its short half-life of only 13.6 years. Therefore, it is necessary to breed the tritium in situ from the transmutation of lithium, driven by the 14.1 MeV neutron ejected from the plasma via:

$$ {}_0^1n + {}_3^{6(7)}Li \rightarrow {}_2^4He + {}_1^3H + \left({}_0^1\mathrm{n}\right). $$

In order to sustain the fuel cycle, it is essential that the tritium breeding ratio, defined as the rate of tritium production to the rate of tritium burned in plasma, must be greater than unity [49]. Tritium breeding will occur in the breeder blanket, situated immediately behind the plasma-facing first wall. In addition to breeding tritium, the blanket is also where the kinetic energy of the neutrons will be converted into heat for electricity generation, while also providing protection for the magnets.

The two leading blanket concepts employ either a lithium ceramic or liquid lithium lead breeder material, with the solid breeder considered a more near-term proposition. Ceramic breeder materials are attractive due to their high lithium densities, enabling higher breeding ratios and their relatively low chemical reactivity with structural materials, their low thermal conductivity offers a lower conversion efficiency than the liquid concepts. In concepts, such as the EU's Helium Cooled Pebble Bed [50], the ceramics will be in pebble form to provide resistance to the high thermal loads.

Selection of candidate materials is constrained by the desire to use low activation materials and avoid elements whose future availability may be limited (e.g., Zn, Ga, Ge, As, and Nb) [51]. Ideal candidate materials must have a high lithium density and be able to avoid excessive fragmentation due to thermal stresses during operation while maintaining consistent heat transfer properties as well as being tolerant to high levels of neutron irradiation and simultaneously displaying good tritium release characteristics [52]. Therefore, a range of candidate materials has been considered, including Li_2O [53], γ-$LiAlO_2$ [54], Li_2ZrO_3 [55], and Li_8PbO_6 [56].

In recent years, lithium orthosilicate (Li_4SiO_4) and lithium metatitanate (Li_2TiO_3) have emerged as the leading candidate materials and have been selected for application in the Test Blanket Modules on ITER [57]. While Li_4SiO_4 has a higher lithium density, leading to improved tritium breeding and better release characteristics at low temperatures, Li_2TiO_3 offers better thermomechanical stability [58] and is insensitive to moisture [59]. Both ceramics have complex crystal structures; the titanate adopts a monoclinic unit cell with alternating metal and oxide ions and the C2c space group [60], while the silicate exhibits the low symmetry p21/m space group [61]. The Li_2O–TiO_2 phase diagram shows that the Li_2TiO_3 phase can accommodate significant deviations from stoichiometry ranging from 47 to 51.5 mol % TiO_2 [62]. This flexibility enables the fabrication of the titanate with excess lithium to enhance the tritium breeding ratio. Further, the structure can accommodate the loss of lithium due to transmutation [63]. By contrast, Li_4SiO_4 is a line compound that cannot accommodate substantial nonstoichiometry [64], therefore, a loss of lithium is accommodated in an Li_2SiO_3 secondary phase.

Fabrication of ceramic breeder pebbles and their subsequent characterization has been studied extensively, as shown in the table below [65].

The melt spraying techniques for the fabrication of Li_4SiO_4 developed in China [66] and the EU [67] offer highly spherical pebbles with a high theoretical density and low porosity. These pebbles demonstrate crush resistances of up to 16 N after heat treatment, although the average is closer to 7 N, as the variation between pebbles is significant. For comparison, the slurry dropping approach to fabrication of Li_4SiO_4 results in pebbles with low theoretical densities and large pores [68].

A summary of approaches to the fabrication of different lithium ceramics and the resulting porosities.

Entity	Material	Fabrication process	Pebble size (mm)	Pebble density (%)	Open/closed porosity	Pebble bed density (%)
China	Li_4SiO_4	Melt-spraying	0.8–1.0	94	<4/<1.5	60.5
	Li_2TiO_3	Extrusion-spheronization	0.8–1.2	85	<11/<3	53
EU	Li_4SiO_4	Melt-spraying	0.25–0.63	95	<1/<5	61
	Li_2TiO_3	Extrusion-spheronization	0.8–1.2	90	5/5	55
India	Li_2TiO_3	Solution combustion or solid state	1–1.25	90	6–9/2–5	60
Japan	Li_2TiO_3	Sol-gel	1.18	89	n.a.	64
Korea	Li_4SiO_4	Slurry dropping	0.9–1.1	77	11/12	48
	Li_2TiO_3	Sol-gel	0.9–1.1	91	5/4	56

As the neutrons ejected from the plasma can penetrate breeder pebbles, the tritium ions will initially appear in the ceramic matrix. From there, they must diffuse to the pebble surface via a number of competing diffusion processes, such as intergranular and intragranular diffusion as well as trapping and detrapping processes involving defects. Once at the surface, tritium will undergo isotope exchange reactions with hydrogen in the flow gas (predominantly He) before being recovered. This whole process is illustrated in Fig. 1.6 [69].

The rate-limiting step to tritium recovery is diffusion in the fuel grains. Multiple peaks in thermal desorption spectra of neutron irradiated lithium ceramics show that there are a number of distinct mechanisms contributing to the tritium release process, with their relative importance dependent on temperature.

Experimental activation energies for tritium diffusion vary from 0.63–1.5 eV [70–72] for Li_2TiO_3 to 0.4–0.88 eV for Li_4SiO_4 [73,74]. The experimental data show a significant variation that is likely due to the different experimental conditions but clearly demonstrates the complexity of the process. Tritium diffusion through the ceramic pebbles has also been explored computationally, using Density Functional Theory (DFT). In Li_2TiO_3, the activation energy for tritium diffusion as an interstitial defect was predicted to be 0.33 eV [75]. However, the introduction of lithium vacancy defects that can act as traps leads to an increase in the activation energy to as high as 1 eV [28,78]. This indicates that the presence of defects, arising from either Li burn-up or irradiation damage, might impact tritium extraction from the pebbles. In fact, it has been observed that irradiation by both neutrons [76] and γ-rays [77] leads to decreasing rates of release of hydrogen isotopes. Any increase in retention in the pebbles during operation may compromise the sustainability of the tritium economy.

In order to overcome the limitations presented by the single-phase breeding pebbles, there have been a number of studies that examine the possibility of using multiple phases. For example, Knitter et al. incorporated a small amount of titania into Li_4SiO_4 pebbles to improve the mechanical properties, through the formation of secondary Li_2TiO_3 phases [19]. Taking this idea further, Xiang et al. developed Li_2TiO_3–Li_4SiO_4 core-shell pebbles that offer significantly increased crush loads [78]. The fabrication of core-shell pebbles enables reconsideration of other candidate ceramics that have been discarded. For example, Li_4TiO_4 was discarded due to its sensitivity to air; however, covering a Li_4TiO_4 core with a protective Li_2TiO_3 shell may offer an attractive breeder material [79]. Some examples of these Li_4SiO_4–Li_2TiO_3 pebbles are presented in Fig. 1.7.

In order to ensure a high tritium breeding ratio, it is also necessary to enrich the ceramics with Li-6 as the majority of tritium will be generated from the transmutation of this isotope. Historically, most isotopic enrichment of lithium has been performed using the column exchange (COLEX) process, where aqueous lithium hydroxide (LiOH) is contacted with a lithium-mercury amalgam. Recent research has suggested an approach to making use of the higher mobility of ^6Li through a porous electrode under the application of a DC voltage [80], with ceramics such as spinel-structured $LiTi_2O_4$ proposed as the porous electrodes [81].

More details on advanced ceramics such as SiC, UHTCs, MAX phases, and metal hydrides for fusion applications are available in Chapter 3.

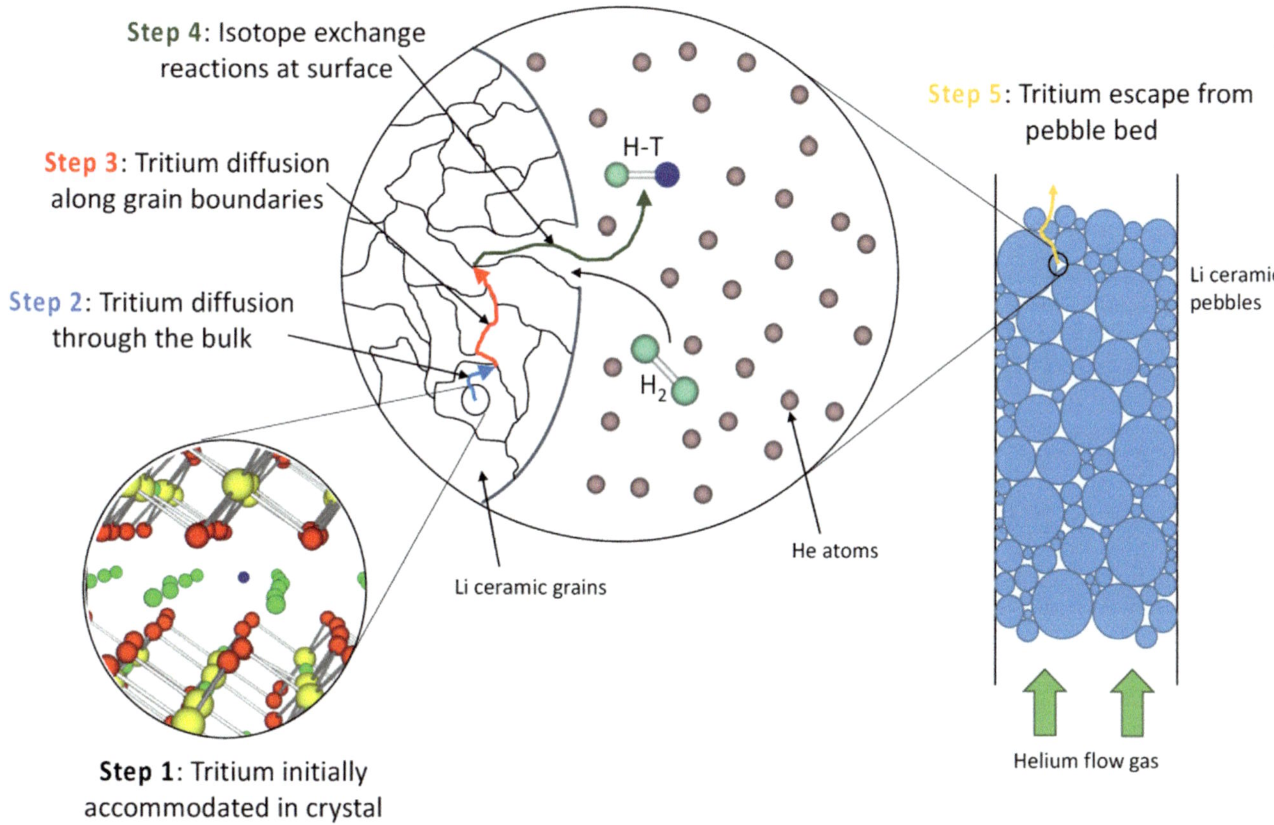

Figure 1.6 The tritium release process from ceramic breeder pebbles. Taken from Goswami and Murphy [69].

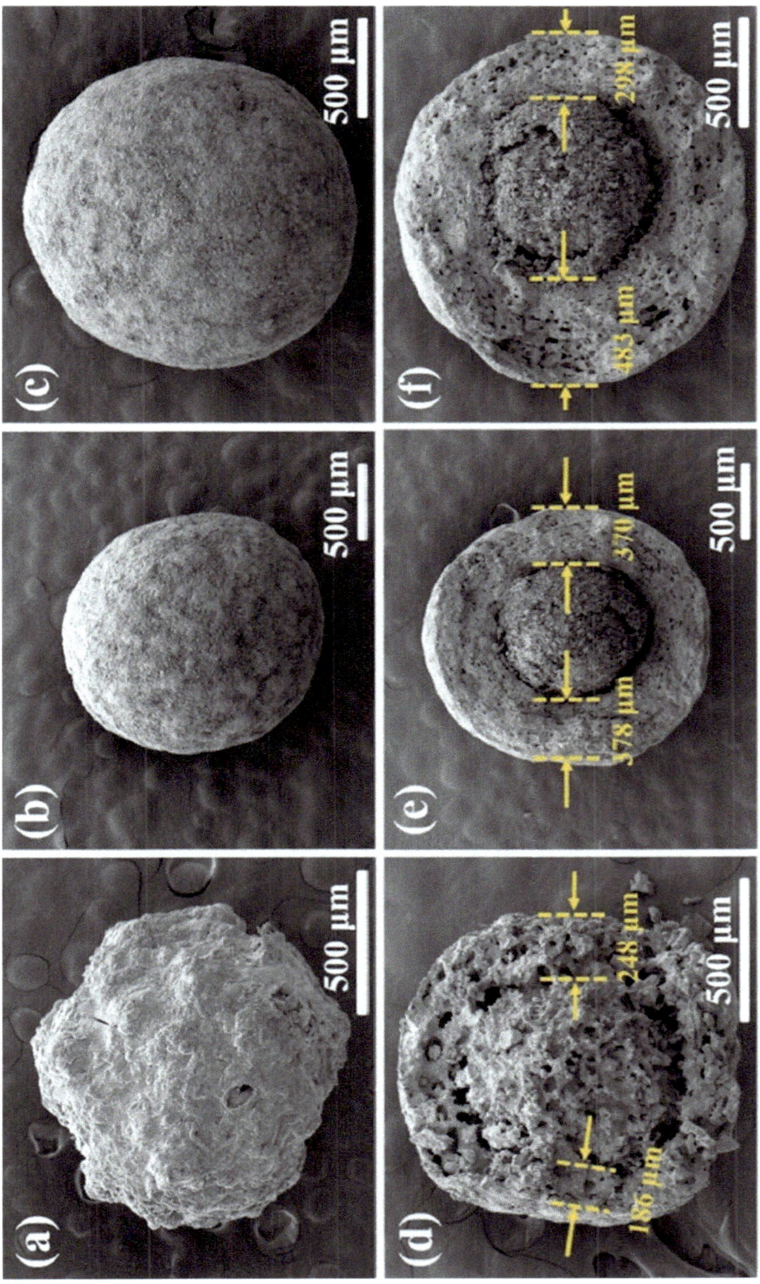

Figure 1.7 Some examples of Li_4TiO_4–Li_2TiO_3 core-shell pebbles.
Taken from Chen et al. [79].

4. Structural materials

Ceramics in the nuclear sector present significant opportunities, particularly as structural materials. Due to their inherent properties, such as high-temperature resistance, durability, and ability to withstand radiation, ceramics like SiC and graphite are increasingly being utilized in advanced reactor systems. The inclusion of ceramics as structural materials in nuclear reactors enhances the overall performance envelope, safety margins, and efficiency, making them advantageous for advancing nuclear technology. More details for the SiC and graphite materials are available in Chapters 4 and 5.

5. Waste forms

The long-term storage of high-level radioactive waste poses a significant challenge. The decay of radioactive isotopes will result in a constantly evolving chemistry of the system, coupled with high temperatures, radiation damage, and the corrosive action of water, which will degrade the waste form over immobilization timescales. Therefore, any waste form must be able to withstand these conditions while also displaying the flexibility to incorporate a wide variety of different cations in different ratios. To address these challenges, significant attention has been given to the concept of immobilizing radioactive elements in an assemblage of mineral phases as first proposed by Hatch [82]. Since then, a wide variety of different mineralogic-ceramic have been proposed, including the Synthetic Rock (SYNROC) [83], Sandia titanate-based ceramic [84], silicate-phosphate supercalcine ceramics [85], alumina-based ceramics and Pu pyrochlores [86]. More details about waste forms will be discussed in Chapter 6.

6. Use of advanced ceramics in other nuclear applications

Ceramics are also used in radiation detection and space exploration. While these topics aren't the focus of this book, they will be briefly covered in this chapter. Interested readers should refer to the references for further reading.

6.1 Radiation detection

The detection of radiation is an essential capability in a wide variety of applications, including scientific research (for example high energy physics, nuclear physics, astrophysics and medical physics), diagnostics in industry (based on X-ray, neutron and ion and electron beams), medical diagnostics (X-rays etc.) and national security (i.e., detecting fissile material) [87]. In a radiation detector, a scintillator acts as the primary radiation sensor that emits light or "scintillates" after absorption of a high-energy photon. In a detector, the light emitted is reflected to photodiodes or photomultipliers for photoelectric conversion [88]. Single crystal and polycrystalline ceramics are two key members of the six-strong scintillator family. In terms of single crystal ceramics, $CdWO_4$, $PbWO_4$, Gd_2SiO_5, and $Bi_4Ge_3O_{12}$ offer excellent efficiency for the detection of X- and gamma-rays [89]. Further, $CdWO_4$ and $Gd_2SiO_5(Ce)$ have also been suggested as scintillators for thermal neutron detection [90]. However, fabrication of large single crystals for use in detectors is a complex process. Therefore, there has been significant effort expended on developing transparent ceramic scintillators that are chemically similar to the oxides discussed above but are polycrystalline. These materials maintain key properties, including high density and high optical phonon yield, but can be fabricated at low temperatures and readily formed in complex shapes [91].

The first ceramic scintillator developed was $Y_{1.34}Gd_{0.6}O_3(Eu_{0.06}, Pr)$ (YGO), which has been deployed commercially in X-ray imaging. In YGO, the scintillation arises from 4f → 4f transitions in the Eu^{3+} dopant ion [92]. Trapping of holes by the Pr codopant reduces the afterglow; however, this naturally reduces light output. Similarly, Gd_2O_2S (GOS) has been a commercial success, although unlike YGO it is translucent as opposed to transparent [53]. An important group of scintillator materials are based on Ce^{3+} activated $Y_3Al_5O_{12}$ (Ce:YAG) garnet. While the light yield in the YAG materials was initially inferior to that of single crystal scintillators, the suppression of antisite defects led to improved performance [93]. Chemical modification of this structure has led to developments including Ce: $(Gd–Y)_3Al_5O_{12}$, which offers better gamma ray stopping power [94] and Ce: $Gd_3(Al,Gd)_5O_{12}$ (Ce:GaGG), which exhibits light yields of 46,000 ph MeV^{-1} [95] (Fig.1.8).

6.2 Radioisotope power sources for space applications

In addition to fission and fusion, it is also possible to obtain useful energy from radioactive decay. The energy can be utilized either as heat in a radioisotope heating unit (RHU) or converted into electricity using a radioisotope power system (RPS) of which the thermoelectric generator (RTG) is the most common. Such devices have been deployed as reliable energy sources for lighthouses, beacons, and weather stations in remote environments [96]. A further important application of RTGs is in the power systems of satellites in Earth orbit and for lunar and deep space exploration [97]. In fact, all missions to Jupiter and beyond (including the Voyager missions, Galileo, Cassini and New Horizons) have relied on RTGs as the solar intensity becomes insufficient to power the spacecraft and associated scientific equipment. The Voyager probes have now both left the Solar System and are reporting the first measurements of the temperature and density of the interstellar medium [98] thanks to their RTGs, which have now powered the spacecraft for nearly 50 years.

A typical radioisotope power system contains a thermal source and a power conversion system that converts a portion of the thermal energy into electricity. In general, the power conversion system can be either static or dynamic. Static systems require no moving parts and can be either thermoelectric, thermionic or

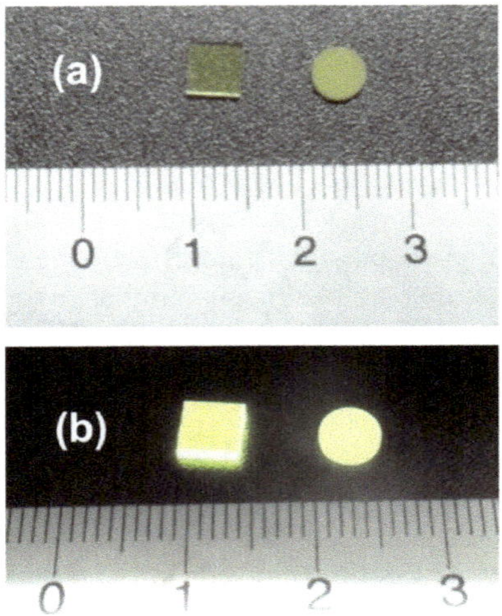

Figure 1.8 Samples of single crystal (square) and ceramic (circle) 1% Ce doped GACC scintillators. (a) Under normal light and (b) emission under UV light.
Image taken from Ref. [93].

thermophotovoltaic. Dynamic systems utilize a working fluid and operate like a typical heat engine to transform heat into electricity. As mentioned above, all of the deep space missions to date have employed RTGs because thermoelectrics have high levels of reliability over long timescales while being rugged and compact. The major limitation associated with thermoelectrics is their low conversion efficiency, which is somewhere below 10%. While dynamic systems may offer higher conversion efficiencies, they employ moving parts that may be subject to mechanical failure [2].

Figure 1.9 A PuO_2 fuel pellet as used in the GPHAS-RTG. Note the pellet growing red hot due to the alpha decay of the ^{238}Pu. image Courtesy of the U.S. Department of Energy / Los Alamos National Laboratory (LANL).

Table 1.2 Potential isotopes for radioisotopes power systems.

Isotope	Half-life	Specific power (W_t/g in elemental form)	Decay mode and radiation emissions	Relevant chemical forms	Production and availability
^{90}Sr	29 years	0.935	β, plus bremsstrahlung photons from β-shielding	$SrTiO_3$, SrO, $SrZrO_3$	Fission products, could potentially be obtained from a nuclear fuel reprocessing plant.
^{144}Ce	285 days	25.5	β, also γ, mainly at 134 keV, plus bremsstrahlung photons from β-shielding	CeO	Fission products, could potentially be obtained from a nuclear fuel reprocessing plant.
^{210}Po	138 days	144	α, some neutrons from (α,n) reaction	HgPo, PbPo, Po	Reactor irradiation of ^{209}Bi
^{238}Pu	88 years	0.568	α, some neutrons from spontaneous fission and (α, n) reaction	PuO_2, Pu_2C_3	Reactor irradiation of ^{237}Np or ^{241}Am
^{241}Am	433 years	0.115	α, some neutrons from spontaneous fission and (α, n) reaction, significant soft γ	AmO_2, Am_2O_3	Chemical separation from aged reactor-grade Pu
^{242}Cm	163 days	122	α, some neutrons from (α,n) reaction	CmO_2, Cm_2O_3	Reactor irradiation of ^{241}Am
^{244}Cm	18 years	2.83	α, strong neutron activity	CmO_2, Cm_2O_3	Reactor irradiation of Pu isotopes

The thermal source generates heat from the α and β-decay of unstable heavy isotopes. The choice of radioisotope is driven by the required operational lifetime and specific power output [99]. Two factors that are inversely related to each other: isotopes with longer half-lives offer longer operational lifetimes at lower power outputs, while isotopes with short half-lives offer higher power outputs but for shorter durations. It is also important to consider the emission of ionizing radiation from the source. While both α and β-decay processes are accompanied by either γ or X-rays, the intensity is typically much greater for a β-emitter. Most α emitters are also subject to spontaneous fission decay. These events are much rarer; however, the highly penetrating neutrons can provide significant doses. Therefore, the choice of radioisotope will also depend on the application and the ability to provide shielding.

Early RTGs experimented with a number of different radioisotopes, the US's Systems for Nuclear Auxiliary Power (SNAP) 1 used ^{144}Ce, while the SNAP 7 and the Russian Beta-M employed ^{90}Sr. However, later SNAP variants, and subsequent Multihundred Watt (MHW) and General Purpose Heat Source (GPHS) RTGs converted to using ^{238}Pu [3]. An example of a PuO_2 thermal source is presented in Fig. 1.9. Due to the limited availability of ^{238}Pu and its associated high costs, the European Space Agency (ESA) has recently been exploring the possibility of using ^{241}Am [100]. ^{241}Am is attractive as it is produced from the decay of ^{241}Pu that can be extracted in a relatively isotopically pure form from plutonium from reprocessed nuclear fuel.

Presented in Table 1.2 is a summary of the different possible radioisotopes for use in RPSs. As illustrated in Table 1.2 most of the radioisotopes' heat sources are in their ceramic oxide form. This is due to the long-term stability of the oxides at the high operating temperatures of up to 1000 K. However, the use of oxygen results in an increase in the neutron yield owing to (α,n)-reaction reactions with ^{17}O and ^{18}O. The neutron yield can be significantly reduced by enriching the concentration of ^{16}O to levels higher than the natural isotopic ratios. A process for the enrichment of PuO_2 has already been demonstrated [101]. Concerns around the stability of the simple AmO_2 and Am_2O_3 under operational conditions [102,103] have prompted an exploration of more complex Am-bearing oxides such as $AmAlO_3$ [104].

References

[1] F. Hund, Fluoritmischphasen der Dioxide von Uran, Thorium, Cer und Zirkonium mit Wismutoxid, Z. Z. Anorg. Allg. Chem. 333 (1964) 248–255.

[2] M.W.D. Cooper, S.T. Murphy, D.A. Andersson, The defect chemistry of $UO_{2\pm x}$ from atomistic simulations, J. Nucl. Mater. 504 (2018) 251–260.

[3] B.T.M. Willis, Positions of the Oxygen Atoms in $UO_{2.13}$, vol 197, 1963, pp. 755–756.

[4] T. Abe, K. Asakura, Uranium oxide and MOX production, in: R. Konings (Ed.), Comprehensive Nuclear Materials, Elsevier, Amsterdam, Holland, 2012, pp. 394–422.

[5] Experiences and trends of manufacturing technology of advanced nuclear fuels, in: IAEA-TECDOC-1686, International Atomic Energy Association, Vienna, Austria, 2012.

[6] W. Wiesenack, Nuclear fuel assembly design and fabrication, in: I. Crossland (Ed.), Nuclear Fuel Cycle Science and Engineering, Woodhead Publishing, Cambridge, UK, 2012, pp. 203–233.
[7] L. Van Brutzel, R. Dingreville, T.J. Bartel, Nuclear Fuel Deformation Phenomena, vol. 5, NEA/NSC/R, 2015.
[8] D. Olander, Nucler fuels – present and future, J. Nucl. Mater. 389 (2009) 1–22.
[9] K. Whittle, Nuclear Materials Science, IOP Publishing, 2016
[10] K. Une, S. Kashibe, Fission gas release during post irradiation annealing of BWR fuels, J. Nucl. Sci. Tech. 27 (1990) 1002.
[11] J. Rest, M.W.D. Cooper, J. Spino, J.A. Turnbull, P. Van Uffelen, C.T. Walker, Fission gas release from UO_2 nuclear fuel: A review, J. Nucl. Mater. 513 (2019) 310–345.
[12] V.V. Rondinella, T. Wiss, The high burn-up structure in nuclear fuel, Mater. Today 13 (2010) 24–32.
[13] J. Arborelius, K. Backman, L. Hallstadius, M. Limback, J. Nilsson, B. Rebensdorff, G. Zhou, K. Kitano, R. Lofstrom, G. Ronnberg, Advanced doped UO_2 pellets in LWR applications, J. Nucl. Sci. Tech. 43 (2006) 967–976.
[14] C. Delafoy, P. Dewes, T. Miles, AREV NO Cr_2O_3 doped fuel development for BWRs. American Nuclear Society – 2007 LWR Fuel Performance/Top Fuel.
[15] J.J. Carbajo, G.L. Yoder, S.G. Popov, V.K. Ivanov, A review of the thermophysical properties of MOX and UO_2 fuels, J. Nucl. Mater. 299 (2001) 181–198.
[16] Managing the UK Plutonium Stockpile. Parliamentary Office for Science and Technology, POSTNOTE 531 2016.
[17] P. Netter, Reprocessing of spent oxide fuel from nuclear power reactors, in: I. Crossland (Ed.), Section 16 in Nuclear Fuel Cycle Science and Engineering, Woodhead Publishing, Cambridge, UK, 2012.
[18] H. Bairiot, P. Deramaix, MOX fuel development: yesterday, today and tomorrow, J. Nucl. Mater. 188 (1992) 10–18.
[19] D. Haas, A. Vandergheynst, J. Van Vliet, R. Lorenzelli, J.-L. Nigon, Mixed-oxide fuel fabrication technology and experience at the Belgonucleaire and CFCA plants and further developments for the MELOX plant, Nucl. Technol. 106 (1994) 60–82.
[20] D. Haas, D.J. Hamilton, Fuel cycle strategies and plutonium management in Europe, Prog. Nucl. Ener. 49 (2007) 574–582.
[21] A.R. Massih, Models for MOX fuel behaviour, SKI Report 10 (2006).
[22] H. Roepenack, F.U. Schlemmer, G.J. Schlosser, Development of thermal plutonium recycling, Nucl. Tech. 77 (1987) 175–186.
[23] C.T. Walker, W. Goll, T. Matsumura, MOX Fuel Irradiation Behavior; Results from X-Ray Microbeam Analysis, International Atomic Energy Agency, Vienna, 1995, pp. 301–318. IAEA-TECDOC-941.
[24] P. Deramaix, D. Haas, J. Van de Velde, In-pile performance of mixed-oxide fuel with particular emphasis on MIMAS fuel, Nucl. Technol. 102 (1993) 47–53.
[25] M. Lippens, Densifications of UO_2-PuO_2 fuel, J. Nucl. Mater. 81 (1979) 99–105.
[26] J. Edwards, R.D. Grimoldby, S.J. Marshall, R.W. Stratton, BNFL Supply of MOX Fuel Assemblies to the Beznau 1 PWR of Nok, International Atomic Energy Association, IAEA-TECDOC-941, Vienna, 1997, pp. 57–67.
[27] P.K. Ivison, P.M.A. Cook, S. Bremier, C.T. Walker, Quantification of the Homogeneity of BNFL SBR MOX Fuel Using Compositional X Ray Mapping, International Atomic Energy Association, IAEA-TECDOC-1233, Vienna, 2001, pp. 19–23.
[28] P.K. Ivision, S.B. Fisher, Quantitative plutonium mapping using energy dispersive X-ray analysis, in: Proceedings of TopFuel99, Avignon, France, September 1999, pp. 13–15.

[29] M.W.D. Cooper, S.C. Middleburgh, R.W. Grimes, Modelling the thermal conductivity of $(U_xTh_{1-x})O_2$ and $(U_xPu_{1-x})O_2$, J. Nucl. Mater. 466 (2015) 29–35.
[30] I.S. Kamimura, Y. Kobayashi, T. Nomata, Helium generation and release in MOX fuels 556, OECD Nuclear Energy Agency, Issy-les-Moulineaux (France), 2000, pp. 263–270. International symposium on MOX fuel cycle technologies for medium and long term deployment; Vienna (Austria); 17–21.
[31] OECDNEA, Accelerator-Driven Systems (ADS) and Fast Reactors (FR) in Advanced Nuclear Fuel Cycles, OECD, Paris, France, 2002.
[32] M. Aslan, I. Leboucher, E. Bouvier, Fuel cycle strategies to optimize the MOX in reactors, Ener. Proc. 39 (2013) 59–68.
[33] M.S. Kazimi, K.R. Czerwinski, M.J. Driscoll, P. Hejzlar, J.E. Meyer, "On the Use of Thorium in Light Water Reactors" MIT-NFC-TR-016, Massachusetts Institute of Technology, Massachusetts, Cambridge, USA, 1999.
[34] M. Lung, O. Gremn, Perspectives of the thorium fuel cycle, Nucl. Eng. Des. 180 (1998) 133–146.
[35] Thorium Fuel Cycle – Potential Benefits and Challenges. IAEA-TECDOC-1450, Vienna (2005).
[36] M. Lung, A Present Overview of the Thorium Nuclear Fuel Cycles, European Commission Report, EUR-17771, Luxembourg, 1997.
[37] S.K. Mukerjee, T.R.G. Kutty, N. Kumar, R.V. Pai, A. Kumar, Fabrication technologies fir ThO_2-based fuel, in: Thoria-based Nuclear Fuels, Green Energy and Technology, Springer-Verlag, London, UK, 2013.
[38] D. Woods, M. Saoudi, C. Mayhew, R. Ham-Su, Characterization of plutonium distribution in the ThO_2-PuO_2 mixed oxides by electron probe microanalysis, CNL Nucl. Rev. 7 (2008) 157–163.
[39] T.R.G. Kutty, P.V. Hedge, K.B. Kahn, T. Jarvis, A.K. Sengupta, S. Majumdar, H. S. Kamath, Characterization and densification studies on ThO_2-UO_2 pellets derived from ThO_2-U_3O_8 powders, J. Nucl. Mater. 335 (2004) 462–470.
[40] T.R.G. Kutty, P.S. Somayajulu, K.B. Khan, A. Kumar, H.S. Kamath, Characterization of $(Th,U)O_2$ pellets made by advanced CAP process, J. Nucl. Mater. 384 (2009) 3030–3310.
[41] T.R.G. Kutty, R.V. Kulkarni, P. Sengupta, K.B. Kahn, K. Bhanumurthy, A.K. Sengupta, J.P. Panakkal, A. Kumar, H.S. Kamath, Development of CAP process for fabrication of ThO_2-UO_2 fuels. Part II characterization and property evaluation, J. Nucl. Mater. 373 (2008) 309–318.
[42] T.R. G Kutty, P.S. Somayajulu, K.B. Kahn, A. Kumar, H.S. Kamath, Characterization of $(Th,U)O_2$ pellets made by advanced CAP process, J. Nucl. Mater. 384 (2009) 303–310.
[43] R. Bohler, A. Quaini, L. Capriotti, L. Cakir, O. Benes, K. Boboridis, A. Guiot, L. Luzzi, R.J.M. Konings, D. Manara, The solidification of behaviour of the UO_2-ThO_2 system in laser heating study, J. Alloy. Compd. 616 (2014) 5–13.
[44] "World Thorium Occurrences, Deposits and Resources", IAEA-TECDOC-1977, International Atomic Energy Agency, Vienna, Austria, 2019.
[45] T.R. Pavlov, T. Wangle, M.R. Wenman, V. Tyrpekl, L. Vlahovic, D. Robba, P. Van Uffelen, R.J.M. Konings, R.W. Grimes, High temperature measurements and condensed matter analysis of the thermo-physical properties of ThO_2, Sci. Reps. 8 (2018) 5038.
[46] J. Park, E. Farfan, K. Mitchell, A. Resnick, C. Enriquez, T. Yee, Sensitivity of thermal transport in thorium dioxide to defects, J. Nucl. Mater. 504 (2018) 198–205.
[47] A. Barry, Experimental measurements to evaluate the effects of fuel microstructure on fission gas release in thoria, A. Nucl. Ener. 151 (2021) 107933.

[48] Y. Park, M. Lee, H. Ann, Y.H. Choi, H. Lee, "A superconducting joint for $GdBa_2Cu_3O_{7-\delta}$-coated conductors", NPG Asia Mater. 6 (2014) e98.
[49] N. Zandi, H. Sadeghi, M. Habibi, I. Jalali, M. Zare, Blanket simulation and tritium breeding ratio calculation for ITER reactor, J. Fus. Ener. 34 (2015) 1365–1368.
[50] M. Dalle Donne, H. Albrecht, L.V. Boccaccini, U. Fischer, S. Gordeev, E. Hutter, K. Kleefeldt, P. Norajitra, G. Reimann, P. Ruatto, K. Schleisiek, H. Schnauder, European helium cooled pebble bed blanket: design of blanket: module to be tested in ITER, Fus. Eng. Des. 39–40 (1998) 825–833.
[51] F.A. Fernandez, P. Pereslavtsev, First principles review of options for tritium breeder and neutron multiplier materials for breeding blankets in fusion reactors, Fus. ENg. Des 137 (2018) 243–256.
[52] M. Zmitko, Y. Poitevin, L. Boccaccini, J.-F. Salavy, R. Knitter, A. Moslang, A. J. Magielsen, J.B.J. Hegeman, R. Lasser, Development and qualification of functional materials for the EU Test Blanket Modules: strategy and R&D activities, J. Nucl. Mater. 417 (2011) 678–683.
[53] C.E. Johnson, G.W. Hollenberg, Recent advances in the development of solid breeder blanket materials, J. Nucl. Mater. 123 (1984) 871–881.
[54] W. Jiang, T. Wang, Y. Wang, Z. Zhu, L. Shao, D. Senor, Deuterium diffusion in γ-$LiAlO_2$ pellets irradiated with He_+ and $D2_+$ ions, J. Nucl. Mater. 538 (2020) 152357.
[55] K. Tobita, S. Nishio, A. Saito, M. Enoeda, C. Liu, H. Tanigawa, S. Sato, D. Tsuru, T. Hirose, Y. Seki, M. Yamada, Water-cooled solid breeding blanket for DEMO, Proceedings of ITC18 P1–43 (2008) 285–288.
[56] I. Palermo, J.M. Gomez-Ros, G. Veredas, J.P. Catalan, F. Ogando, J. Sanz, L. Sedano, Preliminary neutronic assessment of a helium-cooled Li_8PbO_6 breeding blanket design for DEMO, Fus. Eng. Des. 87 (2012) 195–199.
[57] L.M. Giancarli, M.-Y. Ahn, I. Bonnett, C. Boyer, P. Chaudhuri, W. Davis, G. Dell'Orco, M. Iseli, R. Michling, J.-C. Neviere, R. Pascal, Y. Poitevin, I. Ricapito, I. Schneiderova, L. Secton, H. Tanigawa, Y.L. Tonqueze, J.G. van der Laan, X. Wang, R. Yoshino, ITER TBM program and associated system engineering, Fus. Eng. Des. 136 (2018) 815–821.
[58] R. Knitter, M.H.H. Kolb, U. Kaufmann, A.A. Goraieb, Fabrication of modified lithiutm orthosilicate pebbles by addition of titania, J. Nucl. Mater. 442 (2013) S433–S436.
[59] C. Alvani, S. Casadio, V. Contini, A. Di Bartolomeo, J.D. Lulewicz, N. Roux, Li_2TiO_3 pebbles reprocessing, recovery of 6Li as Li_2CO_3, J. Nucl. Mater. 307–311 (2002) 837–841.
[60] K. Kataoka, Y. Takahashi, N. Kijima, H. Nagai, J. Akimoto, Y. Idemoto, K. Ohshima, Crystal growth and structure refinement of monoclinic Li_2TiO_3, Mater. Res. Bull. 44 (2009) 168–172.
[61] D. Tranqui, R.D. Shannon, H.-Y. Chen, S. Iijima, W.H. Baur, Crystal structure of ordered Li_4SiO_4, Acta Cryst B35 (1979) 2479–2487.
[62] H. Kleykamp, Phase equilibria in the Li–Ti–O system and physical properties of Li_2TiO_3, Fus. Eng. Des. 61 (2002) 361–366.
[63] T. Hoshino, Optimization of sintering conditions of advanced tritium breeder pebbles fabricated by the emulsion method, Fus. Eng. Des. 98–99 (2015) 1788–1791.
[64] B. Konar, M.-A. Van Ende, I.-H. Jung, Critical evaluation and thermodynamic optimization of the Li-O, and Li_2O-SiO_2 systems, J. Eur. Ceram. Soc. 37 (2017) 2189–2207.
[65] R. Knitter, P. Chaudhuri, Y.J. Feng, T. Hoshino, I.-K. Yu, Recent developments of solid breeder fabrication, J. Nucl. Mater. 442 (2013) S420–S424.
[66] Y.F. Feng, K.M. Feng, Q.X. Cao, J. Hu, H. Tang, Fabrication and characterization of Li_4SiO_4 pebbles by melt spraying method, Fus. Eng. Des. 87 (2012) 753–756.

[67] M.H.H. Kolb, R. Knitter, U. Kaufman, D. Mundt, Enhanced fabrication process for lithium orthosilicate pebbles as breeding material, Fus Eng. Des 86 (2011) 2148–2151.
[68] Y.-H. Park, I.-K. Yu, M.Y. Ahn, S. Cho, D.Y. Ku, Fabrication of Li_4SiO_4 pebbles using slurry droplet wetting method for solid breeding material, Fus. Sci. Tech. 62 (2011) 185–189.
[69] K.N. Goswami, S.T. Murphy, Influence of lithium vacancy defects on tritium diffusion in β-Li_2TiO_3, J. Phys. Chem. C 14 (2020) 12286–12294.
[70] T. Kinjyo, M. Nishikama, M. Enoeda, S. Fukada, Tritium diffusivity in crystal grain of Li_2TiO_3 and tritium release behavior under several purge gas conditions, Fus. Eng. Des. 83 (2008) 580–587.
[71] T. Tanifuji, D. Yamaki, S. Nasu, K. Noda, Tritium release behaviour from neutron-irradiated Li_2TiO_3 single crystal, J. Nucl. Mater. 258–263 (1998) 543–548.
[72] D. Zhu, T. Oda, Y. Shonon, S. Tanaka, Release Behaviour of Hydrogen Isotopes Thermally Sorbed in Li_2TiO_3 Single Crystal, vol 442, 2013, pp. S437–S441.
[73] C. Xiao, X. Gao, M. Kobayashi, K. Kawasaki, H. Uchimura, K. Toda, C. Kang, X. Chen, H. Wang, S. Peng, X. Wang, Y. Oya, K. Okuno, Tritium release kinetics in lithium orthosilicate pebbles irradiated with low thermal-neutron fluence, J. Nucl. Mater. 438 (2013) 46–50.
[74] T. Kinjyo, M. Nishikawa, Tritium release behaviour from Li_4SiO_4, Fus. Sci. Tech. 46 (2003) 561–570.
[75] Y. Shi, J. Qi, Y. Hun, T. Lu, Anisotropic diffusion of a charged tritium interstitial in Li_2TiO_3 from first principles calculations, Phys. Rev. Appl. 10 (2018) 024021.
[76] M. Kobayashi, Y. Oya, K. Munakata, K. Okuno, Developing a tritium release model for Li_2TiO_3 with irradiation-induced defects, J. Nucl. Mater. 458 (2015) 22–28.
[77] M. Gonzalez, E. Carella, A. Morono, M.H.H. Kolb, R. Knitter, Thermally induced outdiffusion studies of deuterium in ceramic breeder blanket materials after irradiation, Fus. Eng. Des. 98–99 (2015) 1771–1774.
[78] M. Xiang, Y. Zhang, Y. Zhang, S. Liu, H. Liu, C. Wang, C. Gu, Preparation of Li_2TiO_3-Li_4SiO_4 core-shell ceramic pebbles with enhanced crush load by graphite bed process, J. Nucl. Mater. 466 (2015) 477–483.
[79] R. Chen, Z. Liao, Y. Shi, H. Wang, H. Guo, Y. Zeng, J. Qi, Q. Shi, T. Lu, Preparation of Li_4TiO_4-Li_2TiO_3 core-shell ceramic pebbles with thick shells and high strength through an improved granulation method, J. Nucl. Mater. 543 (2021) 152580.
[80] T. Hoshino, T. Terai, Basic technology for 6Li enrichment using an ionic-liquid impregnated organic membrane, J. Nucl. Mater. 416 (2011) 696–699.
[81] K. Kobayashi, T. Hoshino, Numerical investigation for lithium isotope effect in ionic superconductor, Fus. Eng. Des. 136 (2018) 205–209.
[82] L.P. Hatch, Ultimate disposal of radiaoactive wastes, Am. Sci. 41 (1953) 410–421.
[83] A.E. Ringwood, V.M. Oversby, S.E. Kesson, SYNROC: Leaching performance and process technology, in: Proceedings of the Seminar on Chemistry and Process Engineering for High Level Liquid Waste Solidification, Julichw, Germany, 1-5 June 1981, pp. 495–506.
[84] R.O. Schoebel, Stabilization of high level waste in ceramic form, Bull. Am. Ceram. Soc. 54 (1975) 459.
[85] G.J. McCarthy, J.G. Pepin, D.E. Pfoertsch, D.R. Clarke, Crystal Chemistry of the Synthetic Minerals in Tcurrent Supercalcine Ceramics. US DoE Report CONF-790420, Batelle Pacific Northwest Labs, Richland, WA, USA, 1979, pp. 315–320.
[86] B.E. Burakov, M.I. Ojovan, W.E. Lee, Crystalline Materials for Actinide Immobilization, Imperial College Press, London, UK, 2020.

[87] B.D. Milbrath, A.J. Peurrung, M. Blass, W.J. Weber, Radiation detector materials: An overview, J. Mater. Res. 23 (2008) 2561–2581.

[88] C. Greskovich, S. Duclos, Ceramic scintillators, Annu. Mater. Sci. 27 (1997) 69–88.

[89] S.P. Burachas, F.A. Danevich, A.S. Georgadze, H.V. Klapdor-Kleingrothaus, V. V. Kobycheva, B.N. Kropivyanskya, V.N. Kutsa, A. Muller, V.N. Muzalevskya, A. S. Nikolaikoa, O.A. Ponkratenkoa, V.D. Ryzhikovb, A.S. Sai, I.M. Solskyd, V. I. Tretyaka, Y.G. Zdesenkoa, Large volume $CdWO_4$ crystal scintillators, Nucl. Instru. Meth. Phys. Rev. A 369 (1996) 164–168.

[90] V. Ryzhikov, L. Nagornaya, V. Volkov, V. Chernikov, O. Zelenskaya, Thermal neutron detectors based on complex oxide crystals, Nucl. Instru. Meth. Phys. Rev. A 486 (2002) 156–159.

[91] K.J. Wilson, R. Alabd, M. Abolhasan, M. Safavi-Naeini, D.R Franklin, Optimisation of monolithic nanocomposite and transparent ceramic detectors for positron emission tomography, Sci. Rep. 10 (2020) 1409.

[92] C.W.E. van Eijk, Inorganic scintillators in medical imaging, Phys. Med. Biol. 47 (2002) R85–R106.

[93] T. Yanagida, K.K.Y. Fujimoto, H. Yagi, T. Yanagitani, Comparative study of ceramic and single crystal Ce:GAGG scinttillator, Opt. Mater. 35 (2013) 2480–2485.

[94] T. Yanagida, T. Itoh, H. Takahashi, S. Hirakuri, M. Kokbun, K. Makishima, M. Sato, T. Enoto, T. Yanagitani, H. Yagi, T. Shigetad, T. Ito, Improvement of ceramic YAG(Ce) scintillators to $(YGd)_3Al_5O_{12}(Ce)$ for gamma-ray detectors, Nucl. Instru. Meth. Phys. Res. A 579 (2007) 23–26.

[95] K. Kamada, T. Endo, K. Tsutumi, T. Yanagida, Y. Fujimoto, A. Fukabori, A. Yoshikawa, J. Pejchal, M. Nikl, Composition energineering ni Cerium-doped $(Lu, Gd)_3(Ga,Al)_5O_{12}$ single crystal scintillators, Cryst. Growth Des. 11 (2010) 4484–4490.

[96] E. Levnor, I. Linkov, J.M. Proth, Strategic Management of Marine Ecosystems, Springer, 2005, p. 21.

[97] G.R. Schmidt, T.J. Sutcliff, L.A. Dudzinki, Radioisotopes Power: A key technology for deep space exploration, in: Radioisotopes – Applications in Physical Sciences by Nirmal Singh, IntechOpen, 2011, pp. 419–456.

[98] J.D. Richardson, J.W. Belch, P. Garcia-Galindo, L.F. Burlaga, Voyager 2 plasma observations of the heliopause and interstellar medium, Nat. Astro. 3 (2019) 1019–1023.

[99] L. Summerer, K. Stephenson, Nuclear power sources: a key enabling technology for planetary exploration, Proc. IMEchE 225 Part G: J. Aero. Eng. (2010) 129–143.

[100] D. Horlait, R. Caraballo, F. Lebreton, C. Jegou, P. Roussel, T Delahaye, J. Solid, State. Chem 217 (2014) 159–168.

[101] O.L. Kruger, Process of making oxygen enriched plutonium dioxide (PuO_2), United States Patent 7315092, 1972.

[102] E. Epifano, C. Gueneau, R.C. Belin, R. Vauchy, F. Lebreton, J.-C. Richaud, A. Joly, C. Valot, P.M. Martin, Insight into Am-O phase equilibria: a thermodynamics study coupling high-temperature XRD and CALPHAD modelling. Inorg. Chem. 56 (2017) 7416–7432.

[103] F. Lebreton, R.C. Belin, T. Delahaye, P. Blanchart, In-situ X-ray diffraction study of phase transformations in the Am-O system, J. Solid State Chem. 196 (2012) 217–224.

[104] J.-F. Vigier, K. Popa, L. Martel, D. Manara, O.D. Blanco, D. Freis, R.J.M. Konings, Plutonium and Americium aluminate perovskites, Inorg. Chem. 58 (2019) 9118–9126.

Advanced ceramics and composite fuels

2

E. Kardoulaki[1] *and Dong Liu*[2]
[1]Materials Science and Technology Division, Los Alamos National Laboratory, Los Alamos, NM, United States; [2]Department of Engineering Science, University of Oxford, Oxford, United Kingdom

1. Monolithic fuels

1.1 Uranium nitride

Uranium nitride (UN) exhibits a plethora of properties that deem it attractive for use in a variety of nuclear applications and reactors. Early comprehensive reviews of UN and (U,Pu)N properties from 1972 [1] were the first to incite interest in these fuel forms. This led to nitride fuel development programs, such as the Systems for Nuclear Auxiliary Power (SNAP) US program [2]. When SNAP terminated, development of nitride fuels in the US was also paused and UN did not receive much interest until the SP-100 nuclear-electric power space application program, which selected UN, as well as UO_2, as the primary fuel candidates (see Fig. 2.1) [3]. The selection of UN for SP-100 stemmed from the requirements for a compact and lightweight reactor with a high power-to-mass ratio. More specifically, UN has a high uranium density, high thermal

Figure 2.1 UN pellets fabricated for the SP-100 program [3].

Ceramics for Nuclear Applications. https://doi.org/10.1016/B978-0-12-823865-3.00011-0

conductivity, high decomposition temperature, and a good chemical stability, which make it an attractive option for many nuclear applications including the US Advanced Liquid Metal Reactor (ALMR) and lead-cooled small, sealed, transportable, autonomous reactor (SSTAR), and Gen IV fast reactors.

UN synthesis presents a challenge that could create obstacles when considering its commercialization. There are currently three laboratory-scale synthesis techniques for UN with the most established one being carbothermic reduction and nitridation (CTRN) [3]. The precursors for CTRN are UO_2 powder and carbon black powder that are mixed to produce a compact, which when heated reacts to form UC. Subsequently, the UC is heated under an H_2–N_2 atmosphere to form UN. CTRN is the conventional method of forming UN and can produce high-purity material with tight control of impurities, such as carbon and oxygen, albeit in relatively small quantities. The batch quantities in CTRN are dictated partly by the equipment size and the availability of inert atmosphere infrastructure, and party by the nature of the reaction, from carbide to nitride, which is diffusion limited [4]. Even though several techniques, such as using an overpressure of flowing nitrogen or chemical milling, can be employed to promote diffusion of nitrogen into the material, the precursor compacts are still somewhat limited in size. UN synthesized from CTRN typically features impurities on the order of 3000 ppm for carbon and 1000 ppm for oxygen. These impurities, which are very sensitive to batch size [3], can be reduced significantly with further high-temperature processing; however, this is a time consuming and energy intensive step. Carbon and oxygen impurities need to be well controlled, as they have been shown to influence fuel swelling [3].

The other two UN synthesis methods are a direct hydride–dehydride–nitride route that utilizes U metal as the starting material [5], and ammonolysis of uranium fluorite [6,7]. The hydriding and nitriding synthesis route can yield a high purity material (~1000 ppm of carbon and ~1000–2000 ppm of oxygen) with relatively low temperature processing, as the hydriding/dehydriding steps usually require temperatures below 500°C [2,8]. However, it also poses challenges as handling of the formed U metal powders, which feature small particle sizes, under inert atmospheres is required due to their extreme oxygen sensitivity. Despite this, the availability of high enrichment U metal precursor material can make this process attractive. Finally, synthesis from uranium tetrafluorite to uranium nitride can provide an attractive option from an infrastructure perspective, as it requires lower processing temperatures of around 800°C for synthesis of UN_2 [7], and it makes use of the preexisting uranium fluoride infrastructure currently in place for industrial synthesis of UO_2. This synthesis route is, however, still relatively new and further studies need to be conducted to fully understand the thermodynamics and kinetics of the reactions, as well as the consistency and purity of the final product. Table 2.1 below summarizes the key benefits and drawbacks of the three synthesis techniques for UN.

Fabrication of high-density UN pellets is also challenging, as high-temperature processing is required. For example, to achieve higher than 90% TD pellets, sintering temperatures in excess of 1900°C are often required [3,9]. Additionally, high energy ball milling and long milling times are key to achieving high-density, large-grained pellets, as shown in Fig. 2.2 [3]. Sintering of UN is benefited by an overpressure of

Table 2.1 Summary of the three synthesis techniques for UN.

	CTRN	Hydride/ dehydride–nitride	Ammonolysis
Precursor	UO_2 + graphite	U metal	UF_4 or UF_6
Impurities	Carbon, oxygen	Hydrogen, impurities native to the U metal	Fluorine
Advantages	Established process, ability to control impurity levels	Low temperature processing, high purity product	Low temperature processing, use of UF industrial infrastructure
Drawbacks	High temperatures, long processing times, impurity control, inert infrastructure	Oxygen sensitivity, proliferation, inert infrastructure	HF byproducts, limited understanding of reaction
Reference	[3,4]	[2,8]	[7]

N_2 given that this can eliminate decomposition at high temperatures. Additionally, controlling any U_2N_3 impurities in the UN powder is also important as conversion of U_2N_3 to UN and N_2 at the sintering temperatures can have detrimental effects, as reported by Matthews et al. [3]. Arai et al. [10] studied the effect of oxygen impurities on UN sinterability and characteristics of the dense fuel. It was found that, while the oxygen impurities did not have an effect on the sinterability, the average grain size decreased with increased oxide content, as shown in Fig. 2.3. This is due to the formation of oxide precipitates that pin the grain boundaries limiting grain growth.

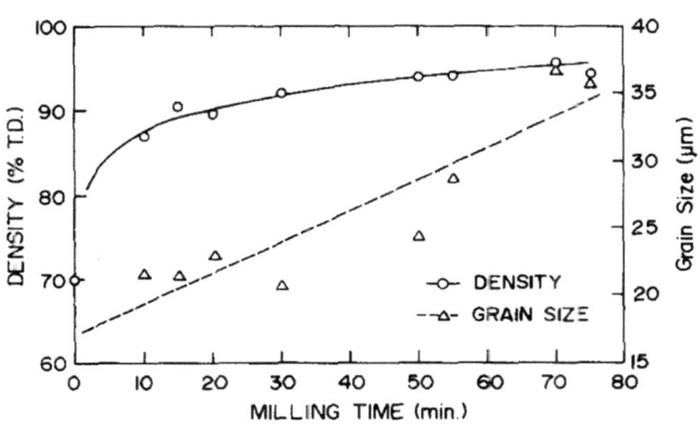

Figure 2.2 Effect of milling time on density and grain size of UN as reported by Matthews et al. [3].

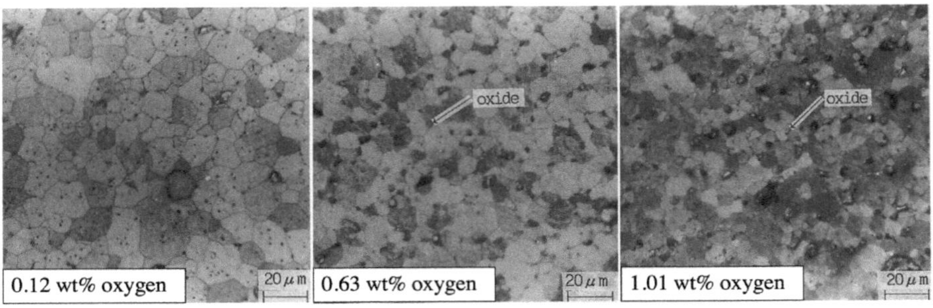

Figure 2.3 Microstructures of UN pellets as a function of oxygen impurities. Adapted from Arai et al. [10].

Arai et al. [10] achieved densities higher than 90% TD at temperatures of 1750°C in an Ar-8%H_2 atmosphere. Since conventional sintering has been shown to require very high temperatures to produce high-density UN pellets, techniques that implement pressure (e.g., hot isostatic pressing (HIP) and spark plasma sintering (SPS) [11,12]) have been utilized for UN pellet fabrication. As with the synthesis infrastructure, inert atmospheres are also required for the handling and fabrication of dense UN pellets.

Irradiations on UN fuel have been performed at thermal reactors with a variety of parameters including various cladding alloys (i.e., tungsten, molybdenum, niobium etc.), temperatures over the 1350–2470°C range, burnups up to 4.5 at.%, fuel densities from 80%–96% TD, and various grain sizes and stoichiometries [3,13,14]. Unfortunately, the data from these irradiations produced substantially varied results for similar conditions. Nonetheless, the following trends were established: fission gas release rates were low, especially for large grained and higher density UN; high-density UN showed no restructuring or cracking; fuel swelling was relatively low for high-density pellets; and niobium cladding alloys (with a tungsten liner) were deemed most viable for operation below 1127°C for minimal fuel/cladding chemical interactions. Overall, density and stoichiometry were established as the parameters that had the most impact on fuel swelling. The role of impurities such as oxygen and carbon were also thought to be key. The results from these irradiations informed a UN specification for the SP-100 program with densities of 96 %TD, grain sizes higher than 30 μm, a N/U ratio of 1.00–1.05, and impurities of less than 3000 and 1000 ppm for carbon and oxygen, respectively [14].

The SP-100 program enabled significant advances in fundamental research of the synthesis, fabrication and irradiation behavior of UN and cemented it as an attractive option for reactors for space propulsion, lead fast reactors, microreactor applications, and light water reactors [3,13]. Despite its attractive attributes, as a fine powder UN is pyrophoric requiring great care when handling, which must take place under inert atmospheres. UN also has poor characteristics against water and steam environments [15], which pose challenges when considering its use in LWRs. Mitigation of these issues using additives such as UO_2, to increase the high-temperature oxidation and corrosion resistance of UN have been proposed but will be discussed in a later section (see Section 4). Finally, the replacement of the naturally occurring isotope ^{14}N for ^{15}N is required to enhance neutron economics and to reduce the unwanted production of long-lived ^{14}C. Due to the impact on the synthesis and fabrication infrastructure, such a modification will have a large impact on the economics of commercial UN and has to be seriously considered.

1.2 Uranium carbide

Uranium carbide, similarly to uranium nitride, has favorable high-temperature characteristics, such as a high thermal conductivity, high melting point (2526°C), and high uranium density. These properties can enable a lightweight reactor core with effective high-temperature behavior. Therefore, UC, as well as (U,Pu)C, have been considered for space applications [16], fast and breeder reactors [17–19], and a generation IV supercritical water-cooled concept reactor [20]. Mixed carbide fuels such as ZrC, NbC, TaC, and HfC mixed with UC had also been considered by the NERVA/ROVER program in the 1960s [21]. However, implementation of carbide or mixed carbide fuel was ultimately hindered due to difficulties associated with the chemistry control of the fuel, interactions of carbon with other nuclear materials in the core, dissatisfactory fuel swelling and fission gas release characteristics, and the high pyrophoricity of the fuel. In recent years, there has been a renewed interest in UC, primarily driven by generation IV design plans. As such, comprehensive reviews on the utilization of carbide fuels [22] and their thermophysical properties [23] have been published.

Carbothermic reduction (CTR) is one of the more established synthesis routes for UC. CTR uses UO_2 powder as the precursor combined with graphite powder to reduce the compact to UC. This process requires inert atmosphere handling, as well as high temperatures and long times to drive the reaction to completion. Studies on small scale samples have shown that temperatures of 1600°C are required to complete the reaction and synthesize impurity-free UC within minutes [24,25]. However, experimental studies targeted at higher yields often require several hours to synthesize UC with minimum impurities. The temperatures and atmospheres selected during CTR are parameters that have to be very well controlled as they define the final O and N impurity contents, which have been shown to significantly affect the properties of the fuel [26,27]. Recently, an innovative CTR route utilizing SPS was presented by Salvato et al. [28], who synthesized phase pure UC at 1200°C under a 5 min vacuum hold, a significant temperature and time reduction compared with traditional CTR methods. Other methods of fabrication include arc melting of metal uranium with carbon [29,30], direct reaction sintering of uranium hydride with graphite [31], and reaction of uranium metal with CH_4 [30]. These methods are often difficult to control and can result in energetic exothermic reactions making it challenging to scale them up to produce higher yields.

Fabrication of high-density UC pellets is also not without challenges due to the high temperatures and long sintering times required. Additionally, all processing has to be performed under an inert atmosphere, which imposes further restrictions. UC with high and low oxygen impurities was fabricated by Gorlé et al. [32] to establish their effect during irradiation. It was found that a ball milling step of 8 h was necessary to obtain reproducible results and that the sintering temperature was the most influential factor for achieving high densities, as shown in Fig. 2.4. On the other hand, longer sintering times did not translate to significant increases. Gorlé et al. [32] were able to achieve densities of 94% TD at 1850°C after 3 h while also achieving low oxygen impurities (300 ppm). Small additions of nickel (0.1% Ni) were also tested as a sintering aid and it was found that a density of 94% TD could be achieved with a reduced sintering temperature of 1600°C after 3 h. However, due to the lower sintering temperature, the oxygen impurities were higher in these samples ranging from 500 to 600 ppm. UC has also been recently sintered through SPS by Salvato et al. [33] at 1700°C with a 10 min dwell at a pressure of 100 MPa under vacuum to obtain a pellet with 95% TD. The study also showed that UC powders that had been synthesized using an SPS modified CTR method at a lower temperature (1200°C) showed increased sinterability and produced higher-density pellets.

Despite the difficulties in synthesis and fabrication of high-density UC or mixed carbide fuels, which make it a commercially unattractive choice, the high fissile density and high thermal conductivity can enable lightweight cores and ultrahigh-temperature performance that would be hard to achieve with other fuel forms. These attributes are particularly important for space applications. What is more, UC has the distinct advantage of not requiring isotope enrichment, as in the case of UN. Irradiation studies have confirmed that the swelling of UC is higher than that of oxide fuel; however, carburization of the cladding, which had been a significant historic concern, has not been attributed as the failure mechanism in any of the studies [34–36]. Continued understanding of the in-pile performance of UC and mixed carbides is an important step for the development of this fuel form.

1.3 Uranium boride

Compounds of the U–B system include uranium diboride, UB_2, uranium tetraboride, UB_4 and uranium dodecaboride, UB_{12}, all of which are line compounds [37]. Limited

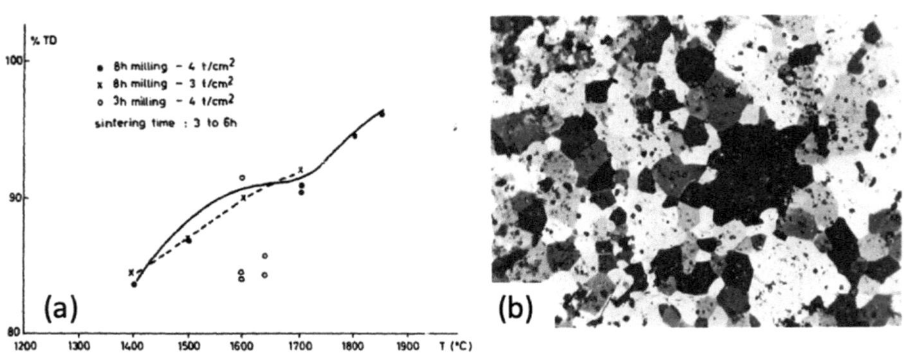

Figure 2.4 (a) Influence of sintering temperature and milling time on the densities of UC and (b) 94% TD UC fabricated with Ni as a sintering aid at 1600°C for 3 h under vacuum. Adapted from Gorlé et al. [32].

research has been performed on these fuel forms primarily due to the large thermal neutron cross-section of ^{10}B (3800 b) [38]. However, enrichment to ^{11}B, an isotope with a significantly lower cross-section (0.005 b), is possible [39,40]. Tailoring the isotopic ratio in uranium borides could then be achieved, leading to a fuel with integrated burnable poison and fissile nature. What is more, uranium borides have high melting points [41], high thermal conductivities due to their metal-like bonding and, in the case of UB_2, higher uranium density that UO_2. They could, therefore, be considered for LWRs provided that their properties would result in substantial operational and economic advantages.

Traditional synthesis of uranium boride compounds has been done through arc melting of metal uranium and metal boron [42–46]. Arc melting of these compounds requires inert atmospheres, due to the volatile nature of boron and high temperatures. Despite arc melting being an established technique, typically only small batches of material can be synthesized rendering it unattractive when it comes to scale-up and commercialization. Recently, a carbo/borothermic reduction synthesis method for UB_2 was carried out by combining UO_2 with B_4C and graphite in low partial pressures of CO with temperatures of up to 1800°C [47]. A novel synthesis technique to form UB_4 from UCl_4 and MgB_2 has also been demonstrated by Lupinetti et al. [48] under vacuum at 850°C. Experiments to determine the effect of time and temperature on the process, revealed that the lowest temperature required was 600°C and the shortest time required to form UB_4 was 1 day. A borothermic reduction of uranium dioxide at 1550°C, followed by sintering of the powders into rods and induction zone melting was also previously used for the synthesis of UB_{12} by Troc et al. [49].

Due to their high melting points, uranium borides require high temperatures and long sintering times for fabrication of high-density pellets. Difficulties in their manufacturing has prevented, in many cases, extensive study of their thermophysical properties. UB_2 has traditionally received more attention, due to its high uranium density, and so some studies assessing its specific heat and thermodynamic properties exist [44,50,51] and have been reported in a review by Chevalier et al. [52]. Matterson and Jones [53] used hot pressing to achieve high densities in UB_2 and UB_4 samples, however, they reported partial decomposition of UB_2 upon contact with the graphite die and identified the reaction byproducts to be UB_4, UO_2 and UBC. They were successful in sintering UB_4 at 2000°C and 28 MPa, which resulted in samples with <1% porosity. Conventional sintering of UB_2 has also been conducted by White [45]; however, high temperatures and long sintering times, in excess of 3 h, were required to reach densities above 90% TD. High-temperature data on the thermal conductivity of UB_2 have been recently published from density functional theory (DFT) simulations up to 1500°C from Jossou et al. [54] and up to 1000°C from Evitts et al. [55]. In both cases, the reported values were significantly higher than those of UO_2, identifying UB_2 as an important high thermal conductivity compound.

Recently, SPS was used to fabricate UB_2 and UB_4 [46], which resulted in pellets with higher than 90% TD within minutes at temperatures as low as 1750°C. During that study it was established that aging of the feedstock resulted limited densification in a subset of the samples. This was also observed through conventional sintering of these materials [45]. Achieving high-density UB_2 and UB_4 pellets enabled high-temperature thermal diffusivity and thermal expansion measurements, as well as the experimental determination of the thermal conductivity of these materials up to 1500°C for the first time, shown in Fig. 2.5 [46]. Establishing experimental values for the high-temperature thermal conductivity of UB_2 and UB_4 was an important milestone for these fuels as it was shown that it is significantly higher than that of UO_2. The thermal conductivity increases linearly with temperature for UB_4. The conductivity of UB_2 decreases from room temperature to 600°C after which point it starts to increase. This is an important characteristic of these systems as it indicates optimized thermal conductivity at the high temperatures relevant to accidents.

Further research into uranium borides is necessary to assess their full potential for deployment in nuclear reactors. In particular, very little is known about their resistance to oxidation and steam attack, as well as their behavior under irradiation. A recent study by Burr et al. [57] used DFT to calculate defect energies and migration energies for UB_2 and identified anisotropy in the migration of boron and uranium vacancies and boron interstitials, indicating that UB_2 might possess lower radiation tolerance than UO_2. Nonetheless, experimental corroboration of these results is required to further assess the irradiation performance of UB_2. No known studies have assessed the irradiation performance of UB_4 or UB_{12}.

2. Composite fuels

Aside from monolithic fuel concepts, such as doped UO_2, UN, and UC, composites provide an alternative approach, whereby different uranium compounds can be selected and combined based on attractive properties, such as high thermal

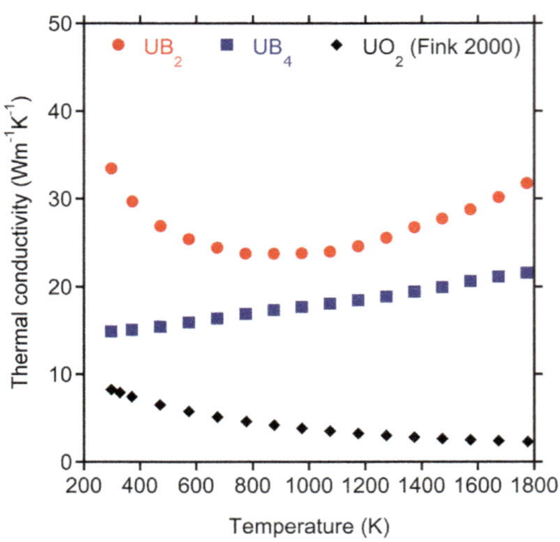

Figure 2.5 Thermal conductivity of UB_2 and UB_4 compared with that of UO_2 [56], from Ref. [46].

conductivity, resistance to oxidation and steam attack, and high radiation tolerance. As such, a composite fuel form with enhanced performance compared with monolithic fuels can be engineered.

A typical approach is to consider a UO_2-based composite with another fissile phase that could potentially improve the thermal conductivity of the fuel. The advantages of this approach are twofold; using UO_2 as the primary constituent phase allows years of experience in its fabrication and thermophysical properties to be leveraged, at the same time a uranium-based compound with improved properties compared with UO_2 can serve as the balance constituent phase and, therefore, enhance the overall properties of the fuel while maintaining or, in some cases, increasing the fissile fuel density.

Alternatively, a composite of UO_2 with a nonfissile phase can be developed, where the nonfissile phase can have benefits, such as high thermal conductivity, high melting point, and good chemical stability. However, the displacement of fissile material for a nonfissile one, despite an improvement in thermophysical properties, can be a major drawback since it means that increased uranium enrichment might be required to reach the same burnup.

2.1 Composites with fissile secondary phases

Due to the many advantages that UN could provide in a composite system, as outlined in Section 3.1, UO_2-UN has been considered as an advanced fuel form. Of course, one of the major drawbacks of UN is the fact that it readily reacts with water and steam [15], a considerable safety concern when it comes to its use for LWRs. To mitigate this issue, encapsulation of the UN phase in UO_2 has been proposed. Yang et al. [58] studied the UO_2-UN system with a particular emphasis on selecting the appropriate phase fraction mixtures to produce pellets that can be fabricated to high densities and exhibit significant increases in thermal conductivity compared with monolithic UO_2. They used hot pressing to fabricate almost 100% dense UO_2-UN pellets with UN weight fraction additions of 30, 50 and 70 wt.%. One of the main findings was that, during sintering performed at temperatures of 1300, 1500 and 1590°C under vacuum for 4 h, the UN fraction reduced and a small amount of U_2N_3 formed. The reported chemical reaction highlighted that larger UN weight fractions are needed to offset the expected phase fraction reduction (from UN to U_2N_3) and achieve a considerable increase in thermal conductivity. Nonetheless, Yang et al. [58] measured a 100% increase with respect to standard UO_2, at 800°C, in the thermal conductivity for a UO_2-UN composite sample with ~40 wt.% UN after sintering at 1590°C for 4 h. Combining the substantial increases in thermal conductivity with the increase in fissile density, UO_2-UN composites seem like an attractive fuel concept. The microstructures from the sintered pellets of this study, shown in Fig. 2.6, showed UN particles well-surrounded by the UO_2 matrix. This microstructure would theoretically provide tolerance to oxidation and steam corrosion; however, further studies of high-temperature oxidation and corrosion are needed to fully support this claim.

Jaques et al. [8], in an effort to overcome the drawbacks of UN with regards to water and steam corrosion, also studied the UO_2-UN system. In this study, UN was the

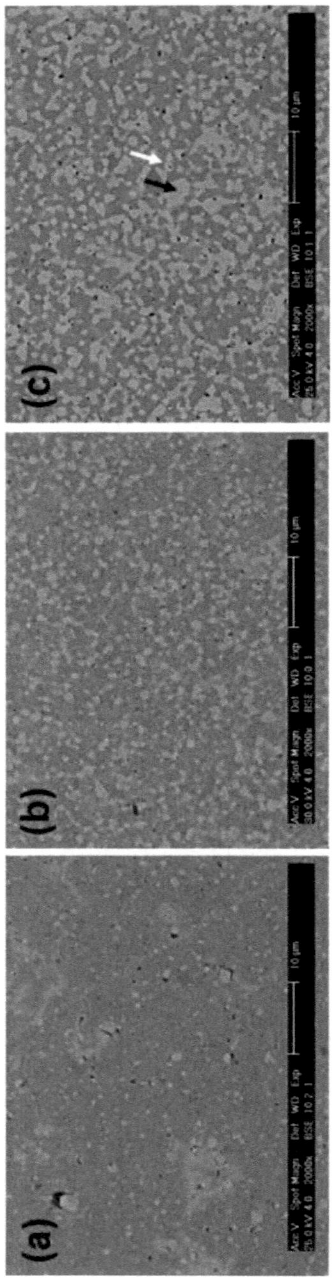

Figure 2.6 Back scattered SEM images of sintered UO_2-based composites with (a) 30, (b) 50, and (c) 70 wt% UN (aimed starting composition) obtained by hot pressing powder mixtures at 1590°C for 4 h.

primary constituent phase and additions of 5 and 10 wt% UO_2 were included to increase oxidation resistance. These composites were sintered at 1900°C for 5 h in UHP-Ar, Ar-100 ppm N_2, and Ar-1at%N_2. Different sintering temperatures, from 1700 to 2000°C were also examined under the Ar-100 ppm N_2 atmosphere. It was found that at higher temperatures, lower densities were recorded, and this was attributed to a preferential formation of oxygen-stabilized nitrides or uranium oxynitrides. The findings of this study further indicate the complex reactions that can be facilitated by high-temperature sintering of UO_2 and UN. These reactions should be studied extensively before a UO_2-UN composite fuel concept can be realized. What is more, very few studies have examined the efficiency of oxide additions, such as UO_2, when it comes to providing oxidation resistance for the fuel.

Recently a study by Costa et al. [59] suggested a variation of a UO_2-UN composite fuel as an ATF, with UN microspheres embedded in a UO_2 matrix (see Fig. 2.7). This approach has the potential to completely shield UN phase from oxidation using the UO_2 matrix and, therefore, substantially increase the safety margins when it comes to an oxidation accident scenario. The work by Costa et al. [59] also provided a comprehensive explanation for the formation of the U_2N_3 phase that is often encountered in these composite fuels. Formation of U_2N_3 during fuel fabrication is undesirable as it can dissociate to form UN and N_2 during high-temperature reactor operation. The N_2 gas can increase the pressure in the pin plenum and risk bursting of the cladding as well as compromising the structural integrity of the fuel pellets due to cracks forming as a result of that reaction. It follows that an in-depth understanding of what prompts the formation of this phase, and therefore how to avoid it, is important when considering UO_2-UN fuel concepts. To probe this, additions of 10, 30 and 50 wt% UN, in the form of microspheres, were introduced to UO_2 and sintered under SPS at various conditions of temperature, pressure, and cooling rate in Costa et al. [59]. It was found that the aforementioned conditions impact the amount, size, and

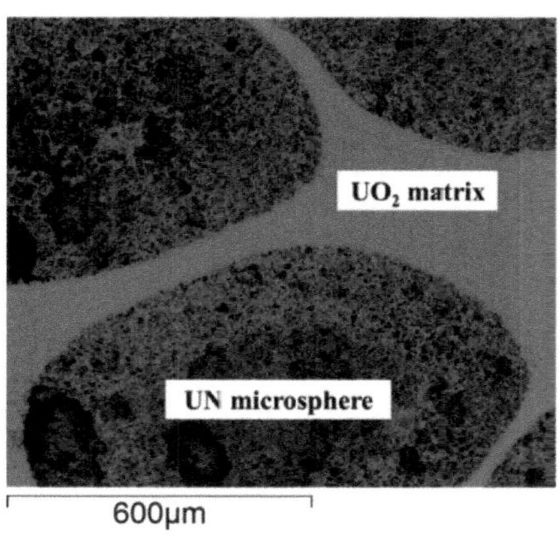

Figure 2.7 SEM-BSE image of a typical microstructure from the UO_2-UN composites from Ref. [59].

morphology of the U_2N_3 phase that formed in all samples. In particular, increases in temperature and pressure, as well as slower cooling rates, contributed to increased formation of U_2N_3. In all cases, changes in the studied SPS parameters resulted in changes to the morphology of the U_2N_3 phase. The proposed mechanism for the formation of U_2N_3 was explained by interdiffusion of O and N into the UN and UO_2 respectively to form a $UO_{2-x}N_x$ phase during sintering. On cooldown the solubility of N_2 in this $UO_{2-x}N_x$ phase decreases, resulting in redistribution and formation of U_2N_3. It should be noted that the UO_2 selected for the study in Ref. [59] was hyper-stoichiometric, which could have affected the kinetics of the proposed mechanism. By optimizing the SPS parameters, 1500°C for 3 min at 80 MPa, UO_2-UN samples of 96% TD were sintered.

U_3Si_2 has also been proposed as a constituent phase for an advanced accident tolerant fuel form. Similarly to UN, U_3Si_2 also has attractive properties, such as high thermal conductivity and high fissile density [60], although a notable disadvantage is the high susceptibility to steam. Previous studies have shown that U_3Si_2 will degrade structurally at temperatures as low as 350°C in contact with steam [61]. It is, therefore, important when considering U_3Si_2 as a constituent phase for LWR composite fuel, that another phase is also incorporated to increase the corrosion resistance of the pellet. An obvious advanced fuel concept that incorporates a silicide phase is a UO2-U_3Si_2 composite, however, there are surprisingly few studies that have tackled such a composite. In fact, the sole study was recently published by Gong et al. [62] who used SPS to sinter pellets with various weight ratios of U_3Si_2 and UO_2. Gong et al. [62] sintered pellets of 95, 75, 50, and 25 wt% U_3Si_2 (Fig. 2.8), the remainder being UO_2, to high densities at various temperatures ranging from 1000 to 1300°C under 2 atmospheres: vacuum and argon. The key findings were that the sintering temperature influenced the final densities, with higher temperatures resulting to higher densities, and that the overall oxidation resistance increased with larger additions of UO_2, although the oxidation onset temperature was reduced. Samples that were sintered at higher temperatures, and, therefore, had high densities, were also more resistant to oxidation as it would be expected. Finally, thermal diffusivity measurements were completed on these samples and recorded increases in diffusivity compared with monolithic UO_2 as the U_3Si_2 phase content increased.

Despite the clear benefits that UO_2 could provide in terms of oxidation resistance, the majority of the studies regarding composites of U_3Si_2 center around UN-U_3Si_2 fuel concepts. This stems primarily from two reasons. Firstly, the interest around UN due to its high fissile density and high thermal conductivity were offset by its poor corrosion resistant, whereas U_3Si_2 was originally thought to be oxidation and corrosion resistant compared with UN. In fact, early studies comparing the degradation of monolithic UN and UN-U_3Si_2 pellets in steam environment showed that indeed the addition of the silicide phase improved the oxidation resistance of the fuel [15]. However, later studies reported that U_3Si_2 has a high affinity for oxygen and steam, and was shown to readily pulverize under steam at temperatures relevant to reactor operation, a major drawback for use in LWRs. The second reason for the conceptualization of a UN-U_3Si_2 composite was that the lower melting point of U_3Si_2 (compared with both UN and UO_2) would facilitate sintering of a UN composite. In the past, UN

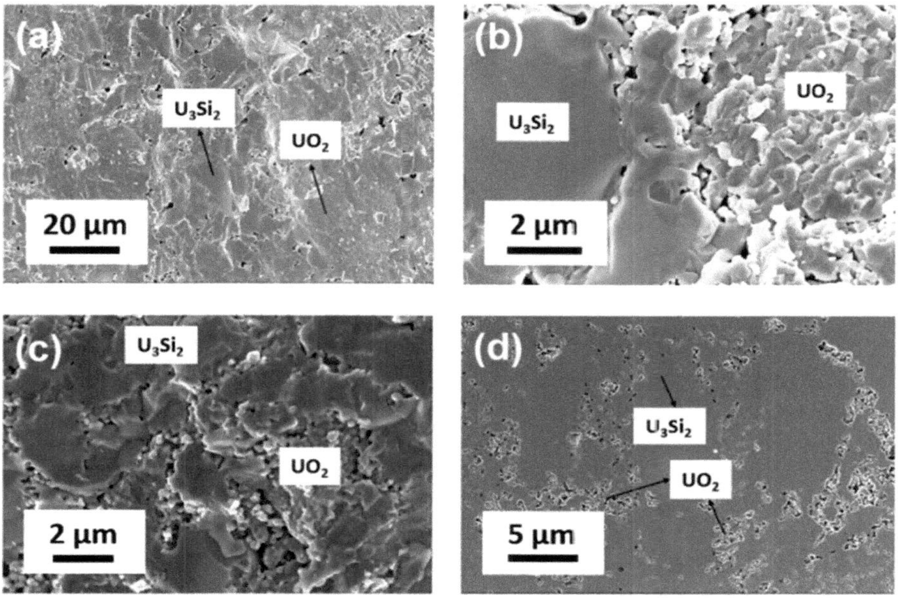

Figure 2.8 SEM images of SPS sintered U_3Si_2 composites with (a) 5 wt%, (b) 25 wt%, (c) 50 wt%, and (d) 75wt% UO_2 additions [62].

was shown to require high sintering temperatures and long sintering times to achieve high densities. The addition of U_3Si_2 as a secondary phase ought, therefore, to reduce the sintering times and temperatures and allow for more economical sintering conditions. Johnson et al. [63] first fabricated a UN-U_3Si_2 composite through SPS and reported the formation of an unknown ternary phase. Following this work, White et al. [64] also investigated the effect of sintering conditions for the fabrication of UN-U_3Si_2 composites and also reported the formation of a U-Si-N ternary phase. White et al. [64] measured the thermal diffusivity and thermal expansion of the composites and calculated their thermal conductivity, which increased as the UN phase fraction increased. Differences in the thermal expansion coefficients of the UN and U_3Si_2 phases led to microcracking upon cooling of the composites and resulted in hysteresis in the thermal diffusivity data. Other composite fuel concepts that incorporate U_3Si_2 include those with additions of UB_2 to act as a burnable absorber [65] and were proposed to increase the overall steam performance of the fuel although limited data were presented to support this claim [66].

Uranium boride phases, such as UB_2 and UB_4, have received attention in recent years primarily due to their high thermal conductivity [46,54,55] and high fissile density, in the case of UB_2. Composites of UO_2-UB_2 and UO_2-UB_4 are therefore of interest as they can enhance performance with respect to UO_2, based on the aforementioned benefits, and can enable incorporation of a burnable absorber phase. A recent study by Kardoulaki et al. [67] used SPS to fabricate high-density UO_2-based composites with additions of 5, 15 and 30 wt% UB_2 and UB_4. The study revealed that above 527°C the boride phase can oxidize to form the higher boride phase and UO_2, resulting in a redistribution of the initial phase fraction within the fuel, which in turn caused lower than expected thermal diffusivity. Fig. 2.9 shows a 30 wt% UB_2-UO_2 composite before and after the oxidizing reaction, whereby the UB_2 phase has reacted to form a mixed UO_2-UB_4 feature. This finding showcases that strict atmosphere control should be exercised for these composites at elevated temperatures in order to limit such reactions. Nonetheless, it was shown that the UO_2-UB_2 and UO_2-UB_4 composites both provided increased thermal diffusivity compared with standard UO_2.

2.2 Composites with nonfissile secondary phases

Non fissile additions in UO_2 fuel such as SiC, carbon nanotubes, diamond, BeO, and ZrB_2 have also been explored to provide benefits such as increased thermal conductivity [68–75], incorporation of a burnable absorber and increased oxidation and corrosion resistance [76]. However, often the gains made through these additions are offset by the fact that fissile material is displaced for nonfissile material, thus, increasing the need for higher enrichment to produce the required burnup. In fact, in order to significantly enhance the thermal conductivity of the fuel, large phase fractions of the nonfissile, high thermal conductivity phase are required, often higher than 10 wt% as shown by Yeo et al. [71] in the case of UO_2-SiC composites (Fig. 2.10), which makes these composite fuels unattractive from an economic viewpoint.

3. Manufacturing techniques of fuel pellets

The traditional route to fuel fabrication is through conventional powder metallurgy and sintering at high temperatures for long times. Typical sintering temperatures can vary from 1700°C for UO_2 to temperatures as high as ~2000°C for UN [9,77,78] and sintering times on the order of 4–6 h are required to reach high densities

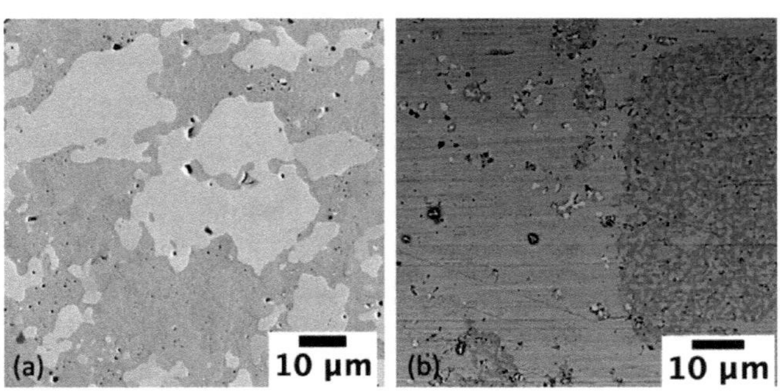

Figure 2.9 A UO_2-UB_2 composite with additions of 30 wt% UB_2 (a) before and (b) after oxidation reaction.

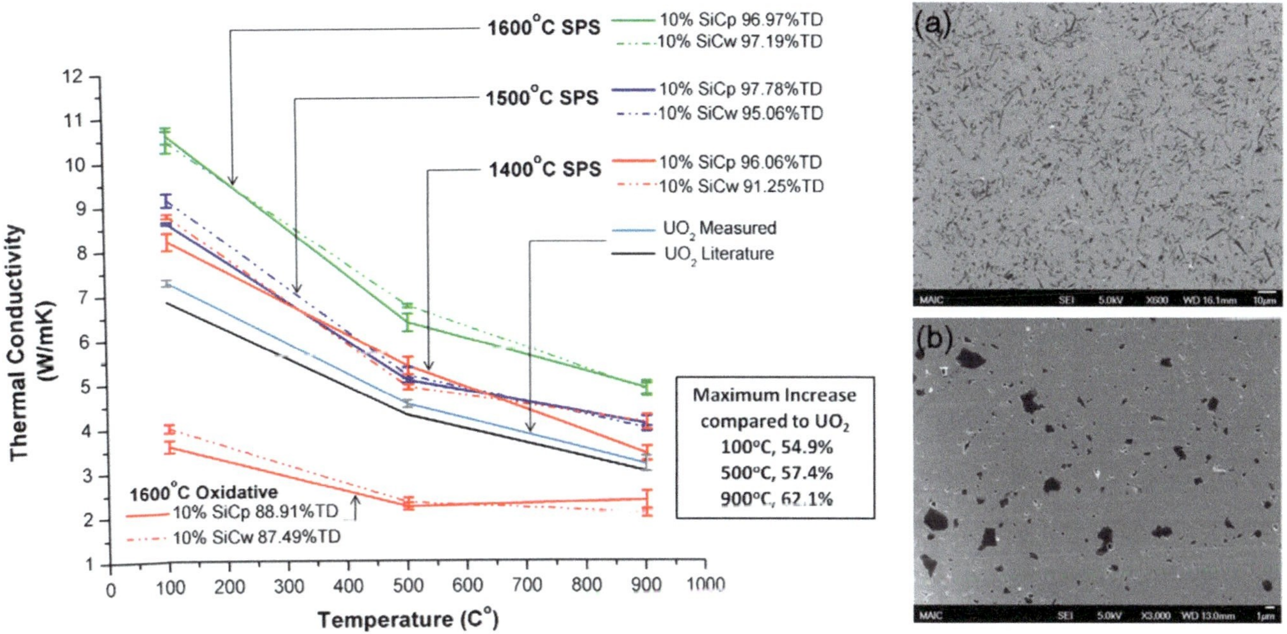

Figure 2.10 Thermal conductivity of UO_2-10 wt% SiC composites sintered through SPS and oxidative sintering at 1600°C along with the micrographs of composites sintered through SPS with SiC (a) whiskers and (b) powder.
Adapted from Ref. [71].

(>95% TD). Conventional sintering for UO_2 is very well established from years of industrial experience and high densities can be readily achieved, however, the high temperatures and times required pose an economic drawback. Various studies have researched the possibility to fabricate dense UO_2 pellets at lower temperatures [79–81] but long sintering times are still required. Field assisted sintering techniques (FAST) such as microwave sintering, SPS, and flash sintering (FS) are relatively new fabrication techniques that utilize internally generated heat through microwaves or through electric fields and currents to densify pellets at lower temperatures and times, thus, significantly reducing the required energy and associated costs. The scale of the potential energy savings is demonstrated in the schematic of Fig. 2.11 below for SPS and FS versus traditional methods. It is apparent by this comparison that FS has by far the highest potential to minimize energy expenditure due to the significantly reduced sintering times and temperatures, even compared with SPS.

3.1 Microwave sintering

Microwave sintering is a technique that has been available since the 1980s [82] but has not received significant attention when it comes to nuclear fuels. Still, SIEMENS has developed and patented a commercial microwave furnace set up for sintering UO_2 [83]. During microwave sintering, the pellet is heated directly by absorbing the applied microwave energy, thus, lowering the overall time required to sinter the material. Microwave sintering has been shown to densify UO_2 pellets to approximately 95%TD at a frequency of 2.45 GHz, powers of 1–1.6 kW, temperatures of 1500–1700°C and dwells of 1–4 h [82,84,85]. However, one of the primary issues with this technique is cracking of the UO_2 pellets, due to the fast heating rates [84,85]. To overcome this issue, Subramarian et al. [85] used UO_2 powder to blanket

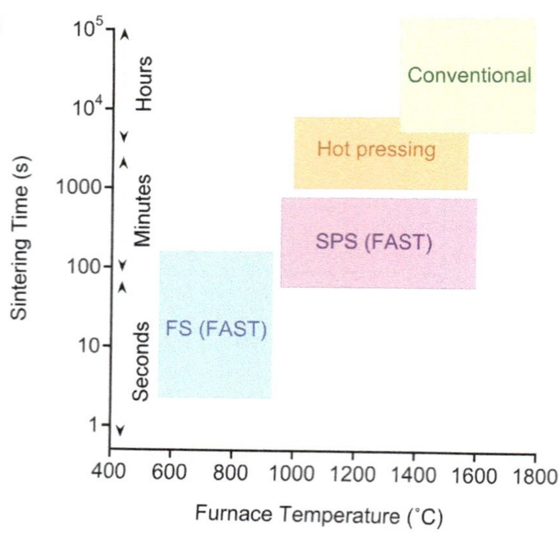

Figure 2.11 Schematic of sintering times and furnace temperatures required for FAST techniques FS and SPS versus traditional methods such as hot pressing and conventional sintering.

the UO_2 pellets during sintering. This resulted in high-density pellets with minimal cracking due to the insulation that the UO_2 powder provided. The resulting microstructures were comparable with those obtained from conventional sintering as show in Fig. 2.12. Despite the limited attention microwave sintering has received for the fabrication of nuclear fuels, it could be a promising technique considering that scale up of appropriately designed microwave furnaces could enable sintering of kilograms of material at a time, requiring significantly less energy and time than conventional sintering does. The scale up of the technique and the feasibility of sintering multiple pellets at a time with the same specification remains to be proven.

3.2 SPS

SPS is another FAST method that has been around since the 1950s [86] but has only recently been applied to the fabrication of nuclear fuels. SPS combines the use of electric fields and currents with uniaxial pressure to sinter materials to high densities at lower temperatures and shorter times, compared with conventional sintering. Due to its significant economic advantages and the fact that the addition of pressure is advantageous for sintering and tailoring the microstructure of many materials [87–90], SPS has been established as a successful tool for nuclear fuel fabrication in recent years [91,92]. In SPS, powder is loaded onto a graphite die, which is heated by a pulsed direct current (DC) while simultaneously applying uniaxial pressure. The typical applied voltages are low (<20 V/cm) and the applied currents flowing through the sample and the die are high (>1000 A). Since the die is also heated through Joule heating, due to the current flowing through the graphite, the sample is heated through both convection and internal heating generated from the flowing current. While this complicates the understanding of the sintering mechanisms that are present during SPS, it provides a relatively fast sintering method that can densify high-temperature nuclear ceramics to high densities effectively. SPS has been used to

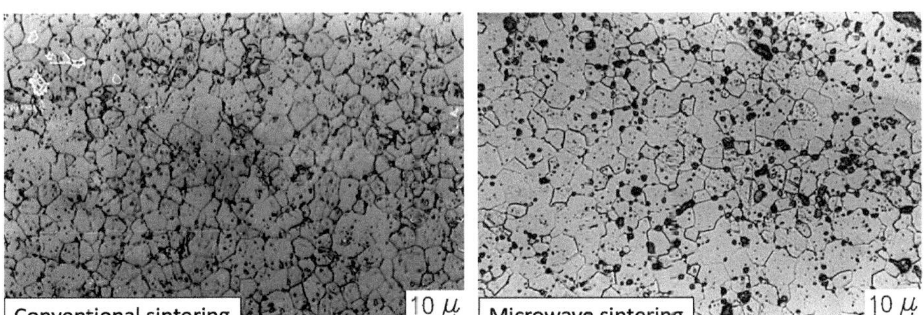

Figure 2.12 Comparison of UO_2 microstructures resulting from conventional sintering (left) and microwave sintering (right) when blanketing the UO_2 pellet with UO_2 powder.
Adapted from Subramarian et al. [85].

densify UN [11,12,93], UC [33], U_3Si_2 [94], UB_2 and UB_4 [46], composite systems such as UN-U_3Si_2 [63], as well as many UO_2-based composites [62,67,68,71,76,95] to high densities. For example, UN, which is notoriously difficult to sinter conventionally and requires temperatures as high as 2000°C for 4 h or longer, has been densified to above 90% TD at 1500°C using pressures of 100 MPa within 10 min with SPS [11].

A potential drawback of SPS is the use of the graphite die since at high temperatures, carbon can diffuse into the fuel. This is particularly important for UN for example where the C impurities need to be precisely controlled as they can affect in-pile performance. Few studies have assessed the diffusion of C into sintered SPS pellets with Mackie et al. [96] having investigate C uptake into Sm(Co, Fe, Cu, Zr) z and Morita et al. [97] having investigated uptake in $MgAl_2O_4$, a-Al_2O_3, and 3Y-TZP. The study by Mackie et al. [96] only showed carbon uptake on the surface of their sample, within 10 μm, however, a relatively low sintering temperature was tested (1100°C) and only for 5 min. For SPS of nuclear fuels, temperatures as high as 1600°C are more typical and these temperatures could certainly enhance carbon diffusion. Morita el al. [97] who performed a more comprehensive study in three different materials found that the heating rate, sintering temperature, and material in question, greatly influence the resulting carbon uptake. For example, they found that high heating rates (i.e., 100°C/min) would result in carbon uptake throughout the whole area of the $MgAl_2O_4$ but would only result in surface contamination for alumina and 3Y-TZP. Further work remains to be carried out to assess the true effect of carbon uptake for fuels such as UN and UC at the appropriate sintering temperatures, nonetheless, the use of a protective foil such as W or Mo foil could provide some protection against this issue.

Despite SPS having a lot of advantages and an established track record for nuclear fuel fabrication, it is still a technique that cannot be readily scaled up. In traditional SPS, only one die is used at a time, making the process very cumbersome for industrial scale fabrication. Ironman et al. [98] studied the feasibility of SPS scale up for LWR sized pellets and reported that while the pellets were reproducible and within specification when fabricated one at a time, efforts to fabricate four or eight pellets in a multiple-pellet die, shown in Fig. 2.13, resulted in pellets with varied densities and microstructural properties. Ultimately SPS is a powerful tool for timely fabrication of advanced ceramic fuels, however, it is not envisioned that it can be easily adapted for commercial-scale fabrication.

3.3 FS

FS is another technique similar to SPS that uses electric voltage and current to sinter green density compacts to high densities. FS was first observed by Cologna et al. [99] who were able to sinter nanograin zirconia at 850°C in less than 5 s, using an electric field of 120 V/cm. Since this breakthrough study, FS has been applied to a plethora of materials including oxides such as 8YSZ [100], cobalt manganese oxide [101], titanium oxide [102,103], tin oxide [104], strontium titanate [105], borides such as ZrB_2 [106], and carbides such as SiC [107]. Typical FS conditions require a higher

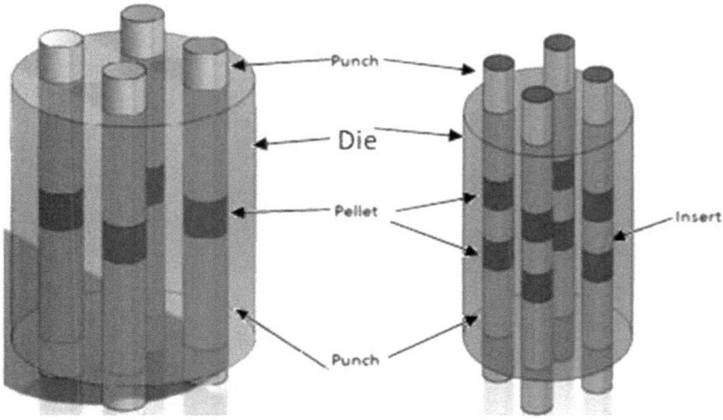

Figure 2.13 Schematic of the die geometry to fabricate four pellets and eight pellets [98].

electric field (>60 V/cm) and a lower current density (<1 A/mm^2) than those required for SPS to sinter a low-density powder compact within minutes. A traditional FS set up only requires a furnace and a power supply that can provide current and voltage to a powder compact in contact with the electrodes. Commonly dog-bone samples are used for FS, which can simplify the set up as the samples can be suspended using Pt wires, which also provide electrical connection [100,105,108–110]. Pellet-shaped samples, which are relevant for nuclear fuel applications, have also been used for FS [111–113] as well as rods [103,114].

Typically, in traditional FS a predefined field is applied to the sample while the power supply operates under voltage control. During this period, often called incubation period, only a very small amount of current is flowing through the sample. Due to Joule heating induced from this flowing current, the temperature of the sample raises, thus, causing its electrical conductivity to increase and permit more current to flow through. This leads to the flash event, which is characterized by a current runaway during which the current exponentially increases until reaching a predefined current limit. The bulk of sintering can occur within seconds under the correct combination of field, current and temperature. Aside from current runaway, events that characterize flash are also a drop in electrical resistivity, bright light emission, and thermal runaway of internally generated Joule heating occurring at low temperatures [115].

FS is a very recent fabrication technique and while it has been applied to a variety of ceramics, it has remained relatively unexplored as a fabrication technique for nuclear fuel. In fact, only a handful of FS studies have been reported in the literature for the fabrication of UO_2 [112–114,116]. Valdez et al. [112] was the first to report densification of UO_2 through FS within seconds at 600°C, a temperature where conventional sintering is not observed, as shown in Fig. 2.14a. While this pioneering study by Valdez et al. [112] highlighted the potential of FS for the fabrication of nuclear fuel, it did also report that sample cracking due to the extremely fast sintering rates the material experiences was unavoidable (Fig. 2.14b).

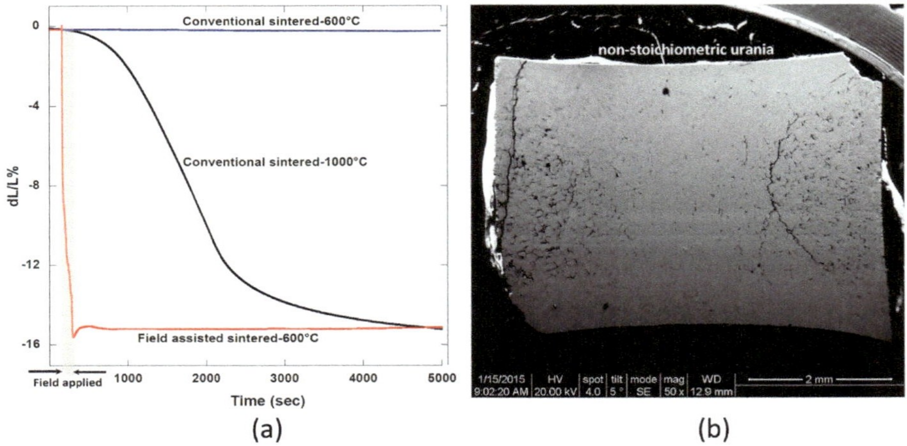

Figure 2.14 (a) Densification profiles versus time for UO_2 samples sintered conventionally at 600 and 1000°C and through FS at 600°C and (b) microstructure of a UO_2 sample sintered through FS at 600°C.
Adapted from Ref. [112].

Recently, a more controlled way to perform FS has been established and has showed promise for materials with low thermal conductivity such as UO_2 [117]. This FS modification from the traditional way is often called controlled current ramp or current-rate FS and it has already been applied to gadolinium doped ceria, zinc oxide and 3YSZ [117–119]. In this variation, the current applied to the sample is increased with respect to time according to a user defined rate and end value while the electric field is applied, usually up to the maximum voltage of the power supply. This results in the reduction or elimination of the power spike seen in traditional FS and consequently in a reduction of the thermal shock the material is subjected to. Controlling the current rate with this method can then allow for controlled sintering rates. This method was recently used to sinter UO_2 samples and it was observed that lower current rates promoted higher densification and eliminated cracking within the pellet [113]. In contrast, the application of higher current rates was found to have a detrimental effect, leading to lower densities. The study also compared the obtained densities from FS and those achieved from conventional sintering and found that moderate current rate FS still produced higher densities than conventional sintering in a fraction of the time, as shown in Fig. 2.15.

Understanding the phenomena present in FS and how they contribute to enhance the sintering rates in such a dramatic way is still a young research topic. In general, the faster sintering rates in FS have been attributed to Joule heating and fast heating rates [120,121]. However, multiple electric field effects may also be taking place, such as grain boundary space charge interactions [122,123], and electric current localization at the particle-particle contact points [115,124,125]. In addition, the nucleation of Frenkel pairs beyond the thermal equilibrium has been suggested as a possible

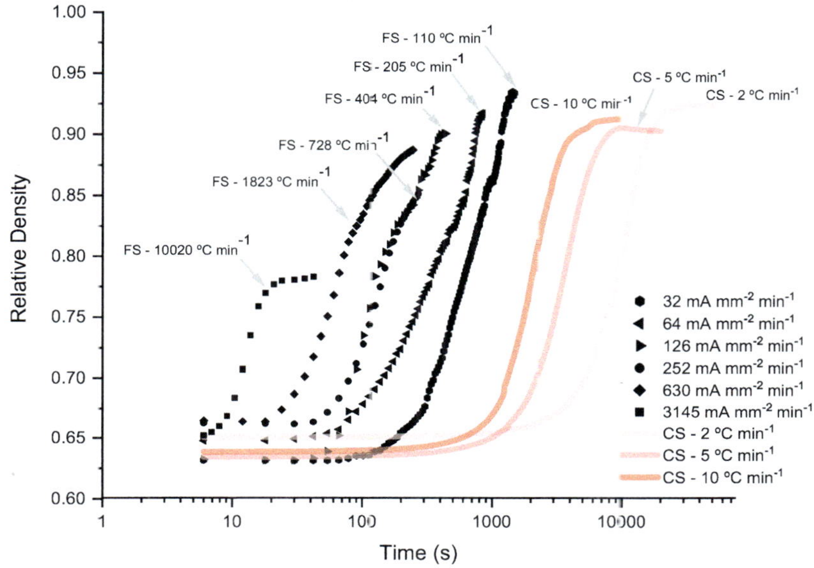

Figure 2.15 Relative density evolution between controlled current rate FS using a variety of current rates and conventional sintering at 2, 5, and 10°C/min.
Adapted from Ref. [113].

enabler of solid-state diffusion in FS [99,117,126]. Nonetheless, the fast heating rates employed in FS can provide multiple benefits including energy savings and the potential to retain nanograin microstructures as the technique lends itself inherently to limiting grain growth [127]. Scale up of FS is also anticipated to be less challenging than that for SPS as the use of a die and the application of pressure are not required.

4. Tristructural-isotropic (TRISO) particle fuel

Tristructural-isotropic particle fuel, known as TRISO, is playing increasingly important role in nuclear energy production worldwide. TRISO has excellent fission product retention capabilities and excellent high-temperature integrity, and it is considered to be "the safest fuel." This section will discuss the history of TRISO fuel and current large development and qualification programs, followed by an introduction to the TRISO structure and typical fabrication processes.

4.1 History and development of TRISO

Historically, tristructural-isotropic particle fuel was invented in the United Kingdom in the 1960s as part of the 20-MWth Dragon Research Reactor (Winfrith in Dorset,

England), which is a high-temperature helium-cooled prismatic-core reactor operated by the UKAEA (United Kingdom Atomic Energy Authority). The Dragon Reactor used coated particles comprise micro-size kernels of a fissile material (such as U-235) surrounded by various layers of ceramic outer coating (e.g., bilayer coating system named as "BISO"). Initially most of the fuel kernels was highly enriched U (e.g., >90% U-235) and later moved to lower enrichment fuel (~20% U-235). Coated particle fuels were further developed and optimized in the 1960–70s and used in other reactors including research reactors such as the pebble bed AVR (Arbeitsgemeinschaft Versuchsreaktor) reactor in Germany, HTTR (high-temperature test reactor) in Japan, HTR-10 in China and Peach Bottom Unit 1 in the USA. TRISO is also used in several commercial reactors including THTR-300 (thorium cycle high-temperature nuclear reactor with pebble bed design, Germany), Fort St. Vrain (prismatic core, USA) and HTR-PM (GenIV HTGR, China). There is a resurgence of TRISO particle fuel in recent years due to its widened applications in microreactors (HTRs or molten salt reactors), as loss of coolant accident tolerant fuel candidate for light water reactors (LWRs) and in space reactor power systems. A comparison between the bilayer coating from the early version of TRISO (named as BISO standing for bistructural isotropic), and modern TRISO particle fuel is compared side by side in Fig. 2.16.

Over the past few decades, TRISO fuel fabrication has been continuously optimized in terms of kernel compositions, particle designs, coating layers, particle compaction and heat treatment. Even since the 1960s, strict process control has already led to very high quality particles (~100 defects in 3.3 million particles) fabricated in Germany at an industrial/production scale benefiting from the improvements made in the fuel production AVR and THTR reactors [17,19]. There are also currently large national/international programs on the development of TRISO fuels.

One of the large TRISO programs is the U.S. Department of Energy Next Generation Nuclear Plant (NGNP) project [20], which aims to license a commercial GenIV VHTR (Very High Temperature Gas Cooled Reactor) that is capable of cogeneration

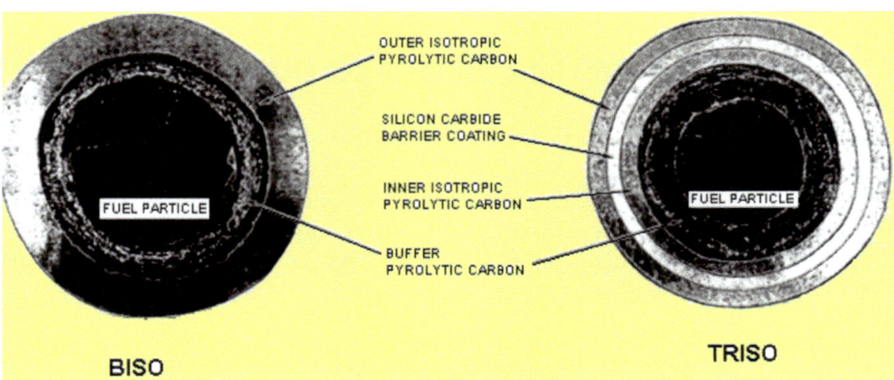

Figure 2.16 Direct comparison of BISO-coated particle fuel (two coating layers) and TRISO particle fuel (four coating layers).

of electricity and industrial process heat. Two types of concepts are under consideration including pebble-bed (TRISO particles with LEU uranium oxide (UO_2) kernel packed to 60 mm diameter pebbles in a graphite matrix) and prismatic-core reactors (TRISO particles with LEU uranium oxycarbide (UCO) kernel embedded in 50-mm long by 12.5-mm diameter cylindrical graphite compacts).

The LEU UO_2 is a very widely used reference type fuel in pebble bed reactors with extensive irradiation and heating experiments/database, in particularly from historical HTGR programs since 1970s. However, it is not the case for UCO kernel TRISO fuel. Historically in 1970s, UC_2 was used in TRISO particles and it was found this kernel type increases the rare earth attack of SiC coating layer. UCO kernel was introduced by adding sufficient amount of UC and UC_2–UO_2 to act as a getter for oxygen to prevent CO gas formation during irradiation. This design is able to retain the fission product as oxide form within the kernel and reduce the kernel migration by maintaining a low CO gas pressure. This is also the reason UCO kernel material was selected for the prismatic core design in NGNP as it has excellent ability in achieving high burnups (>20% FIMA) improving the economics. A final point is that compared with UC_2 kernel, the heavy metal loading remains similar for UCO kernel hence there is no need to increase the buffer thickness/volume to accommodate CO generation during uranium depletion. However, there has been a lack of fuel irradiation and qualification programs for UCO kernel type since the closure of German programs in 1990s. Hence UCO kernel has become the focus of the Advanced Gas Reactor Fuel Development and Qualification Program [21,22] (AGR) initiated in 2002 under NGNP.

Similarly, European-based projects, such as the European Commission's Joint Research Center and the Institute for Energy (JRC-IE), supported an HFR-EU1 irradiation [26] for high burnup of five HTR fuel pebbles to demonstrate its sustainability for conditions beyond that of the current HTR reactor designs with pebble bed cores. In addition, the Korean NHTD Plan (Nuclear-Hydrogen Technology Development) was initiated to support the development of Very High Temperature Reactor (VHTR) in Korea [23–25]. This program includes the experimental irradiation testing of TRISO fuel at the High-flux Advanced Neutron Application Reactor (HANARO). In summary, there is a resurge in TRISO research worldwide as it plays an important role in the deployment of advanced reactors.

During the past few decades of development in TRISO fuels, a large range of the design and property parameters have been investigated. These can be categorized to three primary aspects including the kernel, coating and fuel forms. For the kernel, it is apparent from the descriptions above that various fission and fertile materials were investigated, e.g., U, Th and Pu on their own or mixed, enriched or natural in the form of oxide, carbide or oxycarbide. The kernel size also varied between 100 and 800 μm. The kernel and coating types for several historically important reactors are listed in Table 2.2. Note that in TRISO, the kernel is separated from the denser coatings (PyC and SiC) by a buffer layer, therefore the understanding gained from various kernel designs are translatable at times. The coating system, either in BISO or TRISO, varied combinations of the deposition temperature, deposition gas, layer thicknesses and materials have been studied. For instance, the SiC layer can be replaced by ZrC and the anisotropy of PyC layer can be modified by adjusting

Table 2.2 BISO/TRISO designs/development in different reactors.

Reactor, country (operation period)	Reactor characteristics	Purpose	Kernel type	Coating layers
Dragon Experimental Reactor, UK (1965–1975)	• Helium-cooled, graphite moderated HTR • "Prismatic" design with 37 fuel elements arranged in a hexagonal array • 24 control rods hosted in the holes of the prismatic graphite "inner live" reflector • Outlet temperature 750°C	• To develop and test fuel and materials for the OECD-NEA (The Organisation for Economic Co-operation and Development Nuclear Energy Agency) High-Temperature Reactor Project.	• Driver fuel (six outer fuel element rods): HEU UO_2; kernel diameter: 800 μm; hollow cylinder compact with graphite matrix • Center elements: LEU UO_2; also thorium and plutonium as oxide and carbide. • Tested a range of other fuels	Buffer/SiC Driver fuel: Buffer/SiC/OPyC (total thickness of 150 μm)
AVR (Arbeitsgemeinschaft Versuchsreaktor) Reactor, Germany (1967–88)	• Helium-cooled, graphite-moderated high-temperature reactor • Pebble bed design • Reached core temperature of 950°C.	• Prototype pebble bed; • To develop and test a variety of fuels, materials and machinery, • Act as the basis of the pebble bed technology licensed to China to build HTR-10 and the HTR-PM.	• UC_2 or Th (1974–78) • UO_x (1983–88)	Buffer/PyC (1974–78)
Peach Bottom Unit 1, USA (1966–74)	• Helium-cooled, graphite-moderated high-temperature reactor • Prismatic core	• Commercial • First prototype HTR in the USA	• Core 1: $(Th,U)O_2$ (1966–69) • Core 2: $(Th,U)O_2$ (1970–74)	PyC (core 1) Buffer/PyC (core 2)

THTR-300, Germany (1983–89)	• Helium-cooled, graphite moderated, • Pebble bed design, • Used TRISO pebble fuel developed by the AVR.	• Prototype HTR • Commercial reactor	• $(Th,U)O_2$ (400 μm in diameter)	Buffer(80 pm)/ dense sealing PyC (30 μm)/ outer PyC (80 μm)
HTTR, Japan (1998–2011; restarted in 2021)	• Helium-cooled, graphite-moderated high-temperature reactor • Prismatic block type (hexagonal fuel assemblies) • Annular fuel pellets (10 mm inner diameter; 26 mm outer diameter) • Outlet temperature 850–950°C	• High-temperature engineering test reactor • Produce heat source for hydrogen production systems	• Originally used UO_2 fuel kernel • Later tested fuels with $(Th,U)O_2$ kernel (600 μm in diameter)	Buffer (60 μm)/ IPyC (30 μm)/ SiC(25 μm)/ OPyC (45 μm)
HTR-10, China (2000–present)	• Helium-cooled, graphite-moderated high-temperature reactor • Pebble bed design, • Outlet temperature 700–950°C	• Prototype HTR • Irradiation test for fuel elements • Experimental development of nuclear process heat (cogeneration of electricity/heat) • Scaled-up to two HTR-PM reactors (2021–present)	• UO_2 kernel (500 μm in diameter)	Buffer (90 μm)/ IPyC (40 μm)/ SiC (35 μm)/ OPyC (40 μm)

fabrication parameters. Finally, the final forms of fuels based on these fuel particles include various geometries such as pebbles in spherical form, and cylindrical and annular-shaped compacts. The packing density, matrix material and fabrication methods can all be varied. There are vast amount of open literature (research papers, white papers and technical reports) discussing these developments and experience in detail. Therefore, the basic processes for fabricating TRISO fuels are well established.

4.2 Modern TRISO particle fuel

It is considered that modern TRISO particle fuels are designed with the criterion not to fracture/crack at temperatures up to 1600°C under stresses during processing and operation, e.g., mismatch stress caused by different thermal expansion over heating or cooling, gas pressure generated by fission reactions, and irradiation induced dimensional change. Thus, the fuel can be contained even in the worst of accident scenarios in a properly designed reactor. Specifically, modern TRISO particle fuel is a hardy microencapsulated fuel with a diameter of about 0.8–1.0 mm typically comprises a spherical UO_2 or UCO fuel kernel (~350–500 μm in diameter), Fig. 2.17. The kernel is coated with four concentric layers of ceramics.

(i) A low-density, porous pyrolytic carbon about 100 μm thick (named as the "buffer layer"), which is considered to have a theoretical porosity of ~50 vol. %. This buffer layer accommodates the accumulation of fission gas and fission recoils, and it functions to decouple the kernel and the other denser coatings.

(ii) A high-density inner pyrolytic carbon (IPyC) layer, which is usually about 40 μm thick. The purpose of this layer has twofold: during the fabrication process, this IPyC layer can provide a good surface for SiC layer deposition and in the meantime protect the kernel from corrosive gases (HCl, C12) liberated during the SiC coating process. During service, IPyC contributes to fission gas retention.

(iii) A dense silicon carbide (SiC) layer, which is usually about 35 μm thick. The SiC layer is considered a very important coating contributing to the structural integrity of the particle and as a dense diffusion barrier layer to help retaining metallic fission products and stop them from releasing to the compact matrix and joining the reactor coolant stream.

(iv) The high-density outer pyrolytic carbon (OPyC) layer, which will protect the SiC layer during particle handling as well provides a bonding surface to the fuel matrix. The OPyC layer also contributes to fission gas retention and the thickness of this layer is similar to the IPyC and is about 40 μm. Both IPyC and OPyC layers have comprised ability in retaining metallic fission products at high temperatures. Another function of OPyC is to apply compressive stress on SiC to further enhance the integrity of the particle.

Note the thickness/diameter numbers quoted here refer to approximate mean values. These coating layers are all fabricated using fluidized bed chemical vapor

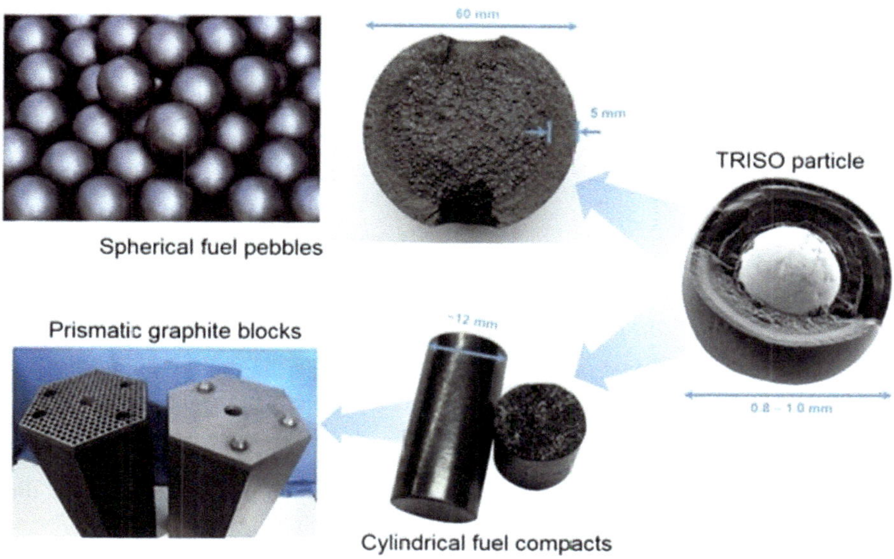

Figure 2.17 This diagram of pictures are produced by the U.S. Department of Energy (DOE) demonstrating the different forms of TRi-structural ISOtropic (TRISO) particle fuel forms. On the right is graph showing a fractured TRISO particle revealing the kernel and coated PyC and SiC layers. These particles were then compacted to spherical pebbles with graphite matrix and cylindrical configuration hosted by prismatic graphite blocks.
Source: DOE.

deposition (FCVD); the gas mixture used and temperature will be elaborated in Section 5.3.

Together, these coating layers maintain the particle's overall integrity in response to the extreme environments experienced under irradiation. After fabrication, these particles are either embedded in a graphite or SiC matrix to form pellets (for light water reactor designs) or so-called pebbles (for pebble bed designs) and subject to irradiation damage as well as thermal/mechanical stresses. A graphic of this is shown in Fig. 2.17. Note that while there has been significant developmental and optimization work on the manufacturing of TRISO particles internationally in the past 60 years, in-depth understanding of the materials dimensional change, irradiation creep and property data is still lacking. This particularly prohibits the mechanistic modeling efforts that seek to improve fuel performance. This is an area that requires additional effort.

Other exotic kernel compositions and different matrix materials have been proposed in addition to the conventional designs, for instance, the Ultra Safe Nuclear Corporation (USNC) proposed "Fully Ceramic Microencapsulated" (FCM) fuel embedding TRISO in a dense and fully sealed SiC matrix as an additional layer of protection against radionuclide release. FCM pellets will act as accident tolerant fuel for

LWRs, as well as fuels in many other reactors including their own Micro-Modular Reactors (MMR), gas-cooled reactors, CANDU reactors and molten salt reactors. In addition, they suggested TRISO fuels with larger kernel size for added benefit in many applications.

4.3 TRISO particle fuel fabrication

This section will provide a very brief description of the typical processing methods for the kernel, TRISO coating layers and the compacting matrix.

4.3.1 TRISO kernel

The process of the kernel varies slightly with the raw materials, desired kernel composition and the intended application. Here a generalized method for production of UCO kernel will be described based on the so-called "internal gelation" process. A schematic is shown in Fig. 2.18 to illustrate to key steps. The first step is to dissolve the uranium oxide power in nitric acid to produce uranyl nitrate solution. By adding urea, carbon and hexamethylenetetramine to the solution, a gelatin solution is formed to produce spherical droplets through a vibrating nozzle. The droplets will be "aged" in a liquid to facilitate gelation reaction; the "aged" particles will then be washed to

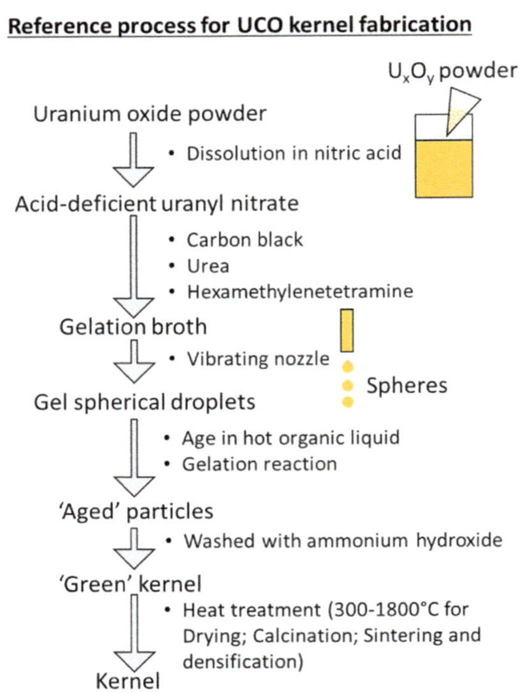

Figure 2.18 Typical fabrication process for UCO kernel.

remove excessive urea, formaldehyde and ammonium nitrate to produce "green" kernels. The final steps involve applying a range of heat treatments to the "green" kernels over temperatures ranging from a few hundred degree Celsius to 1600–1800°C for drying, calcination, sintering and densification.

The exact details of fabrication may vary from different manufacturers, which offers the potential to increase the quality of the kernel and reduce defects.

4.3.2 TRISO coatings

The coating processes of the kernels take place in a fluidized-bed furnace using chemical vapor deposition method (FCVD). The deposition temperature, rate and gas mixture will all have impact on the final nano-/microstructures and properties. For instance, in studies as early as the UK Dragon Project, it was already found the deposition rate of PyC has a direct relation to coating failure, and recommended a deposition rate higher than 5 μm per minute. In modern TRISOs, the coatings will be deposited sequentially in an uninterrupted process, i.e., without unloading, cooling or reheating. A typical deposition temperature for the porous buffer PyC is about 1350°C. The gas used are argon and acetylene. For the dense IPyC and OPyC layers, the temperatures adopted are similar to the porous buffer layer and are both deposited by thermally decomposing acetylene with propylene gas. The fluidizing gas used for all PyC layers are argon gas. For SiC layer, the deposition temperature is higher, at about 1560°C; the carrier gas used is either hydrogen or a mix of hydrogen and argon. SiC is formed by decomposing methyl trichlorosilane gas.

The exact coating condition could be tailored to produce the desired coating microstructure and properties, for instance the AGR program tested TRISO fuels by varying many parameters such as the deposition temperature, mass flow rate and coating gas fractions. These are well reported in various literature and will not be repeated here.

4.3.3 TRISO graphite matrix

The most common matrix material for TRISO compacts or pebbles is graphite. Note the USNC SiC matrix is another very important design, but the fabrication processes contain proprietary information, which will not be elaborated here. The fabrication of graphite matrix is very different from the manufacturing of nuclear graphite moderator. First of all, the matrix graphite processing temperature cannot be as high as that used for nuclear graphite (2700–3000°C)—temperatures lower than 1800°C is preferred to protect the fuel kernel in the TRISO particles. Secondly, the forming methods are very limited as the pressure applied has to be sufficiently low to avoid causing cracks in the TRISO particles. A common practice is to coat the TRISO particles with matrix precursor before the pressing stage to reduce the particle-to-particle contact and unnecessary particle damage.

A typical method adopted in modern TRISO compact/pebble production is shown in Fig. 2.19. The graphite powder is mixed with a thermoplastic resin binder, which will be used to coat the TRISO particles. There are different recipes of the powder mixture, for instance the AGR5/6/7 experiment used a derivative of the German A3-27 matrix formulation to achieve a higher matrix density while earlier AGR experiments used A3-3. This coating step takes places in a drum-shaped rotating vessel where an appropriate number of TRISO particles are loaded. A liquid media, e.g., methanol is added during the process to wet both the TRISO particle surface as well as the powder mixture. The coated TRISO particles will go through a drying process at low temperature about 80°C before being poured to a mold for hot pressing. Depends on the amount of coating on TRISO particles, additional matrix material may or may not be added. This forms "green" compacts in cylindrical or annular shapes or spherical pebbles. The follow-up processes include carbonization in an inert gas environment, usually argon gas, at 900°C, and graphitization at temperatures around 1750–1800°C. This fabrication process is called "thermosetting-matrix-based compacting process" and it is used as the reference process for graphite matrix in the NGNP/AGR Fuel Program.

There are strict quality assurance processes in place to exclude defected particles and compacts in nonstandard shapes, or with missing layers (e.g., missing buffer layer, PyC or SiC layers). These are all to reduce the possibility of particle failure, which are typically marked by the fracture of TRISO coatings (pressure vessels) causing the release of fission products. The reasons causing TRISO particle failure are very diverse and can arise from many different steps during fabrication and handling.

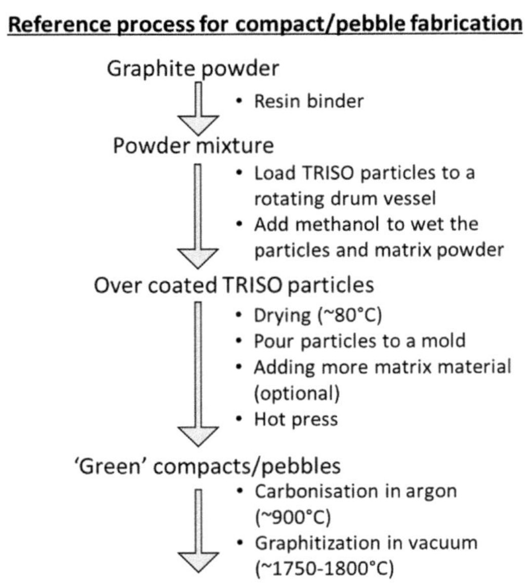

Figure 2.19 Typical process for fabrication of TRISO compacts/pebbles with a graphite matrix.

5. Concluding comments

There are many exciting and very important topics in nuclear TRISO fuels that are not covered in this chapter. For instance, local thermal and mechanical properties in TRISO particle and fuel compacts. There are challenges in terms of measuring the local thermal properties in general but there are recent developments of high-resolution techniques to overcome this problem. Another aspect is the materials response to irradiation damage. SiC layer is a most studied coating in TRISO due to its role in structural integrity, however, it has become increasingly recognized that much more research is needed on the PyC layers due to the stark lack of experimental data. For instance, it is known that the PyC layers in TRISO fuels is not perfect isotropic where the deposition processes are adjusted to reduce the degree of anisotropy. However, the dimensional changes of the PyC layers are still not well understood. This behavior can be further modified by the presence of residual stresses and in-service stresses due to a mechanism called "irradiation creep." The interactions between the PyC layers and the SiC layer sandwiched between them play an important role in keeping the SiC layer under compressive stress during irradiation. Inappropriate design of the PyC-SiC interfacial strength can result in coating fracture hence particle failure.

References

[1] A.A. Bauer, Nitride fuels: properties and potentials, React. Technol. 15 (1972) 87–104.

[2] M.A. DeCrescente, M.S. Freed, S.D. Caplow, Uranium nitride fuel development, SNAP-50 (1963).

[3] R.B. Matthews, K.M. Chidester, C.W. Hoth, R.E. Mason, R.L. Petty, Fabrication and testing of uranium nitride fuel for space power reactors, J. Nucl. Mater. 151 (3) (1988) 345.

[4] A.J. Parkison, S.S. Parker, A.T. Nelson, Fabrication of ThN using a carbothermic reduction to nitridation process, J. Am. Ceram. Soc. 99 (12) (2016) 3909–3914.

[5] F. Anselin, Etude des nitrures d'uranium, de plutonium et de leurs solutions solides, J. Nucl. Mater. 10 (1963) 301–320.

[6] V.H. Funk, H. Bohland, Zur darstellung von metallnitriden aus ammonium-fluorometallaten und ammoniak, Zeitschrift fur Anorg. und Allg. Chemie. Band 282 (1939) 155–162.

[7] C.B. Yeamans, et al., Oxidative ammonolysis of uranium (IV) fluorides to uranium (VI) nitride, J. Nucl. Mater. 374 (2008) 75–78.

[8] B.J. Jaques, et al., Synthesis and sintering of UN-UO_2 fuel composites, J. Nucl. Mater. 466 (2015) 745–754.

[9] V.J. Tennery, T.G. Godfrey, R.A. Potter, Sintering of UN as a function of temperature and N_2 pressure, J. Am. Ceram. Soc. 54 (7) (1971) 327–331.

[10] Y. Arai, M. Morihira, T. Ohmichi, The effect of oxygen impurity on the characteristics of uranium and uranium-plutonium mixed nitride fuels, J. Nucl. Mater. 202 (1–2) (1993) 70–78.

[11] H. Muta, K. Kurosaki, M. Uno, S. Yamanaka, Thermal and mechanical properties of uranium nitride prepared by SPS technique, J. Mater. Sci. 43 (19) (2008) 6429–6434.

[12] P. Malkki, M. Jolkkonen, T. Hollmer, J. Wallenius, Manufacture of fully dense uranium nitride pellets using hydride derived powders with spark plasma sintering, J. Nucl. Mater. 452 (1–3) (2014) 548–551.

[13] S.B. Ross, M.S. El-Genk, R.B. Matthews, Uranium nitride fuel swelling correlation, J. Nucl. Mater. 170 (2) (1990) 169–177.

[14] R.B. Matthews, Irradiation performance of nitride fuels, in: Specialist Conference on Space Nuclear Power and Propulsion Technologies—Materials and Fuel, Podolsk-Moscow, Russia, 1993.

[15] D.A. Lopes, S. Uygur, K. Johnson, Degradation of UN and UN–U_3Si_2 pellets in steam environment, J. Nucl. Sci. Technol. 54 (4) (2017) 405–413.

[16] W.J. Carmack, Status of Fuel Development and Manufacturing for Space Nuclear Reactors at BWX Technologies, Space Technology and Applications International Forum, 2004, pp. 426–431.

[17] P. Rodriguez, Mixed plutonium-uranium carbide fuel in fast breeder test reactor, Bull. Mater. Sci. 22 (3) (1999) 215–220.

[18] M.K. Meyer, R. Fielding, J. Gan, Fuel development for gas-cooled fast reactors, J. Nucl. Mater. 371 (1–3) (2007) 281–287.

[19] R.B. Matthews, R.J. Herbst, Uranium-plutonium carbide fuel for fast breeder reactors, Nucl. Technol. 63 (1982).

[20] L. Grande, B. Villamere, L. Allison, S. Mikhael, A. Rodriguez-Prado, I. Pioro, Thermal aspects of uranium carbide and uranium dicarbide fuels in supercritical water-cooled nuclear reactors, J. Eng. Gas Turbines Power 133 (2) (2011) 1–7.

[21] R.A. Haslett, Space Nuclear Thermal Propulsion Program Final Report (1995).

[22] A.K. Sengupta, R. Agarwal, H.S. Kamath, Carbide Fuel, first ed., vol. 3, Elsevier Inc., 2012.

[23] D. Manara, F. De Bruycker, A.K. Sengupta, R. Agarwal, H.S. Kamath, Thermodynamic and Thermophysical Properties of the Actinide Carbides, vol. 2, Elsevier Inc., 2012.

[24] L. Biasetto, et al., Developing uranium dicarbide-graphite porous materials for the SPES project, J. Nucl. Mater. 404 (1) (2010) 68–76.

[25] H.M. Reiche, S.C. Vogel, M. Tang, In situ synthesis and characterization of uranium carbide using high temperature neutron diffraction, J. Nucl. Mater. 471 (2016) 308–316.

[26] M. Saibaba, S.V. Varamban, C.K. Mathews, Effect of oxygen and nitrogen impurities on the thermodynamic properties of uranium-plutonium mixed carbide fuel, J. Nucl. Mater. 144 (1–2) (1987) 56–64.

[27] D. Srivastava, S.P. Garg, G.L. Goswami, Thermodynamic analysis of mixed carbide, carbonitride and nitride fuels for fast breeder reactors, J. Nucl. Mater. 161 (1) (1989) 44–56.

[28] D. Salvato, et al., Innovative preparation route for uranium carbide using citric acid as a carbon source, Ceram. Int. 42 (15) (2016) 16710–16717.

[29] R.J. Gray, W.C. Thurber, C.K.H. DuBose, Preparation and Metallography of Arc-Melted Uranium Carbides (1958).

[30] A.E. Austin et al., Progress on the Development of Uranium Carbide-Type Fuels (1959).

[31] G. Vasudevamurthy, T.W. Knight, Production of high-density uranium carbide compacts for use in composite nuclear fuels, Nucl. Technol. 163 (2) (2008) 321–327.

[32] F. Gorlé, L. Coheur, W. Timmermans, Uranium carbide fuel technology, J. Nucl. Mater. 51 (3) (1974) 343–353.

[33] D. Salvato, J.F. Vigier, M. Cologna, L. Luzzi, J. Somers, V. Tyrpekl, Spark plasma sintering of fine uranium carbide powder, Ceram. Int. 43 (1) (2017) 866–869.
[34] A.W. Hare, S. Alfant, F.A. Rough, D.I. Sinizer, Further results of irradiation of uranium carbide, Nucl. Sci. Eng. 10 (1) (1961) 24–30.
[35] H. Blank, Fabrication of Carbide and Nitride Pellets and the Nitride Irradiations Niloc 1 and Niloc 2 (1991).
[36] D.C. Crawford, D.L. Porter, S.L. Hayes, Fuels for sodium-cooled fast reactors: US perspective, J. Nucl. Mater. 371 (1–3) (2007) 202–231.
[37] H. Okamoto, B-U (Boron-Uranium), Bin. Alloy Phase Diagrams 1 (1990) 551–552.
[38] V.F. Sears, Neutron scattering lengths and cross sections, Neutron News 3 (3) (1992) 26–37.
[39] P.P. Pronko, P.A. Vanrompay, Z. Zhang, J.A. Nees, Isotope enrichment in laser-ablation plumes and commensurately deposited thin films, Phys. Rev. Lett. 83 (1999) 2596–2599.
[40] M. Joseph, P. Manoravi, Boron isotope enrichment in nanosecond pulsed laser-ablation plume, Appl. Phys. A 76 (2003) 153–156.
[41] B.W. Howlett, A note on the uranium-boron alloy system, J. Inst. Met. 88 (1959) 91–92.
[42] J.P. Dancausse, et al., Compression study of uranium borides UB_2, UB_4 and UB_{12} by synchrotron X-ray diffraction, J. Alloys Compd. 189 (2) (1992) 205–208.
[43] L. Brewer, D.L. Sawyer, D.H. Templeton, C.H. Dauben, A study of the refractory borides, J. Am. Ceram. Soc. 34 (6) (1951) 173–179.
[44] H.E. Flotow, D.W. Osborne, P.A.G. O'Hare, J.L. Settle, F.C Mrazek, W.N. Hubbard, Uranium diboride: preparation, enthalpy of formation at 298.15°K, heat capacity from 1° to 350°K, and some derived thermodynamic properties, J. Chem. Phys. 51 (2) (1969) 583–592.
[45] J.T. White, Report on the Basic Chemistry, Microstructure and Physical Properties of High Uranium Density Boride Compounds (2015).
[46] E. Kardoulaki, et al., Thermophysical and mechanical property assessment of UB_2 and UB_4 sintered via spark plasma sintering, J. Alloys Compd. 818 (2020) 1–14.
[47] J. Turner, F. Martini, J. Buckley, G. Phillips, S.C. Middleburgh, T.J. Abram, Synthesis of candidate advanced technology fuel: uranium diboride (UB_2) via borothermic reduction of UO_2, J. Nucl. Mater. (2020) 152388.
[48] A.J. Lupinetti, J.L. Fife, E. Garcia, P.K. Dorhout, K.D. Abney, Low-temperature synthesis of uranium tetraboride by solid-state metathesis reactions, Inorg. Chem. 41 (9) (2002) 2316–2318.
[49] R. Troć, R. Wawryk, A. Pikul, N. Shitsevalova, Physical properties of cage-like compound UB12, Philos. Mag. 95 (21) (2015) 2343–2363.
[50] D.R. Fredrickson, M.G. Barner, R.D. Chasanov, R.L. Nuttall, R. Kleb, W.N. Hubbard, The enthalpy of uranium diboride from 600 to 1500 ° K by drop calorimetry, High Temp. Sci. 1 (80) (1969) 373–380.
[51] D.R. Fredrickson, R.D. Barnes, M.G. Chasanov, Enthalpy of uranium diboride from 1300 to 2300°K by drop calorimetry, High Temp. Sci. 2 (1970) 299–301.
[52] P.Y. Chevalier, E. Fischer, Thermodynamic Modelling of the C ± U and B ± U Binary Systems, vol 288, 2001.
[53] K.J. Matterson, H. Jones, A study of the tetraborides of uranium and thorium, Trans. Br. Ceram Soc 60 (1961) 475–493.
[54] E. Jossou, L. Malakkal, B. Szpunar, D. Oladimeji, J.A. Szpunar, A first principles study of the electronic structure, elastic and thermal properties of UB_2, J. Nucl. Mater. 490 (2017) 41–48.

[55] L.J. Evitts, S.C. Middleburgh, E. Kardoulaki, I. Ipatova, M.J.D. Rushton, W.E. Lee, Influence of boron isotope ratio on the thermal conductivity of uranium diboride (UB_2) and zirconium diboride (ZrB_2), J. Nucl. Mater. 528 (2020) 1–7.
[56] J. Fink, Thermophysical properties of uranium dioxide, J. Nucl. Mater. 279 (1) (2000) 1–18.
[57] P.A.A. Burr, E. Kardoulaki, R. Holmes, S.C.C. Middleburgh, Defect evolution in burnable absorber candidate material: Uranium diboride, UB_2, J. Nucl. Mater. 513 (2019) 45–55.
[58] J.H. Yang, D.J. Kim, K.S. Kim, Y.H. Koo, UO_2-UN composites with enhanced uranium density and thermal conductivity, J. Nucl. Mater. 465 (June) (2015) 509–515.
[59] D. Costa, M. Hedberg, S. Middleburgh, J. Wallenius, P. Olsson, D.A. Lopes, UN microspheres embedded in UO2 matrix: an innovative accident tolerant fuel, J. Nucl. Mater. 540 (2020) 152355.
[60] J.T. White, A.T. Nelson, J.T. Dunwoody, D.D. Byler, D.J. Safarik, K.J. Mcclellan, Thermophysical properties of U_3Si_2 to 1773 K, J. Nucl. Mater. 464 (2015) 275–280.
[61] E.S. Wood, J.T. White, C.J. Grote, A.T. Nelson, U_3Si_2 behavior in H_2O: part I, flowing steam and the effect of hydrogen, J. Nucl. Mater. 501 (2018) 404–412.
[62] B. Gong, et al., U_3Si_2 and UO_2 composites densified by spark plasma sintering for accident-tolerant fuels, J. Nucl. Mater. 534 (2020) 1–12.
[63] K.D. Johnson, A.M. Raftery, D.A. Lopes, J. Wallenius, Fabrication and microstructural analysis of UN-U_3Si_2 composites for accident tolerant fuel applications, J. Nucl. Mater. 477 (2016) 18–23.
[64] J.T. White, A.W. Travis, J.T. Dunwoody, A.T. Nelson, Fabrication and thermophysical property characterization of UN/U_3Si_2 composite fuel forms, J. Nucl. Mater. 495 (2017) 463–474.
[65] J. Turner, S. Middleburgh, T. Abram, A high density composite fuel with integrated burnable absorber: U_3Si_2-UB_2, J. Nucl. Mater. 529 (2020) 1–5.
[66] J. Turner, T. Abram, Steam performance of UB_2/U_3Si_2 composite fuel pellets, compared to U_3Si_2 reference behaviour, J. Nucl. Mater. 529 (2020) 151919.
[67] E. Kardoulaki, D.M. Frazer, J.T. White, U. Carvajal, A.T. Nelson, D.D. Byler, Fabrication and thermophysical properties of UO_2-UB_2 and UO_2-UB_4 composites sintered via spark plasma sintering, J. Nucl. Mater. (2020).
[68] Z. Chen, G. Subhash, J.S. Tulenko, Raman spectroscopic investigation of graphitization of diamond during spark plasma sintering of UO_2-diamond composite nuclear fuel, J. Nucl. Mater. 475 (2016) 1–5.
[69] Z. Chen, G. Subhash, J.S. Tulenko, Spark plasma sintering of diamond-reinforced uranium dioxide composite fuel pellets, Nucl. Eng. Des. 294 (2015) 52–59.
[70] J. Tulenko, Development of Innovative Accident Tolerant High Thermal Conductivity UO2-Diamond Composite Fuel Pellets (2016).
[71] S. Yeo, E. McKenna, R. Baney, G. Subhash, J. Tulenko, Enhanced thermal conductivity of uranium dioxide-silicon carbide composite fuel pellets prepared by Spark Plasma Sintering (SPS), J. Nucl. Mater. 433 (1–3) (2013) 66–73.
[72] S. Yeo, R. Baney, G. Subhash, J. Tulenko, The influence of SiC particle size and volume fraction on the thermal conductivity of spark plasma sintered UO_2-SiC composites, J. Nucl. Mater. 442 (1–3) (2013) 245–252.
[73] S. Ishimoto, M. Hirai, K. Ito, Y. Korei, Thermal conductivity of UO_2-BeO pellet, J. Nucl. Sci. Technol. 33 (2) (1996) 134–140.

[74] W. Zhou, R. Liu, S.T. Revankar, Fabrication methods and thermal hydraulics analysis of enhanced thermal conductivity UO_2-BeO fuel in light water reactors, Ann. Nucl. Energy 81 (2015) 240–248.
[75] A. Cartas, H. Wang, G. Subhash, R. Baney, J. Tulenko, Influence of carbon nanotube dispersion in UO_2-carbon nanotube ceramic matrix composites utilizing spark plasma sintering, Nucl. Technol. 189 (3) (2015) 258–267.
[76] T. Yao, et al., UO2 + 5 vol% ZrB2 nano composite nuclear fuels with full boron retention and enhanced oxidation resistance, Ceram. Int. 46 (2020) 26486.
[77] T. Takahashi, T. Kido, T. Kikuchi, Sintering of Uranium Mononitride (1975).
[78] R. Metroka, Fabrication of Uranium Mononitride Compacts, (1970).
[79] N. Fuhrman, L.D. Hower, R.B. Holden, Low-temperature sintering of uranium dioxide, J. Am. Ceram. Soc. 46 (3) (1963).
[80] J. Gao, X. Yang, R. Li, Y. Wang, F. Zhong, Low-temperature sintering mechanism on uranium dioxide, J. Mater. Sci. 42 (15) (2007) 5936–5940.
[81] X. Yang, J. Gao, Y. Wang, X. Chang, Low-temperature sintering process for UO_2 pellets in partially-oxidative atmosphere, Trans. Nonferrous Met. Soc. China (English Ed.) 18 (1) (2008) 171–177.
[82] T.A. Thornton, V.D. Holaday Jr., Sintering UO2 and oxidation of UO2 with microwave radiation, US Patent 4389355, 1983.
[83] W. Dörr, T. Gerdes, G. Gradel, B. Schmitt, W.-P. Monika Method and furnace for microwave sintering of nuclear fuel, US patent 6437303B1, 2002.
[84] J.H. Yang, et al., Microwave process for sintering of uranium dioxide, J. Nucl. Mater. 325 (2–3) (2004) 210–216.
[85] T. Subramanian, P. Venkatesh, K. Nagarajan, P.R. Vasudeva Rao, Novel method of sintering UO_2 pellets by microwave heating, Mater. Lett. 46 (2–3) (2000) 120–124.
[86] F.V. Lenel, Resistance sintering under pressure, JMET (J. Med. Eng. Technol.) 7 (1) (1955) 158–167.
[87] N. Tamari, T. Tanaka, K. Tanaka, I. Kondoh, M. Kawahara, M. Tokita, Effect of spark plasma sintering on densification and mechanical properties of silicon carbide, J. Ceram. Soc. Japan 103 (1199) (1995) 740–742.
[88] L. Gao, J.S. Hong, H. Miyamoto, S.D.D.L. Torre, Bending strength and microstructure of Al_2O_3 ceramics densified by spark plasma sintering, J. Eur. Ceram. Soc. 20 (12) (2000) 2149–2152.
[89] K.D. Johnson, D.A. Lopes, Grain growth in uranium nitride prepared by spark plasma sintering, J. Nucl. Mater. 503 (2018) 75–80.
[90] V. Tyrpekl, M. Cologna, J.F. Vigier, A. Cambriani, W. De Weerd, J. Somers, Preparation of bulk-nanostructured UO_2 pellets using high-pressure spark plasma sintering for LWR fuel safety assessment, J. Am. Ceram. Soc. 100 (4) (2017) 1269–1274.
[91] L. Ge, G. Subhash, R.H. Baney, J.S. Tulenko, E. McKenna, Densification of uranium dioxide fuel pellets prepared by spark plasma sintering (SPS), J. Nucl. Mater. 435 (1–3) (2013) 1–9.
[92] V. Tyrpekl, et al., On the role of the electrical field in spark plasma sintering of UO_{2+x}, Sci. Rep. 7 (April) (2017) 1–9.
[93] H. Muta, K. Kurosaki, M. Uno, S. Yamanaka, Thermophysical properties of several nitrides prepared by spark plasma sintering, J. Nucl. Mater. 389 (1) (2009) 186–190.
[94] A. Mohamad, Y. Ohishi, H. Muta, K. Kurosaki, S. Yamanaka, Thermal and mechanical properties of polycrystalline U_3Si_2 synthesized by spark plasma sintering, J. Nucl. Sci. Technol. 55 (10) (2018) 1141–1150.

[95] R.C. O'Brien, N.D. Jerred, Spark Plasma Sintering of W-UO_2 cermets, J. Nucl. Mater. 433 (1–3) (2013) 50–54.
[96] A.J. Mackie, G.D. Hatton, H.G.C. Hamilton, J.S. Dean, R. Goodall, Carbon uptake and distribution in Spark Plasma Sintering (SPS) processed Sm(Co, Fe, Cu, Zr)z, Mater. Lett. 171 (2016) 14–17.
[97] K. Morita, B. Kim, H. Yoshida, K. Hiraga, Y. Sakka, Distribution of carbon contamination in oxide ceramics occurring during spark-plasma-sintering (SPS) processing: II—effect of SPS and loading temperatures, J. Eur. Ceram. Soc. 38 (6) (2018) 2596–2604.
[98] T. Ironman, J. Tulenko, G. Subhash, Exploration of viability of spark plasma sintering for commercial fabrication of nuclear fuel pellets, Nucl. Technol. 200 (2) (2017) 144–158.
[99] M. Cologna, B. Rashkova, R. Raj, Flash sintering of nanograin zirconia in <5 s at 850°C, J. Am. Ceram. Soc. 93 (11) (2010) 3556–3559.
[100] M. Cologna, A.L.G. Prette, R. Raj, Flash-sintering of cubic yttria-stabilized zirconia at 750°C for possible use in SOFC manufacturing, J. Am. Ceram. Soc. 94 (2) (2011) 316–319.
[101] A.L.G. Prette, M. Cologna, V. Sglavo, R. Raj, Flash-sintering of Co2MnO4 spinel for solid oxide fuel cell applications, J. Power Sources 196 (4) (2011) 2061–2065.
[102] Y. Zhang, J. Nie, J. Luo, Effects of phase and doping on flash sintering of TiO_2, J. Ceram. Soc. Japan 124 (4) (2016) 296–300.
[103] B. Yoon, et al., Measurement of O and Ti atom displacements in TiO2 during flash sintering experiments, J. Am. Ceram. Soc. 101 (5) (2018) 1811–1817.
[104] R. Muccillo, E.N.S. Muccillo, Electric field-assisted flash sintering of tin dioxide, J. Eur. Ceram. Soc. 34 (4) (2014) 915–923.
[105] A. Karakuscu, et al., Defect structure of flash-sintered strontium titanate, J. Am. Ceram. Soc. 95 (8) (2012) 2531–2536.
[106] S. Grasso, et al., Flash spark plasma sintering (FSPS) of pure ZrB2, J. Am. Ceram. Soc. 97 (8) (2014) 2405–2408.
[107] E. Zapata-Solvas, S. Bonilla, P.R. Wilshaw, R.I. Todd, Preliminary investigation of flash sintering of SiC, J. Eur. Ceram. Soc. 33 (13–14) (2013) 2811–2816.
[108] S.K. Jha, R. Raj, The effect of electric field on sintering and electrical conductivity of Titania, J. Am. Ceram. Soc. 97 (2) (2014) 527–534.
[109] M. Cologna, J.S.C. Francis, R. Raj, Field assisted and flash sintering of alumina and its relationship to conductivity and MgO-doping, J. Eur. Ceram. Soc. 31 (15) (2011) 2827–2837.
[110] H. Yoshida, Y. Sakka, T. Yamamoto, J.M. Lebrun, R. Raj, Densification behaviour and microstructural development in undoped yttria prepared by flash-sintering, J. Eur. Ceram. Soc. 34 (4) (2014) 991–1000.
[111] J.S.C. Francis, R. Raj, Flash-sinterforging of nanograin zirconia: field assisted sintering and superplasticity, J. Am. Ceram. Soc. 95 (1) (2012) 138–146.
[112] J.A. Valdez, D.D. Byler, E. Kardoulaki, J.S.C. Francis, K.J. McClellan, Flash sintering of stoichiometric and hyper-stoichiometric urania, J. Nucl. Mater. 505 (2018) 85–93.
[113] R.R.I. Neto, K.J. Mcclellan, D.D. Byler, E. Kardoulaki, Controlled current-rate AC flash sintering of uranium dioxide, J. Nucl. Mater. 2 (2020).
[114] D.J. Sprouster, et al., In situ X-ray characterization of uranium dioxide during flash sintering, Materialia 2 (August) (2018) 176–182.
[115] M. Biesuz, V.M. Sglavo, Flash sintering of ceramics, J. Eur. Ceram. Soc. 39 (2–3) (2019) 115–143.

[116] A.M. Raftery, et al., Onset conditions for flash sintering of UO2, J. Nucl. Mater. 493 (2017) 264–270.
[117] T.P. Mishra, R.R.I. Neto, R. Raj, O. Guillon, M. Bram, Current-rate flash sintering of gadolinium doped ceria: microstructure and defect generation, Acta Mater. 189 (2020) 145–153.
[118] P. Kumar, D. Yadav, J.-M. Lebrun, R. Raj, Flash sintering with current rate: a different approach, J. Am. Ceram. Soc. (October 2017, 2018).
[119] H. Charalambous, S.K. Jha, K.H. Christian, R.T. Lay, T. Tsakalakos, Flash sintering using controlled current ramp, J. Eur. Ceram. Soc. 38 (10) (2018) 3689–3693.
[120] F. Lemke, W. Rheinheimer, M.J. Hoffmann, A comparison of power controlled flash sintering and conventional sintering of strontium titanate, Scr. Mater. 130 (2017) 187–190.
[121] W. Ji, B. Parker, S. Falco, J.Y. Zhang, Z.Y. Fu, R.I. Todd, Ultra-fast firing: effect of heating rate on sintering of 3YSZ, with and without an electric field, J. Eur. Ceram. Soc. 37 (2017) 2547–2551.
[122] T.B. Holland, T.B. Tran, D. V Quach, U. Anselmi-tamburini, J.R. Groza, A. K. Mukherjee, Athermal and thermal mechanisms of sintering at high heating rates in the presence and absence of an externally applied field, J. Eur. Ceram. Soc. 32 (14) (2012) 3675–3683.
[123] W. Rheinheimer, X. Li, H. Wang, F. Lemke, M.J. Hoffmann, H. Wang, The role of point defects and defect gradients in flash sintering of perovskite oxides, Acta Mater. 165 (2019) 398–408.
[124] C. Manière, G. Lee, E.A. Olevsky, All-materials-inclusive flash spark plasma sintering, Sci. Rep. (October) (2017) 1–8.
[125] O. Guillon, et al., Field-assisted sintering technology/spark plasma sintering: mechanisms, materials, and technology developments, Adv. Eng. Mater. (2014) 830–849.
[126] M. Jongmanns, R. Raj, D.E. Wolf, Generation of Frenkel defects above the Debye temperature by proliferation of phonons near the Brillouin zone edge, New J. Phys. 20 (2018) 1–8.
[127] E. Kardoulaki, Report on the Effects of Dopants on FLASH Sintering of Enhanced UO2 (2019).

Ceramics in magnetic confinement fusion reactors

3

Samuel Humphry-Baker [1], *Thomas P. Davis* [2,3] *and Tyler Abrams* [4]
[1]Department of Materials, Imperial College London, London, United Kingdom; [2]Oxford Sigma, Oxford, United Kingdom; [3]Nuclear Futures Institute, Bangor University, Bangor, United Kingdom; [4]General Atomics, San Diego, CA, United States

1. Introduction

For long-term, baseload energy production, nuclear fusion energy could be the ultimate terrestrial energy source [1,2]. Fusion has the potential of producing decarbonized, near-limitless energy from the abundance of fuel with no long-term ($\gtrsim$500,000 years) radioactive material [3]. Achieving fusion requires generating a high-temperature gas on the order of 100 millions of degrees to overcome the electrostatic repulsion of nuclei [4]. At these extreme temperatures, the electrons in the gas are stripped away from the nuclei, creating a plasma, the fourth state of matter. Two main plasma confinement methods are currently being pursued: magnetic confinement and inertial confinement [5]. This chapter is concerned with magnetic confinement, with a particular focus on tokamak (toroidal) reactors, which are widely considered the fastest route toward fusion energy production [6].

As fusion energy transitions toward its engineering/implementation phase, the power density of devices is rapidly increasing, as are durations for exposure to harsh conditions and expected component lifetimes. This poses more extreme conditions on the reactor's structural components, radiation shielding, plasma facing surfaces, and other functional components, meaning that significant improvements over conventional nuclear materials are needed [7]. Due to their high melting points, creep resistance, and corrosion resistance, ceramics are emerging as promising candidates for many of these applications. This chapter addresses three such emerging trends: (i) high temperature structural ceramics; (ii) neutron/gamma-ray shielding materials; and (iii) new first-wall materials with enhanced high temperature capability. More exotic ceramics may also be good candidates for other fusion applications, such as breeder materials [8] and electrical insulators [9], but these topics are not addressed here for brevity. We first introduce some key fusion engineering concepts and recent trends in reactor design, which naturally form a set of materials criteria to guide the selection of advanced ceramics.

1.1 Introduction to fusion energy

Nuclear fusion converts two lighter elements into a heavy element. The basic fusion process powers the Sun in our Solar System by fusing two hydrogen (H) atoms (i.e.,

Ceramics for Nuclear Applications. https://doi.org/10.1016/B978-0-12-823865-3.00010-9

$p + p \rightarrow {}^2\text{He}$. The reaction with the highest cross section is the fusion of Deuterium (${}^2_1\text{H}$, or D) and Tritium (${}^3_1\text{H}$, or T). It is this reaction that will be considered within this chapter, as it is the preferred reaction for terrestrial fusion energy. Other fusion reactions are being pursued, some of which have the benefit of being aneutronic [10], but these are at a significantly lower readiness level and thus beyond the scope of this chapter. The D-T fusion reaction can be written:

$$\text{D} + \text{T} \rightarrow {}^1_0n + {}^4_2\alpha + 17.6 \text{ MeV}$$

with the neutron's, n, kinetic energy equal to 14.1 MeV and the alpha particle, α, kinetic energy as 3.4 MeV. To achieve fusion of D and T the nuclei must overcome the electrostatic Coulomb barrier, which is approximately 0.4 MeV and corresponds to a temperature of

$$E = kT; T = \frac{0.4 \text{ MeV}}{0.862 \times 10^{-4} \text{ eV K}^{-1}} \approx 4.64 \times 10^9 \text{K}.$$

The Coulomb repulsion suggests the temperature of the fusion plasma must be on the order of a 10^{9}°C. However, due to quantum tunneling, the temperature required actually is decreased by a factor of ≈ 10, to 10^{8}°C [4].

Tritium (T) is radioactive and decays through the beta minus mechanism with a half-life of 12.32 years (5.5% per year). However, tritium can be bred in nuclear fission reactors (through the use of tritium-producing burnable absorber rods (TPBARs) [11,12] or as a by-product with transmutation of deuterium in the CANDU heavy water coolant. While these methods have been explored to temporarily supply fusion fuel [13]), real-time breeding of T within fusion reactors will ultimately be necessary to sustain fusion energy. The most likely nuclear reaction to breed T is:

$${}^6\text{Li} + n \rightarrow \text{T} + \alpha$$

$${}^7\text{Li} + n \rightarrow \text{T} + \alpha + n$$

with the natural relative abundance of ^{6}Li and ^{7}Li as 7.4% and 92.6%, respectively [2]. Lithium extracted from terrestrial supplies could supply future fusion power plants for ≈ 1000 years [14]. Moreover, lithium could be extracted from sea-water with the latest reserves estimated to last ≈ 6 million years [14]. The tritium supply for fusion is a key challenge to its success [15].

The "break-even" parameter to describe the performance of a fusion reactor is called the Q-value, defined as:

$$Q = \frac{E_{\text{fusion}}}{E_{\text{heat-in}}}$$

where E_{fusion} is the energy provided from the fusion reaction and $E_{heat-in}$ is the energy provided by the equipment to heat the plasma. For scientific break-even, $Q = 1$, and for ignition (where the reaction is self-sustaining) then $Q = \infty$ because $E_{heat-in} \rightarrow 0$. The ignition criterion was described in more detail by Lawson in 1958 [16]. His *Lawson Criterion*, defines a minimum value for the triple product of the plasma density, *n*, the ion temperature, *T*, and the confinement time, *τ*, in order to be a self-sustaining fusion reaction. For a typical D-T reaction, the Lawson Criterion is:

$$nTt \geq 3 \times 10^{21} \text{keV.s/m}^3.$$

While a fusion reactor will not necessarily have to be fully self-sustaining, obtaining core plasma conditions close to the Lawson Criterion are essential to demonstrate the cost-effective generation of electricity.

1.2 Magnetic confinement fusion

Magnetic confinement fusion (MCF), the most developed method to control a fusion plasma on earth, uses strong magnetic fields to confine the plasma to the required density and temperature. The ороидальная каМера с МагнитныМи катушкаМи (toroidal'naya kamera magnitnymi katushkami), or *tokamak*, fusion reactor concept has been the most successful at containing and controlling a fusion plasma [1]. The Eurofusion program [17], National Strategic Plan for US Plasma Research [18], Japan [19], South Korea [20] and China [21] all have decided that the tokamak device configuration is the highest-priority MCF concept. Due to the significant investment through governments, industry, universities, and national laboratories, only the tokamak-based fusion reactor will be explicitly considered within this book chapter. Many of the material challenges for D-T fusion energy, however, are largely agnostic to the particular magnetic configuration considered.

The basic principles behind the tokamak reactor are:

- A fusion plasma is a fully ionized gas that is dominated by long-range electric and magnetic fields and is a very good electrical conductor. Charged particles experience Lorentz forces and follow magnetic field lines.
- In a tokamak, a series of metallic coils are placed around a donut-shaped plasma chamber (torus) that provides the toroidal magnetic fields. The basic schematic of a tokamak-based reactor is shown in Fig. 3.1. Very strong fields (up to 10 T) are needed to contain the high-temperature plasma.
- The plasma can be considered the secondary windings of a transformer. A central solenoid is used to induce current within the plasma, which provides the initial plasma breakdown and temperature increase by ohmic heating.
- Because a pure toroidal magnetic field will induce a charge separation through the ΔB drift and curvature drift forces, superimposed onto the toroidal field is a weak poloidal field generated by the plasma flowing around the torus. It is the combination of these magnetic fields that result in a helical magnetic field that constrains the fusion plasma inside the torus.
- To maintain the high temperature of the plasma core a combination of ohmic heating, neutral beam injection heating, and radio-frequency heating can be used. Once the plasma

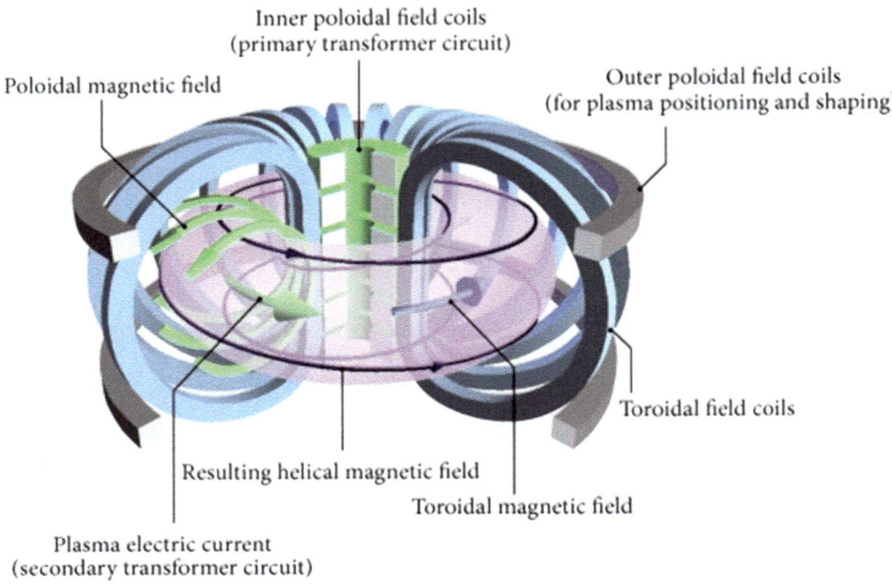

Figure 3.1 Schematic of a tokamak reactor and magnetic field profile. The combination of toroidal and poloidal magnetic fields results in a helical magnetic field.
Reproduced with permission from Ref. [22].

reaches a state where α particles also contribute significantly to the volumetric heating of the plasma (potentially still with some amount of external heating), it is this referred to as a "burning plasma."

The physics principles of tokamak plasma are described further in many textbooks, such as those by Wesson [23], Harms [5], and Friedberg [24].

The largest tokamak fusion reactors ever constructed are the Joint European Torus (JET) [1] and the Tokamak Fusion Test Reactor (TFTR) [25], which first demonstrated fusion power production with D-T fuel in 1997 and 1994, respectively. The fusion power as a function of time for some of these experimental reactors are found in Fig. 3.2. In addition to JET (now known as JET-ITER-Like-Wall, JET-ILW), other major operating tokamaks include DIII-D [26] in the United States, ASDEX-Upgrade in Germany [27], EAST in China [28], and JT-60U in Japan [29].

A new era of fusion energy started in 2005 with the international agreement for the construction site of ITER in Cadarache, France signed by USA, European Union, Russia, India, China, Japan, and South Korea [30]. Construction started in 2010 and is now in the 5-year assembly phase [31]. The goal for ITER ("Iter" means "the way" in Latin) is to demonstrate steady state plasma (>500 s), produce 500 MW(thermal) through fusion energy with only 50 MW(thermal) input to heat the plasma (i.e., $Q > 10$) [31,32]. The fusion reactor should begin with the first plasma shots (H/He) in 2025 and commission with D-T in 2035 [33].

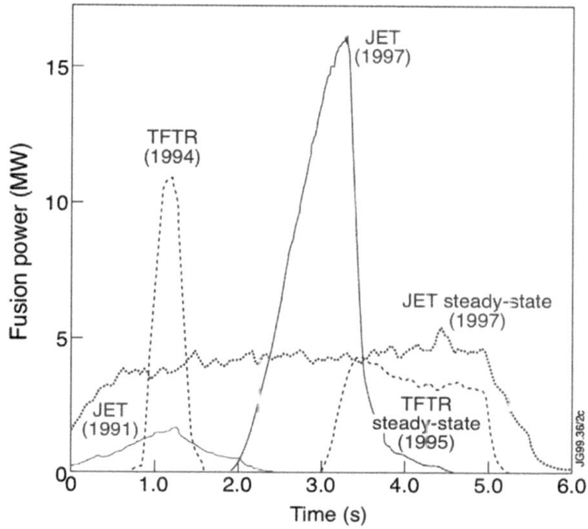

Figure 3.2 Fusion power produced by the D-T reaction campaigns for the JET and TFTR reactors.
Reproduced with permission from Ref. [6].

For the nuclear materials selection in and similar devices, the environment is relatively benign regarding creation of radiation damage, corrosion, high temperature creep, mechanical properties, high temperatures, and steady-state plasma operation (which creates plasma erosion). Therefore traditional nuclear materials have been used for the construction and operation of ITER, such as 316 stainless steel for the vacuum vessel, first wall support, and divertor supports [30,34]. The PFCs and first-wall materials for ITER will be tungsten and beryllium, and the breeder blanket designs will contain the first low activation structural material EUROFER97 [30,34,35]. However, these materials do not offer the performance needed and cannot be relied upon for power producing devices, which will operate at higher temperatures, over a longer lifetime with increased cumulative neutron irradiation and fusion plasma exposure. The ceramics that are discussed here are of interest to the future fusion power station configuration designs, such as DEMO, where the environment could be too extreme for these traditional materials.

1.3 Nuclear fusion power stations

The EU fusion roadmap published in 2018 [36] describes the "Demonstration Reactor" (DEMO), the successor to ITER, that will be capable of generating net electricity and operating with a closed tritium fuel cycle through the use of breeder blankets. It is this key component for the future of DEMO to sustain nuclear fusion as a possible energy source [6,36]. The roadmap predicts that the decision for construction will be in the early 2030s following from the ITER D-T campaign [36].

Table 3.1 Common expected radiation damage values and temperatures of fission and fusion reactors. DEMO values are highly dependent on the design and are subject to change.

Component	Temperature (C)	Radiation damage (dpa)	Ref
Reactor pressure vessel steel (40 years)	290	0.1–0.4	[39]
Fuel cladding of light water reactor	320	20–30	[40]
SiC–SiC fuel cladding for gas-cooled fast reactor (GFR)	800–1000	>100	[41]
JET divertor title	1000–1200	<0.001	[42]
ITER W divertor tile	800–1000	0.1–0.51	[42–44]
ITER stainless steel divertor structure	100–175	1.28–1.52	[44–46]
ITER first wall Be tile	600–1000	1.6	[42,47]
EU-DEMO1 W first wall	1000–1500	50–100	[42,48]
EU-DEMO1 first wall steel structure	400–500	50–100	[48]
EU-DEMO1 W divertor tile	1000–1800	30–50	[38,49]
ARIES-AT first wall silicon carbide (SiC)	700–996	200	[45,50]
ARIES-AT SiC divertor tile	700–996	200	[45]

The predicted levels of radiation (neutron) damage in DEMO materials are highly dependent on the tokamak design, breeder blanket design choice, and plasma parameters. EU-DEMO1 is the nearer-term DEMO variant, which uses a more conservative baseline design. The specifications for EU-DEMO1 will be used to fix these parameters [37] for this chapter. The predicted neutron flux for the first wall surface is 8.25×10^{14} n/cm^2/s, which corresponds to a ~18 dpa per full power year (fpy) into tungsten. W PFCs will be subjected to ~9 dpa/fpy and the W divertor tiles will be subjected to ~7 dpa/fpy [38]. A comparison of the common radiation damage values and temperatures for components within fission and fusion reactors is shown in Table 3.1. In addition, for all the types of applications listed, the cost of R&D, raw material sources, and manufacturing must be considered. The increase in temperature and neutron damage predicted in DEMO-like fusion power station concepts could eliminate traditional structural, radiation shielding, and functional materials from consideration for use [7,38,51]. The next section will discuss the criteria that must be considered when developing new materials for fusion energy.

1.4 Material selection criteria

The properties that must be considered for any new structural and functional materials in DEMO-like fusion reactors can be categorized in two groups: general and nuclear properties.

- General properties include factors such as mechanical strength, ductility, dimensional stability, fabricability, fracture toughness, maximum operating temperature, cost, availability, and thermal conductivity.

- Nuclear properties include factors such as neutron absorption, moderation and reflectivity characteristics, susceptibility to induced radioactivity, resistance to radiation damage, helium embrittlement, and shielding performance.

The materials selection criteria must not only consider the material and nuclear properties, but also the overall system parameters. These parameters could impact the acceptability and performance limits of the material (e.g., the material surpasses the acceptable limit for residual radioactivity after use). Advanced structural and functional materials can improve reactor performance through increased safety margins and design flexibility, in particular by increasing the tensile strength, thermal creep resistance and providing superior neutron radiation damage resistance. A balance must be obtained between the maturity of the material (standardizations, nuclear history, known limits), safety margins, functionality, material properties, and existing knowledge gaps.

Major criteria for selecting existing and new ceramic materials for fusion reactions are summarized Table 3.2. The list is not exhaustive but instead provides a high-level understanding of the needs and desired properties of materials for each type of component. These parameters have been separated into the component system because each system has different general and nuclear requirements. The implementation of new ceramic and ceramic composite materials in any of these application areas would also involve additional investments in R&D, raw material sourcing, and manufacturing scale-up.

Some ceramic components have been proposed for the structural materials of the breeder blanket (see Section 2 for further details). The structural material within these components could be considered as a primary safety barrier and their structural integrity must be guaranteed throughout life. Thus, the applicability of using nuclear design codes might be required in the future. There are three main codes that are used for the design and construction of components for nuclear power plants; American Society of Mechanical Engineers (ASME) Boiler and Pressure Vessel Code (BPVC) [52], Japanese Society of Mechanical Engineers (JSME) Code for Nuclear Power Generation Facilities [53], and the European RCC-MRx code (which has been extended for ITER) [54]. The purpose of these nuclear design codes is to establish national or international standards that consist of a set of rules based on state-of-the-art knowledge, experience, and experimental feedback from nuclear facilities. The codes provide the bridge between different suppliers, participants, researchers, designers, manufacturers, and regulators. The ASME BPVC Section 3 (nuclear installations) Division 4 "Fusion Energy Devices" has formed and, at the time of writing, is compiling the nuclear code design that will cover in-vessel components [55,56]. If such ceramic-based structural materials should be used in the breeder blankets of future fusion power stations, the use of Division 4 could be required, at least in the USA.

One of the major advantages of fusion over fission energy is the lack of high-level nuclear waste production, or sometimes referred to as "long-lived" radioactive waste [57]. In addition, the fusion community discussed methods to reduce the volume of radioactive waste generated by neutron activation. The U.S. Department of Energy in 1982 set out a criterion to limit the radioactive lifetime of fusion waste materials

Table 3.2 General criteria for selecting existing and new materials for fusion reactors.

Criterion	Structural	Radiation shielding	First-wall
Location	*Divertor, breeder blankets, first wall, flow channel inserts*	*Central column*	*Plasma-facing component*
Mechanical properties under irradiation	Required	Not necessary if not also a structural component	Limited
Thermal properties	Important to limited, depending on location	Limited	Critical
Joining	Required	Not necessary	Not necessary[a]
Primary radiation safety barrier	Possibly	No	Yes
Manufacturing standardization and nuclear code qualifications	Yes, if safety barrier	Not necessary due to not being a safety barrier for radiation release	Yes
Low activation	Required	Required	Required
Plasma-facing performance	Minimal	Not necessary	Required
Manufacturability (joining, inspection, forming, etc.)	Very important	Not necessary, depending on design	Important
Tritium barrier	Yes	No	Potentially
Coolant medium interaction	Yes	Depending on design	No
Ferromagnetic properties	Need to be considered	Need to be considered	Need to be considered
Neutron shielding	Not necessary	Yes	No (want PFCs to be neutron transparent)

[a]Dependent on the design—ITER tiles, for example, are bolted onto the structure.

compared with fission materials [58]. This criterion implemented in 1982 introduced the concept of "reduced-activation" or "low-activation" fusion materials, and the definition is provided by Gorley [57]:

> *The materials selection for fusion energy's nuclear waste production, after an initial ~100 years removal from the reactor, can be disposed of in low-level waste repositories.*

This criterion has been the focal point of all fusion structural and functional material development [59]. For example, the fusion materials community have determined that the reduced-activation structural steels would be based on Fe–9Cr materials, such as EUROFER97 reduced-activation steel [60]. For this purpose reduced-activation (or low-activation) structural steels were designed, and neutronics modeling concluded these steels would meet the low-activation criterion and hence only be classified as low-level waste (LLW) [61–64]. Any proposed ceramic-based material that would be exposed to the fusion neutrons, and, thus, subsequent activation, implies that any existing or new ceramic-based material for fusion should meet this criterion. This is one criterion that sets apart the development of fusion materials compared with fission materials.

1.5 Motivation for new structural and functional materials

The motivation for this chapter is to update the fusion and fission community on the latest developments of ceramics for use as the structural, radiation shielding, and plasma-facing components materials. The high radiation damage and temperature conditions created in DEMO-like fusion reactors creates an environment that requires new structural and functional materials. The remainder of this chapter is organized as follows: Section 2 discusses the latest developments in the structural ceramics for breeder blankets and divertors. Section 3 discusses the latest developments in the radiation shielding field for fusion reactors (mainly the central column to protect the magnet core). Finally, Section 4 discusses the developments in ceramics for plasma-facing components. Established and emerging ceramic concepts for each component are discussed and areas of current and future work are identified.

2. Structural components

2.1 Introduction

Strict constraints are placed on the selection of structural materials for fusion due to the harsh electromagnetic, neutronic, thermal, and chemical conditions they withstand. The exact requirements for such materials vary depending on the exact function and location within the vessel, but all will be required to maintain temperatures of >600°C without notable degradation of bulk properties for the duration of one maintenance cycle—likely several years of >90% duty cycle operation. Higher temperatures in any thermodynamic conversion cycle are preferable in the interest of thermodynamic efficiency, but tensions exist between component robustness and operating temperature. Near the high heat flux regions of the plasma-facing surface–the diverter armor–thermal conductivities of 80–100 W/mK are also required, and 15–40 W/mK is required for materials on the remainder of the first wall. These high thermal conductivity values can be a challenge for some ceramics, such as SiC, which undergo irradiation-induced thermal conductivity degradation. In contrast, some components deeper in the reactor structure (e.g., flow-channel inserts in breeder blankets) will require very low thermal conductivities, of the order 1–2 W/mK. This section focuses on materials that serve a structural purpose, i.e., those that experience significant mechanical loading during normal operation. Materials that function primarily as radiation shielding are discussed in Section 3, and materials that function to shield the remainder of the structure from energetic ion bombardment are considered separately in Section 4.

Structural materials must maintain their thermal and/or mechanical properties under the fusion neutron spectrum, which peaks near 14 MeV for components close to the surface but softens slightly moving deeper into the breeder blanket. This high fusion neutron energy spectrum causes additional material performance challenges, such as transmutation [65], which are not encountered in the thermal neutron spectrum found in a light water fission reactor. All in-vessel components will receive an integrated dose of at least 10 dpa between maintenance cycles, although the exact duty cycle of reactor maintenance can vary significantly depending on the chosen reactor design-for example, whether the design includes demountable high-temperature superconducting (HTS) magnetic field coils. The materials must also lie below acceptable levels of He generation through transmutation (<100 appm) to prevent significant embrittlement, creep, or other mechanical deficiencies.

Other considerations manifest for more specific areas of the reactor structure. For materials inside the vacuum vessel, only low sublimation levels and low outgassing rates are acceptable. Inside the breeding blanket chemical compatibility with corrosive lithium compounds comprising the breeding material are essential. Waste classification ratings, while not affecting the technical feasibility of the reactor design, can also be an important cost driver in the overall engineering design, as discussed in Section 1.4. Finally, the manufacturing capability for various components must be considered; for example, it is preferable to leverage existing technologies, or moderate extrapolations thereof, for large components that must be fabricated at scale. For ceramics, this often implies that innovative joining methods must be developed, which is discussed in more detail in Section 2.2.2 below.

Three general classes of materials are considered relevant structural materials for fusion applications. The operating range for each of these material classes is summarized in Fig. 3.3. The first candidates are the nuclear-grade steels: reduced activation ferritic/martensitic (RAFM) high-Cr steels and the oxide dispersion strengthened (ODS) ferritic steel variants. These alloys are available at relatively high Technology Readiness Level (TRL) due to synergistic developments for aerospace and Generation-IV fission applications but are limited by their creep strength to relatively low temperature design point (<700°C) [67]. The second category is refractory metals, of which tungsten is the most widely considered. Machining of refractory metals is substantially more difficult than steels, with specialized tooling required. However, their high melting points offer a higher-temperature operational window (>800°C) that is not considered feasible with steels. Mitigating the loss of strength and ductility of refractory metals under neutron irradiation remains a key challenge. The third category, discussed in the present section, is ceramic matrix composites (CMCs). Ceramics generally maintain mechanical strength and thermal properties under high irradiation levels [68] but are at the lowest TRL of any of these three material

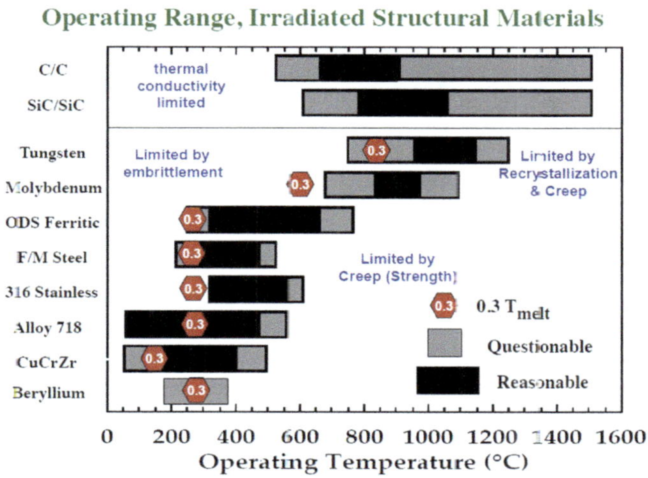

Figure 3.3 Summary of the thermal operating range of various classes of structural materials proposed for fusion applications. Operating temperatures ranges for certain Ultra-High Temperature Ceramics will exceed those of C/C and SiC/SiC composites.
Reproduced with permission from Ref. [66].

classes. Complementary Gen-IV fission development work and work on CMC applications in LWRs is also ongoing and often highly leverageable for fusion applications. Recent advances, current challenges, and emerging opportunities for ceramic structural materials are the focus of the present section.

2.2 Silicon carbide

Silicon carbide (SiC) has been recognized as a promising refractory nuclear material since the development of high-temperature gas-cooled fission reactors in the 1960s [69]. SiC demonstrates excellent strength, creep resistance and oxidation resistance at high temperatures [70,71], and is relatively inert with a wide range of liquid metals proposed as coolants and breeder materials. Like most ceramics, it has very low permeability to hydrogenic species, including tritium [72]. Importantly, SiC also demonstrates high radiation resistance with low activation and decay heat relative to the metallic options for fusion structural materials [73]. The history of SiC development for nuclear applications has recently been reviewed [74] and performance of bulk SiC in irradiation environments is also well documented [75].

2.2.1 SiC$_f$/SiC fiber-reinforced composites

The low fracture toughness and brittle failure of bulk (monolithic) ceramics, including SiC, prevent their use in nuclear components that undergo any serious mechanical loading. Ceramics are thus commonly impregnated with continuous fibers to improve toughness when used for structural purposes. This process introduces a mechanism for dissipation of local stresses in the material matrix, thereby reducing microcrack formation and propagation, and increasing overall toughness, effectively turning a brittle material into a quasi-ductile one. Continuous silicon carbide fiber was first developed in the 1970s [76] and the potential for high-strength, radiation-resistant SiC-based fiber composites for fusion and fusion structural applications was quickly recognized.

SiC/SiC technology development has progressed substantially in the intervening years to the point at which modern, nuclear grade SiC fiber composites are being investigated for accident-tolerant fuel cladding for existing LWRs and Generation-IV fission designs [77]. They are also given serious consideration for high-temperature fusion reactor designs [50,78–80], see Fig. 3.4 below. 2D woven SiC/SiC composites with very predictable mechanical properties that far surpass those of the pure ceramic have been commercially available for over a decade. For example, SiC composites are routinely fabricated with a fracture resistance ~3 kJ/m^2 [68], relative to a value of 30–50 J/m^2 for monolithic SiC material [75]. Detailed properties of

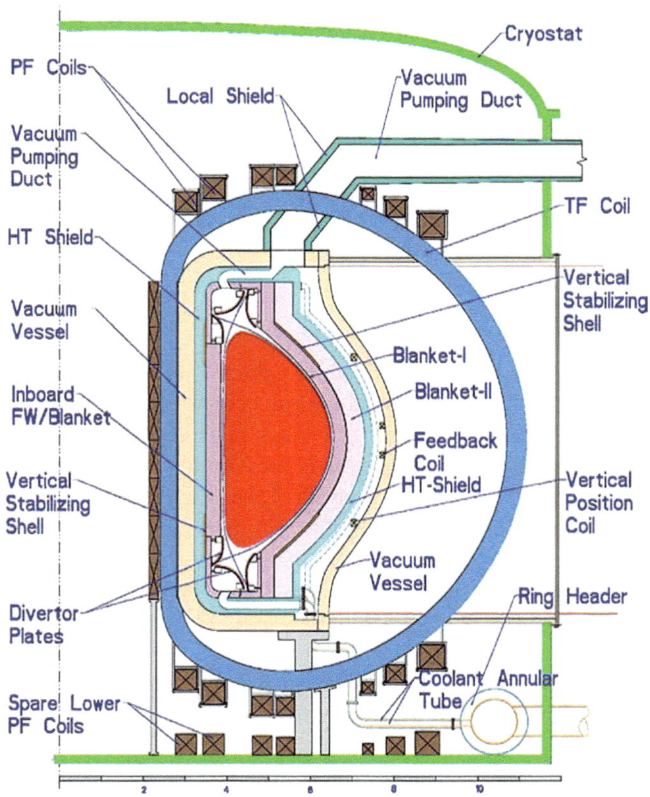

Figure 3.4 Cross section of the ARIES-AT fusion reactor core design.
Reprinted with permission from Ref. [50]. SiC components ("Blanket-I" and "Blanket-II") are illustrated in pink.

SiC/SiC composites and their evolution under irradiation damage are discussed in several recent reviews [68,81], and in Chapter 4 of this book.

2.2.2 Recent advances and outstanding research needs

Recent reviews of the outstanding design issues and research needs for SiC fiber composites can be found elsewhere [82,83] but are briefly reviewed here. Development of joining technologies, both to join SiC/SiC to itself and other materials, is essential to develop integrated blanket structures in the fusion reactor core. A variety of like–like and like–unlike joining technologies have been invented for SiC composites, including diffusion bonding, displacement reaction bonding, pressure-assisted liquid-phase sintering, transient eutectic-phase joining, hybrid preceramic polymer and chemical vapor deposition (HSiC), and glass-ceramic tailored bonding layers [81,84,85]. Databases on the robustness of these joints under irradiation conditions are more limited. Initial results from the High Fluence Isotope Reactor (HFIR) at Oak Ridge National Laboratory, TN, USA indicate that SiC/SiC specimens joined using some of the formulations listed above and then irradiated to 5 dpa show notable microstructural evolution in their joints but limited degradation of thermo-mechanical properties [84]. More recent SiC diffusion-bonded samples irradiated up to 20 dpa at 530°C also indicated minimal degradation of the torsional strength of the joints [85].

Chemical compatibility of the SiC-coolant system within the breeder blanket is another area of active research. Corrosion of silicon carbide by pure molten lithium tends to degrade the strength of SiC_f/SiC at temperatures above 600°C, but an increase of fracture toughness is also noted due to the stress relief mechanisms introduced by Li intercalation [86]. Chemical Vapor Deposition (CVD) SiC surfaces and SiC_f/SiC appear to be more robust to corrosion by static lithium-lead eutectics (LLEs) at temperatures as high as 900°C after 1000s of hours of exposure [87,88]. Some corrosion-namely, formation of a SiO_2 oxide layer-is observed during exposures to flowing LLE systems but a saturation of this effect occurs after 2000–3000 h [87]. The potential for enhanced corrosion effects due to radiation-induced defect formation is being further investigated [89].

Evaluation of potential degradation of the thermo-mechanical properties of SiC_f/SiC under high-fluence irradiations (up to 100 dpa) remains in progress. Hi-Nicalon Type S SiC composites irradiated at $\leq$40 dpa at temperatures ranging from 300–800°C have generally demonstrated minimal degradation of flexural strength or limit stresses [68,81]. More recent evaluations of these same composites at 70–100 dpa, however, have found "severe" degradation of mechanical properties of these materials at irradiation temperatures of 300°C and "moderate" degradation at 800°C. Saturation of swelling and thermal conductivity evolution were found to occur at significantly lower damage levels [90]. Substantial microstructural evolution occurs at irradiation temperatures of 300°C but the effect becomes significantly less pronounced at 800°C [91]. The mechanical properties of the composites can be recovered to some extent by annealing beyond the irradiation temperature [90].

Other potential challenges facing monolithic SiC and SiC-CMCs for fusion include resistance against H/He permeation [92], transmutation effects [93], and irradiation-induced degradation of thermal conductivity [75]. In addition, a more fundamental understanding of how SiC microstructural evolution affects the mechanical properties of the composite material has only recently begun to be explored [94].

2.3 SiC-tungsten alloys

A SiC-tungsten composite alloy represents a possible pathway to leverage the strength under fusion neutron irradiation of SiC composites with the thermal conductivity and erosion resistance of refractory metals. Si–C–W systems are essentially chemically inert below temperatures of 1000°C, but above 1200°C formation of tungsten silicides and carbides occurs through diffusion coupling. An eventual four-phase equilibrium ($W/W_5Si_3/WC/SiC$) occurs between the two materials around 1400°C [95]. SiC–W composites have been successfully fabricated using various sintering techniques including plasma spraying [96], hot pressing [97], and diffusion bonding [98,99]. Most recent work has focused on the latter approach, which demonstrates promising mechanical properties approaching the constituent materials. The formation of native oxide layers of low apparent thermal conductivity at the SiC–W interface, however, has proven difficult to control and leads to unacceptable thermal stresses at heat fluxes as low as several MW/m^2 [100].

2.4 Emerging ultra-high temperature ceramics (UHTCs)

Due to their high strength and creep resistance at extreme temperatures, the ultrahigh-temperature ceramics (UHTCs) show promise for structural (and possibly plasma-facing, Section 4.2.1) applications in fusion [101]. UHTCs are commonly defined as those with melting points exceeding 3000°C [102]. The only materials that meet this 3000°C threshold are the carbides, borides and nitrides of the group IV and V transition metals (Ti, Zr, Hf, V, Nb, Ta) [103]. However, the nitrides are not included here due to their relatively high vapor pressures, concerns about neutron activation to C-14 [104], and relatively poor thermal conductivity. The focus is on the monocarbide (MC) and diboride (M_2B) compounds, which are the most refractory in their respective binary systems.

A thermal stress figure of merit, equal to $M = \sigma k_{th}(1 - \nu)/\alpha E$, (where σ is the material's strength, k_{th} is the thermal conductivity, ν is Poisson's ratio, α is the linear coefficient of thermal expansion, and E is the modulus), is plotted as a function of temperature for SiC and various UHTCs in Fig. 3.5 along with RAFM steels and tungsten for comparison. UTHCs have the potential to dramatically expand the thermal operating range for nuclear materials. In addition, SiC and MAX-phase-based materials may be able to bridge the temperature gap between pure W and RAFM steels, allowing for production of more robust functional components with appreciable thermal gradients, such as will be present in the diverter region of a fusion reactor.

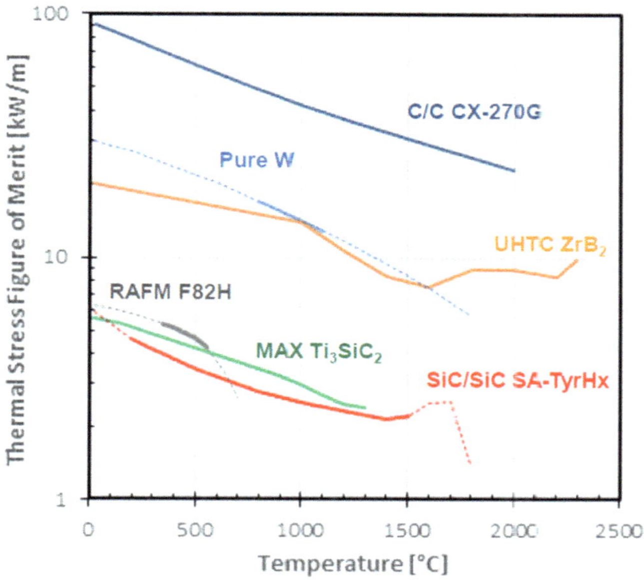

Figure 3.5 Thermal stress figure of merit as a function of material temperature for various UHTCs (including SiC), RAFM steel, and tungsten. The solid lines represent the likely operational space for each material and dotted lines represent questionable operational space. Figure reproduced with permission from Ref. [105].

2.4.1 Thermophysical properties

The two classes of material differ in thermophysical properties, due to their different crystal structures and defect behavior. In the carbides, the carbon atoms sit at the octahedral interstitial sites of a face-centered cubic (FCC) metallic sublattice, with Fm-3m space group. Appreciable substoichiometry is possible, e.g., 15–20 at.% in the case of TiC, which is accommodated by carbon vacancies. Thermophysical properties thus vary strongly with carbon content. By contrast, the borides show a hexagonal P6/mmm space group and accept little deviation from stoichiometry.

Table 3.3 shows some typical thermophysical properties for both the diborides [102,109] and carbides [106–108]. The melting points, T_m, of both carbides and borides increase with increasing period, with the most refractory compounds being Hf and Ta (3928–3950°C for carbides and 3200–3250°C for borides, respectively). The T_m of carbides are generally higher.

For the thermal conductivity, the group IV metal diborides show higher values than the carbides (in the range 60–120 W/m K). These materials show good thermal conductivity owing to their highly metallic bonding character. For the carbides, thermal conductivity decreases monotonically with temperature and increases with C content in the substoichiometric range, peaking at 50 at.% C. Table 3.3 shows a range of values for the carbides [107,108] while for the borides the maximum literature values are given. The measured thermal conductivity values in Table 3.3 should be regarded

Table 3.3 Properties of the group IV and V carbides and diborides.

	Carbides						Diborides					
	Group IV			Group V			Group IV			Group V		
Metallic element	**Ti**	**Zr**	**Hf**	**V**	**Nb**	**Ta**	**Ti**	**Zr**	**Hf**	**V**	**Nb**	**Ta**
Density (g/cc)	4.91	6.59	12.67	5.65	7.79	14.5	4.52	6.09	11.2	5.1	7.21	12.6
Melting point (∘C)	3067	3420	3928	2830	3600	3950	2980	3040	3250	2100	3050	3200
Expansivity ($\times 10^6$/K)	7.4	6.7	6.6	7.2	6.6	6.3	6.6	6.6	6.8	7.6	8	8.2
Conductivity (W/m-K)	6–30	15–40	5–30	10–40	15–40	15–35	120	127	104		16.7	10.9

Density, melting point and expansivity from Ref. [106]. Carbide conductivities from Refs. [107,108]. Diboride conductivities from Ref. [103] and [106].

as lower-bounds, due to the strongly negative influence of contamination and porosity. For example, Lonergan et al. demonstrated that the conductivity of ZrB_2 could be increased by ~30%–127 W/m-K by reaction hot-pressing ZrH_2 and B, thus removing porosity and hafnium/oxygen impurities [110]. Significant improvements in thermal conductivity values in Table 3.3 may be possible in other systems following a similar approach. The UHTCs also show relatively low thermal expansion coefficients ($6–8.2 \times 10^{-6}$/K). The best combination of high thermal conductivity and low thermal expansion is shown for the group IV diborides.

2.4.2 *Manufacturability*

The UHTCs have been little considered as fusion reactor structural materials due to their relatively low technology readiness levels in manufacturing and joining. Some of these issues are being addressed by development in aerospace applications [102]. The primary challenge is the high sintering temperature required for UHTC fabrication. Consolidation of powders into fully dense parts often requires advanced sintering techniques such as Spark-Plasma-Sintering (SPS) or vacuum Hot Pressing (HP). HP is capable of manufacturing more complex geometries, e.g., curved plates, and larger samples, e.g., up to 700 mm in linear dimension. By contrast, SPS machines are generally limited to cylindrical plates and smaller (tens of mm) sized samples [78], although 400 mm sized plates of monolithic carbides (WC) have been reported using a hybrid SPS/HP process [111], and an electric field assisted sintering (EFAS) system with a 2-foot diameter part capacity was recently installed at Idaho National Laboratory [111a].

Brazing of UHTCs is limited by their poor wettability and the relatively low liquidus temperature of typical brazes. This has been addressed by the development of refractory Pd-based brazes with liquid temperatures of ~1500K, leading to sound, defect free joints in ZrB_2-based ceramics [112]. Irradiation stability of braze compositions also must be more thoroughly demonstrated. For very high temperature applications, brazeless joining technology may be necessary. Although welding is not possible for oxide-based UHTCs, due to their typically low electrical conductivities, the welding of carbides and diborides is possible due to their high electrical conductivities. Plasma-arc welding (PAW) has been demonstrated in ZrB_2–ZrC ceramics [113]. Preheating of the parts to ~1500°C was required to reduce thermal shock-induced cracking, however reasonable joint strengths of 150–250 MPa have been achieved [114].

2.4.3 *Mechanical properties*

The UHTCs are hard and brittle at ambient conditions, which poses a challenge for low temperature operation. The Young's modulus of the carbides range from ~250 to 580 GPa [106]. Tabulated values are not provided here as the exact modulus is sensitive to compositional changes. For example, the Young's modulus of ZrC_{1-x} increases with C content between 270 and 400 GPa in the range 40–50 at.% C [107]. Hardness values are similarly sensitive, with ZrC_{1-x} varying from 16–31 GPa in the same range [107]. Fracture toughness values are typical for monolithic ceramics, being in the range 3–5 MPa $m^{1/2}$, although values of 8.6 have been reported for nanostructured NbC [115] and as high as 12.7 MPa $m^{1/2}$, in two phase TaC–Ta_4C_3 ceramics [116]. Flexural strength also increases with carbon content. For example, in ZrC, room temperature values are typically ~200–300 MPa. The carbides undergo a brittle-to-ductile transition around 2000°C, whereupon strength locally peaks and elongation to failure monotonically increases [117].

The properties of diborides are similar to the carbides but show less spread from stoichiometry variation. For example, in ZrB_2, the range for the Young's modulus values are 489–493 GPa, for hardness 21–23 GPa, and for toughness 3.5–4.3 MPa $m^{1/2}$ [103]. The flexural strength is ~565 MPa in pure form but with SiC or $MoSi_2$ additions can exceed 1 GPa [103].

The high temperature strength and creep resistance of the UHTCs is one of their principal advantages. Fig. 3.6 shows a typical deformation behavior of ZrC and ZrB_2, compared with two structural metallic alloys: 316 stainless steel and pure W. The solid lines show the limiting creep life for ~1% strain over 3 years (10^{-10} s^{-1}) and the dotted lines show the yield stress contours (10^{-2} s^{-1}). For components operating at a constant stress of 50 MPa, UHTCs have higher maximum operating temperatures (e.g., the creep life is in the range 1100–1500°C, which is significantly higher than W (~900°C) or 316 stainless (~500°C). We note that ODS steels would show improved temperature capability over unreinforced steels, although the improvement in ODS creep life is generally limited to the order of ~100°C (at least when HT9 steel and ODS-9Cr are compared in the range 650–750°C [118]). ZrB_2 shows better creep resistance than ZrC up to ~1500°C. The softening of carbides at more modest

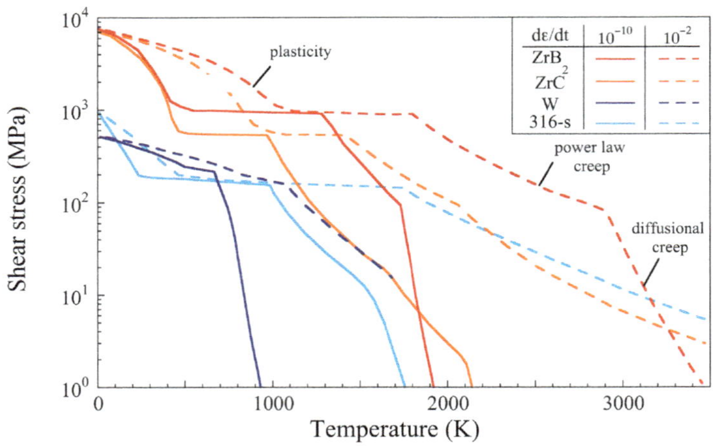

Figure 3.6 Deformation mechanism map for tungsten and 316 stainless steel [120] versus ZrC and ZrB_2 [119], showing two strain rate contours: 10^{-10} and 10^{-2}, which can be used to predict creep stresses (1% over 3 years) and yield strengths respectively.

temperatures may be a more common feature across all the group IV and V compounds, as demonstrated by high temperature indentations [119].

2.4.4 Irradiation damage

Zirconium carbide is the most widely studied UTHC in terms of irradiation damage properties due to its potential replacement of SiC fuel cladding and/or TRISO particles [121]. The available irradiation damage data has been reviewed [122,123]. The material shows good tolerance to swelling. Below 1000°C, the hardness, modulus, and fracture toughness all increase and thermal conductivity decreases, however above 1000°C there is no change according to the limited available data [123]. The main transmutation product in the carbides is Be and He, which form through the (n,α) reaction. Swelling was negligible at stoichiometric compositions, although it increases dramatically with increasing carbon deficiency, e.g., a factor of 8 increase between $ZrC_{0.94}$ and $ZrC_{0.73}$, which may be connected with the ease of carbon being displaced into vacant octahedral sites in substoichiometric compositions. Irradiation data on other transition metal carbides is less well established [124]. However, neutron irradiations of TiC to ~2 dpa have recently been performed at 220–1150°C [124].

Irradiation damage of metal diborides is less well understood. Due to the high neutron absorption cross section for boron-10 (natural abundance ~19.9%), significant irradiation damage occurs due to the neutron absorption reaction $n + {}^{10}B \rightarrow {}^{4}He + {}^{7}Li$. There is little experimental neutron-irradiation data for diborides. The most comprehensive study was performed by Grinik et al., who neutron irradiated (E > 0.1 MeV, i. e., fast neutron spectrum) all the group IV, V and VI compounds (M = Ti, Zr, Hf, V, Nb, Cr, Mo, Nb, W) [125]. The materials were highly porous (generally 5%–10%). Below 1200°C, samples fractured catastrophically at a dose that increased with temperature (for example, at 100°C, this was ~1×10^{24} neutrons/m^2). Above 1200°C the materials swelled but did not fracture.

The transmutation product accommodation in ZrB_2 has been studied from first principles [126]. Helium is able to diffuse with a low activation energy in interstitial sites, and is therefore likely to exit the material, while Li is stable in the vacant B sites [126]. Irradiation damage tolerance is improved in the diborides with boron-11 enrichment due to its lower cross section for neutron capture. ZrB_2 with natural abundancy shows improved He retention compared with B-10 enriched [127]. Helium release in both samples accelerated when irradiated above 540°C. He accommodation in ZrB_2 at 650–850°C has been recently studied, showing similar nanostructured helium bubbles as compared with tungsten in the same temperature range [128]. Similarly, B-11 enriched TiB_2 shows good irradiation damage tolerance under fast neutron irradiation at 200 and 600°C to 2.4×10^{25} n/m^2 (~4.2 dpa) [129]. Swelling was minimal (~1%) with unexpectedly greater expansion in the a-axis than c-axis through XRD. TEM studies show that the anisotropic swelling is due to prism plane loops being two orders of magnitude higher than basal loops [130].

2.4.5 Other considerations

UHTCs also have good tritium permeation barrier properties [131]. For example, a 2.5 μm TiC coating on steel reduced permeation by 5–6 orders of magnitude [132]. A much lower reduction was reported by Forcey et al. [133], indicating film quality is important. The tritium barrier properties of the diborides are also good, as demonstrated in VB_2 [134]. Hydrogen behavior on TiB_2 coatings has been studied in terms of transport rates [134,135] and retention [136].

2.4.6 High-entropy ceramics

High-entropy ceramics (HECs), comprising four or more metal species of equal or near-equal stoichiometry, may have improved irradiation damage tolerance as compared with single component ceramics [137]. While fusion engineering-related research on HECs remains in its infancy, the general properties of these materials are well established [138]. Irradiation damage has been investigated in the Hf-Zr-Ta-Nb-Ti carbide system [137]. The extent and origin of the proposed irradiation damage tolerance of HECs is not yet well understood. Mechanisms similar to high-entropy alloys, such as reduced defect mobility and energy dissipation, have been proposed [137].

HECs could have improved mechanical properties over single-component ceramics. For example, the hardness is higher than the rule-of-mixtures prediction in high entropy boride ceramics (HEBCs) [139–141] and high entropy carbide ceramics (HECCs) [142]. The creep resistance of HECCs is also superior to single-component ceramics by a factor of 10 or more in the range 1400–1600°C [143]. An advantage in terms of fracture toughness is not so clear; values for HEBCs are relatively low at 3.6–4.5 MPa-$m^{1/2}$ [141], and more work is needed to determine the advantage (if any) over single component borides. HECCs [144] and HEBCs [139] also show improved oxidation resistance.

Despite some promise, there are clear disadvantages of the high entropy approach. Thermal stability is often highlighted as a benefit, but the melting points of HECs will likely be lower than their single component counterparts because the entropy of the liquid will also be increased. Furthermore, thermal conductivity degrades due to solid-solution scattering effects; for HEBCs, thermal conductivity falls to 10%–20% that of single-component borides [139,145]. There are also challenges in fabrication of HECs. Due to their reduced defect mobility, full density is only achievable using advanced processing methods such as spark-plasma-sintering and joining techniques are yet to be addressed. More research is needed to assess their suitability for fusion structural materials.

2.5 Summary

Ceramics show promise as structural materials within a fusion reactor in locations expected to reach high temperatures (~1000°C). The ceramic at the highest TRL is currently silicon carbide, which is undergoing active, targeted R&D for fission and fusion applications at many different institutions around the world [74,75,81,82]. Current key areas of study include understanding corrosion effects with Li-based breeder materials, high-fluence irradiation effects on mechanical properties, and radiation-resistant joining technologies. Some other UHTCs also show promise but these ceramics require significant additional study to understand, for example, the effect of neutron damage on their physical properties and the capability for lower-temperature operation (<1000°C), before they can reach commercial manufacturing maturity.

3. Radiation shielding

3.1 Radiation shielding in fusion reactors

Radiation is present in all nuclear fission and fusion reactors; nuclear engineers must design radiation shielding to (a) mitigate damage to critical components, (b) protect operational and maintenance personnel, and (c) protect the public at large. Biological shielding is used in the latter two protection methods, however, in this chapter the former will be discussed; radiation shielding materials that are used to mitigate the damage to critical components within a fusion reactor.

As discussed in Section 1.1, the fusion D-T reaction produces highly energetic neutrons (14.1 MeV on average but there is a spectrum from eV to MeV due to the moderation from the plasma and tokamak materials), which penetrate the walls of the tokamak device. Tokamak reactor designs have a central column to maintain the plasma aspect ratio. Neutrons are known to degrade the performance of superconducting materials, necessitating the use of shielding [146,147]. The tight geometry of tokamak reactors (especially spherical tokamaks, which have smaller central columns) concentrates two technical challenges that must be overcome for an energy source to be viable on the commercial scale:

(1) The challenge to mitigate neutron and gamma-ray radiation damage to the superconducting central column solenoid.
(2) The challenge to mitigate heat deposition into the superconducting central column solenoid to maintain sufficient cryogen temperature cooling.

The operational lifetime of the center column superconducting coil could determine the fusion reactor's overall operational lifetime and economic viability [148–150]. The volumetric constraints unique to spherical tokamaks demands maximum effectiveness of the radiation shield and therefore defines the minimum device size. A schematic view of the radiation shielding for a tokamak device is shown in Fig. 3.7. Fusion creates a unique radiation shielding challenge: highly penetrating neutrons, high neutron flux, and tight volumetric constraints that force a reevaluation of the traditional fission approach to radiation shielding.

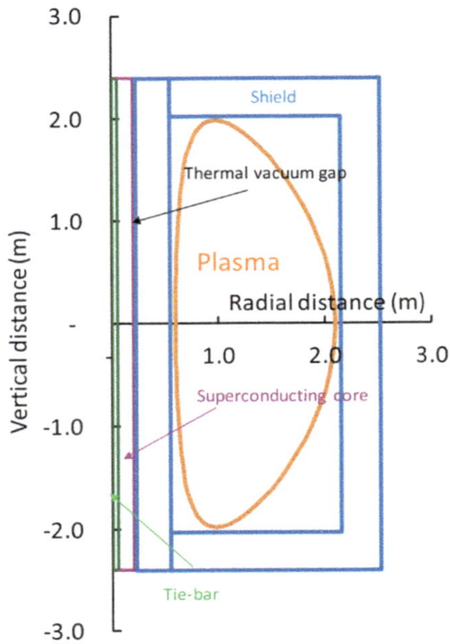

Figure 3.7 A schematic layout of a tokamak reactor and the central column (at radial distance 0 m).
Reproduced with permission from Ref. [151].

3.2 Introduction to radiation shielding

Radiation, in this context, is the product of nuclear processes and can be categorized into charged and uncharged particles. The former includes electrons, protons, and ions, and the latter is electromagnetic radiation (X-ray, gamma-rays) and neutrons. It is this latter that will be of concern in this chapter due to these two radiation forms being prominent within tokamak radiation shielding materials. The energy of the radiation expected is generally within the ranges from eV to MeV. Two excellent textbooks on the fundamentals of radiation shielding and radiation detection are "Radiation Shielding" by Shultis and Faw [152] and "Radiation Detection and Measurement" by Knoll [153], respectively.

Here we are concerned with both neutron and gamma-ray radiation shielding. To shield against neutrons, the material must attenuate the neutron energy and absorb the particle through nuclear processes. The performance measurements of these two factors are characterized as the moderating ability and macroscopic cross section for neutron absorption, respectively. For gamma-ray shielding, the material should possess good gamma-ray scattering and absorption coefficients. The attenuation of gamma-rays is by the interaction between the rays and the shielding material's electrons and electromagnetic field surrounding the nucleus. Thus, heavy elements have large quantities of electrons possessing good gamma-ray scattering and absorption

Table 3.4 The main features neutron and gamma-ray interactions with matter.

Property	Neutrons (thermal energy, 0.025 eV)	Gamma-rays (1 MeV)
Interaction	Nuclear	Electromagnetic
Interaction strength	Weak	Moderate
Penetration depth	cm	mm
μ function of Z	Irregular	Regular (near linear)
Low Z sensitivity	Yes	No
Shield thickness	~0.1–1 m	~1–10 mm

coefficients. The main features of both these radiation types with matter are found in Table 3.4.

3.2.1 Gamma ray shielding

Gamma-rays interact with matter by three main mechanisms: photoelectric absorption, Compton scattering, and pair production. The details of these mechanisms are not required for radiation shielding and are provided in Ref. [153]. Rather, the gamma-ray shielding material's coefficient, called mass attenuation coefficient, is used for design and is defined as the following:

$$\text{Mass attenuation coefficient} = \frac{\mu}{\rho_m}$$

Where μ is the attenuation coefficient in units of cm^{-1} and ρ_m is the radiation shield material density in g/cm^3. μ is defined as the probability per unit path length that the gamma-ray photon is removed from the beam, and is as the following:

$$\mu = \mathscr{T}(\text{Photoelectric}) + \sigma(\text{Compton}) + K(\text{pair} - \text{production})$$

Thus, the attenuation of gamma-rays (and X-rays) can be written for a monoenergetic photon energy as:

$$I = I_0 \exp\left[-\frac{\mu}{\rho_m} x\right]$$

Where I_0 is the incident gamma-ray energy, x is the shield's thickness (in cm), and I is the intensity of the gamma-ray beam after attenuation. The mass attenuation coefficient is approximately proportional to the atomic mass, Z [154–156]. The standard database for the mass attenuation coefficients for each element for incident energies ranging from 1 keV to 20 MeV can be found in the NIST Standard Reference Database 126 [157]. The typical 1 MeV gamma-ray shielding mass attenuation coefficients (cm^2/g) as a function of Z is found in Fig. 3.8 (along with the neutron absorption cross section, which is discussed in the next section). As discussed in later

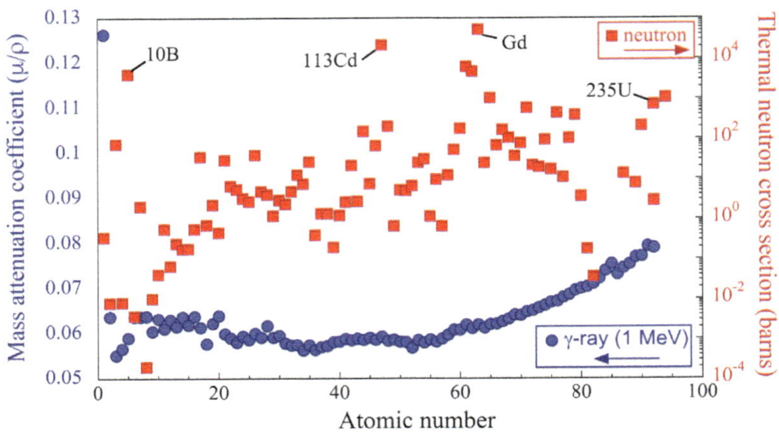

Figure 3.8 The mass attenuation coefficients for 1 MeV gamma-ray incident energies as a function of atomic number, and thermal neutron cross section of all natural elements except the ones highlighted. The gamma-ray data source was gathered from the NIST Standard Reference Database 126 [157]. The neutron cross section data can be found in Ref. [158] and is readily available in the ENDF/B-VIII.0 nuclear database [159].

sections, the average gamma-ray energy expected within a tokamak central column is approximately 1 MeV [149]. However, it should be noted that the gamma-ray shielding effectiveness is highly dependent on three factors: energy of the radiation, the atomic mass, Z, and the atomic density. For example, at 1 MeV gamma-ray, the mass attenuation coefficient of $Z = 1$ (hydrogen) and $Z = 82$ (lead), the respective values are 0.126 cm^2/g and 0.071 cm^2/g. The electron density per nuclei is ~60% greater for hydrogen compared with lead, thus, increasing the coefficient due to Compton scattering domination at these energies. However, the overall atomic density and practicality of lead is much greater than hydrogen (either in gaseous, water, or hydride form), thus, outweighing the benefit of increased gamma-ray absorption performance. To produce an effective gamma-ray shielding material requires the balance of these three factors.

3.2.2 Neutron shielding

To remove a neutron from the system it must be absorbed by a shielding material's atomic nuclei through the nuclear absorption process. As a result of this process, either the neutron is removed, changed direction (scattering), or replaced by one or more secondary radiations (charged particles, electromagnetic radiation or more neutrons). The probability of neutron absorption is called the absorption microscopic cross section, σ_a, and it is the key parameter that is used when designing neutron shields.

σ_a is defined as the number of neutron absorptions per second with one nucleus per unit intensity of the beam and is a function of the cross section for fission, σ_f and (n,p/ γ ...) production σ_p/γ. However, the cross section of scattering, σ_s, competes with σ_a.

The unit of cross section is expressed as a barn (b), where $1b = 10^{-24}$ cm^2. Furthermore, the more common unit in nuclear engineering is the macroscopic absorption cross section:

$$\sum_a = N\sigma_a$$

where N is the density of the material.

All neutron cross sections are a function of the incident neutron energy and target nucleus, and the most common shielding materials σ_a is shown in Fig. 3.9, which shows the irregular dependence on incident energy for the selected nuclei. To reduce the incident neutron energy is called neutron *moderation* and is generally desired to increase the probability of neutron absorption, which is shown in Fig. 3.9. Light atoms, such as the hydrogen within water, enable a neutron to lose much of its energy in this way. Fig. 3.8 provides the σ_a at thermal neutron energies (~0.025 eV).

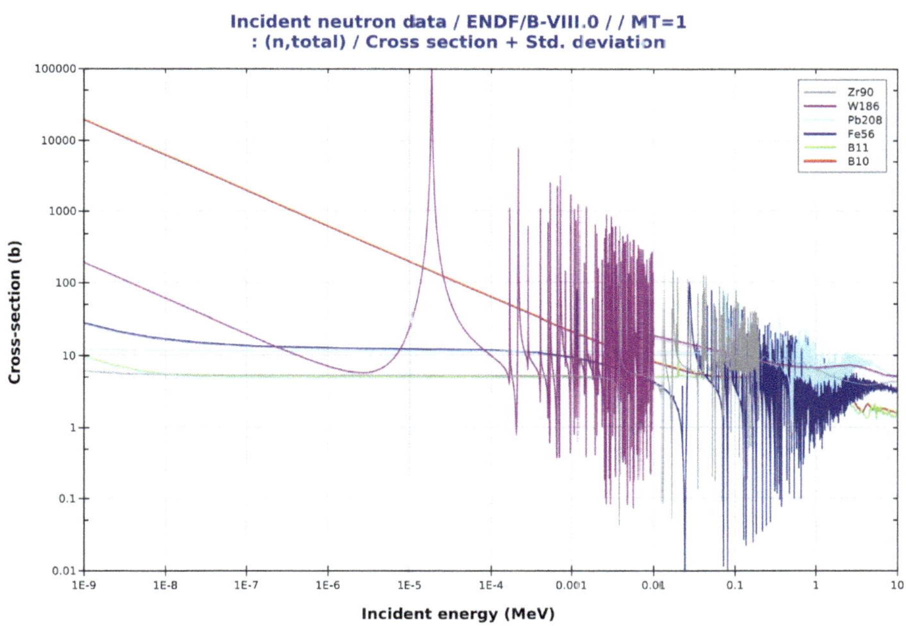

Figure 3.9 The log-log plot of neutron cross section of absorption (σ_a) against incident neutron energy (MeV) for various materials that are used commonly in fission and fusion reactors. Isotopes have been plotted rather than the natural abundances as some have special properties, such as B-10. The cross section data can be found in the ENDF/B-VIII.0 nuclear database [159].

3.3 Radiation shielding materials for tokamak devices

Radiation shielding materials for both gamma-ray and neutron incident radiation should consist of

- Highly dense, high Z materials (due to the space constraints in tokamak devices);
- Neutron moderator materials;
- Neutron absorber materials.

An optimal radiation shield for a tokamak central column needs to perform several functions at the same time: moderation, absorption, energy dissipation, and daughter product removal/accommodation. Some atoms such as boron have high cross sections for neutron absorption (producing Li and He), but only at low "thermalized" neutron energies and therefore require a moderator to be present to function efficiently. Other isotopes, such as those of tungsten used in the first wall, have high (n, γ) cross sections. However, the high-energy gammas created cause their own heating and subsequent damage to the magnetic core and therefore must be absorbed in the shield as much as possible.

Several shielding approaches have been proposed in different tokamak fusion reactor concept designs. Tokamak Energy Ltd is a UK-based privately funded fusion energy company whose approach is based on the merging of two primary technology paths: the first being the spherical tokamak geometry and the second is the use of high-temperature superconducting (HTS) tape, also known as ReBCO (Rare-earth barium copper oxide), to deliver the necessary magnetic field strengths to confine the plasma. For the purpose of reducing the gamma-ray and neutron radiation in the HTS core, the concept radiation shield design for the central column is primarily based on alternating layers of WC-based material and water coolant (to provide both cooling and neutron moderation) [150,151].

Commonwealth Fusion Systems is a US-based privately funded fusion energy company whose approach is based on using HTS tape as the magnet material in a tokamak plasma design, with a molten-salt blanket design, which couples as both a tritium breeding blanket, cooling material, and means of energy extraction [160,161]. The ARC fusion reactor design central column neutron shielding material was selected to be zirconium hydride, ZrH_2, based on the high hydrogen density and extensive use as the moderating material in TRIGA test reactors [162].

A list of possible radiation shielding ceramic materials for fusion devices can be found in Table 3.5. The list has been derived from the radiation shielding materials selection criteria mentioned above and also the criteria set out in Section 1.4. The rationale behind each selection is as follows:

- 316 and 304B7 stainless steel have seen extensive use in the nuclear industry and provides a good benchmark for comparison with ceramic-based shielding materials.
- Tungsten is a high density, reasonable neutron absorption cross section, and gamma-ray absorbing material. The element has a 3 MeV threshold (n,2n) nuclear reaction that enables W to become an effective neutron moderator in fusion only neutron spectrums. Pure W also provides a good benchmark for comparison with ceramic-based shielding materials.

Table 3.5 A list of possible ceramic radiation shielding materials for the central column of fusion reactors.

Material	ρ (g/cc)	T_m	λ (W/mK) at 20–25C	Composition (wt%)	Ref
316 stainless steel	7.90	1390	15	16–18 Cr, 10–12 Ni, 0.08 C, 2 Mn, 1 Mo, Fe – Bal.	[163]
Boronated stainless steel (304B7)	7.80	1390	12.5	16–18 Cr, 10–12 Ni, 0.08 C, 2 Mn, 1.75–2.25 B, Fe – Bal.	[164]
Tungsten	19.25	3422	160	100 W	[149,150,165–167]
Boron carbide, B_4C	2.52	2763	36	78 B, 22 C	[164,168,169]
Tungsten carbide (WC)	15.6	3143	179	50 C 50 W	[170]
Tungsten boride (W_2B)	17.09	2920	52 (for W_2B_5 [171])	33 B 67 W	[151,172]
Zirconium hydride, ZrH_2	5.6	>300	9–10	Zr 97.8, H 2.2	[162,173–175]
Europium oxide (Eu_2O_3)	7.28	2233	3.42	Eu 86.4, O 13.6	[176–178]
Dysprosium titanate (Dy_2TiO_5)	6.20	1870	0.2–0.3	Dy2O3 78.1, TiO2, 19.7, Mo 1.8	[179–181]
Gadolinium oxide (III) (Gd_2O_3)	7.47	2420	0.068	Gd 86.8, O 13.2	[182–184]

The density ρ, melting temperature (T_m) and thermal conductivity λ is provided for a comparison. High level reasons for their selection are described above, and references provided in the table. Three metal systems are provided for comparison to the ceramic alternatives, and these are 316 and 304B7 stainless steels and tungsten.

- Boron Carbide, B_4C, has extensive use as control rod materials in pressurized water reactors and enrichment of B to 10B is well established practice in the industry.
- Tungsten Carbide (WC)-based materials provide a good combination of neutron absorption (W), good gamma-ray shielding material (W), and good neutron moderation (C).
- Tungsten Boride (WB)-based materials provide an excellent combination of neutron absorption (B and W) and good gamma-ray shielding material (W).

- Metal-based hydrides are efficient neutron moderators due to their high hydrogen density (order 10^{22}–10^{23} hydrogen atoms per cm^3). ZrH_2 has been extensively used as the moderating material in TIGRA reactors and is an effective radiation shield in tokamak reactors.
- Europium Oxide (Eu_2O_3) is equivalent to B_4C neutron shielding performance without the He production. The decay daughters of Eu are good neutron poisons. This material has been used in sodium-cooled fast reactor BN-600, nuclear powered icebreakers and test reactors.
- Dysprosium titanate (Dy_2TiO_5) is the control rod material of the Russian VVER and RMBK reactors. Dysprosium has seven stable isotopes and a high neutron absorption cross section.
- Gadolinium Oxide (III) (Gd_2O_3) is used as a nuclear fuel burnable poison, which greatly extends the cycle length of commercial nuclear power plants. ^{157}Gd has a high thermal neutron absorption cross section.

A more efficient radiation shield design for tokamak reactors (efficient in maximizing the reduction of radiation (neutrons and gamma) flux per thickness of radiation shielding material) could provide a few advantageous parameters for tokamak design engineers. For example, a more efficient shield could reduce the volume of shield material, reduce the cryogenic cooling requirements, increase the superconducting material's operational lifetime, and reduce the volume of radioactive shielding material after final shutdown. The novel radiation shielding materials for fusion reactors are based on metal hydride and tungsten-based ceramics, such as tungsten carbide and tungsten boride. The next sections will discuss the development of these material systems for fusion energy.

3.4 Borides and carbides of tungsten

Calculations by several research groups show that alloying tungsten with carbon and/or boron to form carbides and borides can improve its neutron shielding characteristics [45,149,185,186]. This is because while tungsten is an effective gamma-ray attenuator, neutron scatterer (and a partial neutron reflector), carbon and boron can provide effective neutron moderation and absorption. As demonstrated below, a major disadvantage of these materials is their relatively low fracture toughness, which poses a challenge if they are to operate in any structural capacity. A second disadvantage is their high sintering temperatures and relatively poor joinability. Recent developments in ceramic-metal composites, with improved toughness and fabricability have addressed these issues to some extent [172,187,188].

The properties of several candidate shielding materials are shown in Table 3.6. As well as monolithic WC and WB, three ceramic-metal composites, or cermets, are shown: (i) a WC-10FeCr composite [188], (ii) a W_2B–W composite [172], and (iii) a "reactively-sintered boride" (RSB), which is based on the reactive sintering of W + FeCr + B_4C to form various binary and ternary boride and carbide phases [187]. Pure tungsten is also shown for comparison purposes.

The cermets show some advantages over the monolithic compounds in terms of processing and mechanical properties. Their flexural strength and fracture toughness values are generally a factor of two or more greater. Second, they are easier to manufacture, in that they can either be vacuum-hot pressed (in the case of W_2B–W) or

Table 3.6 Thermal-mechanical properties of shielding candidates.

	Fabrication	T_m	Density	$\sigma_{flexural}$	H	K_{1c}	E	α
	Method	(K)	(g/cc)	(GPa)	(GPa)	(MPa $\sqrt{m}$)	(GPa)	10^{-6}
W [67]	HP	3695	19.3	0.4	2.4–3.4	7–45(1)	380	4.4
WB [189]	SPS	2938	15.6	/	23–36	3.8(1)	480–550	6.9
WC [190]	SPS	3143	15.3	1.6	24–26	~5	705	3.4
WC-10FeCr [191]	Vac-sinter	1473*	14.3	3.9	10.4	9.2	580	6
W_2B-0.57W [172]	Vac-HP	2859	17.8	0.65	7	>10	/	/
B5T5WW2 [151,192,193]	Vac sinter	/	12.6	/	14.7	8.4	/	/

"/" means value is not available.

pressurelessly sintered (WC–FeCr and the RSBs). Due to the pressureless nature of the latter, more complicated geometries can be made at relatively low cost. By contrast, monolithic WC and WB must be spark-plasma-sintered to reach full density, which limits the part complexity, and confines part thickness to ~1 cm.

The cermets do have some disadvantages; first, their melting points are relatively modest. For example, WC-FeCr forms a partially liquid eutectic phase at ~1473 K, which is ~1500 K lower than WC monoliths and may preclude their application at very high temperature. Second, their density is significantly lower, due to the presence of the relatively neutron-transparent FeCr additions. Thus the neutron attenuation performance will be reduced, as demonstrated by Windsor et al., who showed that the power attenuation half-distance increased by ~0.5% for every 1% binder addition [151].

A key consideration when designing high temperature fusion reactor components containing tungsten is their accident tolerance. In a loss-of-coolant accident scenario it is possible that the temperature of tungsten will reach 1000–1200°C due to decay heating effects [194]. The problem may be more severe in the case of carbides or borides due to the additional decay heating from the carbon/boron species, however this is yet to be quantified. WC oxidizes at a comparable rate to pure tungsten and may show some advantage about 1000°C [195]. Furthermore, the evaporation of volatile oxides is slowed by the addition of a Fe-based metallic binder [196]. Si-based coatings have proved very effective for improving oxidation resistance of WC-FeCr cermets [197]. The oxidation resistance of tungsten boride is good [198], however the evaporation of the scale is yet to be quantitatively addressed.

Fast neutron irradiations have been performed between 300 and 700°C [199]. WC showed better swelling and cracking resistance compared with Ta, Ti, Zr, and Nb carbides. WC has been studied experimentally by ion-irradiations, mostly with nitrogen ions, due to the possibility to improve the surface hardness [200]. Burnett and Page suggest that WC has a relatively high dose threshold before amorphization compared with other ceramics, due to its strongly metallic-like bonding [201]. According to first-principles studies, displacement-induced defects are mostly accommodated on the carbon sublattice [202] and Ta, Re and Os transmutation products will be accommodated on the tungsten sublattice [203], however more work is needed to establish these elements' solubility limits. The irradiation damage tolerance and transmutation product accumulation of tungsten borides have been addressed by first-principles studies [204]. However experimental verification is needed.

3.5 Hydrides for shielding materials

3.5.1 Introduction to the metal hydride system

Thomas Graham in 1866 discovered the first metal hydride, palladium hydride (PdH_X) [205]. Metal hydrides were not explored outside the scientific community until their selection as the neutron moderator of nuclear fission reactors during the 1950s. The development of zirconium-based hydrides started with the System for Nuclear Auxiliary Power (SNAP) program by the National Aeronautics and Space Administration

(NASA). The hydride provided two functions: neutron moderation and matrix material for uranium fuel [206,207]. The gas-cooled Aircraft Nuclear Propulsion (ANP) reactor used ZrH_x as the moderating material [208]. Low-neutron cross section transition metal hydrides, such as Zr and Y, were chosen for their highly efficient moderation (due to the high hydrogen density) and neutron reflecting performance [174]. These hydride materials are suitable for thermal reactor systems in which core weight and volume must be minimized. YH_x is the candidate for the Transformational Challenge Reactor's moderating material that started development at Oak Ridge National Laboratory in 2019 [209,210]. Hafnium-based hydrides (HfH_x) have been selected as a potential replacement of boron carbide control rod materials in fast reactors [211,212]. Outside of the nuclear technology field, metal hydrides are in development for the use of hydrogen storage, thermal storage, actuation systems, semiconductors, and biomedical systems [213].

The definition of a "hydride" used here describes the binary combination of hydrogen and a metal atom with a hydrogen bond (covalent, metallic, or ionic, or a combination of) [205]. The ordered crystal structures of metal hydrides distinguish them as ceramics, as opposed to disordered solid solutions of hydrogen within a metal lattice. Metal hydrides that exist across a compositional range, MH_x, where $0 > x > 2$, are technically metal-ceramic composites that contain both ordered (ceramic) metal hydrides and disordered metals with dissolved hydrogen.

The main reaction between a metal and hydrogen atoms is:

$$M(s) + H_2(g) > MH_2(s)$$

The metal-hydrogen affinity is relatively weak and will therefore decompose at an elevated temperature (that is dependent on the metal species and metal-hydrogen stoichiometry), releasing hydrogen gas from the material. It is worth noting that reaction with deuterium and tritium with metals has near similar reaction paths to hydrogen [205,214]. The major difference between the isotopes of hydrogen is that the physical properties, such as the density and lattice parameters, are different. This could be important when considering the use of metal hydrides in a fusion environment where all three hydrogen isotopes will be present in the reactor.

There are two main hydride fabrication methods: direct hydrogen reaction with bulk metal pieces (direct hydriding) and powder metallurgy.

For direct hydriding, the general method is to heat a bulk metal in a vacuum and admit (slowly) pure hydrogen gas into the chamber until it is absorbed. To complete the hydride cycle, the partially hydrided metal should then be slowly cooled in an excess of hydrogen until room temperature is reached to avoid cracking and void formation [174,215].

The powder metallurgy route allows for pressureless sintering of green bodies forming the metal hydride into a final or near-final product. The powder metallurgy route has advantages due to the reduced fabrication time, common feedstock supplies and simpler fabrication method compared with hydriding. However, the disadvantage is that sintering less stable hydrides may not be possible if dehydriding occurs before adequate density is reached. For further reading on the manufacturing process of metal hydrides, seek the following references [210,216,217]. This section will review suitable metal hydride systems for radiation shielding in the central column of tokamak reactors.

3.5.2 *Metal hydride systems for radiation shielding*

The group III and IV hydrides are the most promising candidates due to their combination of high hydrogen density, potential/experience use within nuclear reactors, and relatively good thermal stability. The focus of metal hydrides for radiation shielding is ZrH_2, YH_2, TiH_2, and HfH_2.

3.5.2.1 Zirconium hydride

The development of zirconium-based hydrides started with the SNAP program by NASA as the matrix for uranium fuel and moderator material [206,207] and further developed for the ANP reactor [208]. This was further advanced with the use of uranium fuel embedded within the ZrH_x matrix for the TRIGA reactor's fuel [218], and these reactors are still in operation worldwide as of 2023. In part due to this combination of reactor experience and dense hydrogen, ZrH_2 hydride has been proposed as the radiation shielding for the central column in fusion tokamak reactors [162,173]. The radiation shielding effectiveness can be shown with the reduction of neutron flux across the shielding material in a fusion reactor as shown in Fig. 3.10.

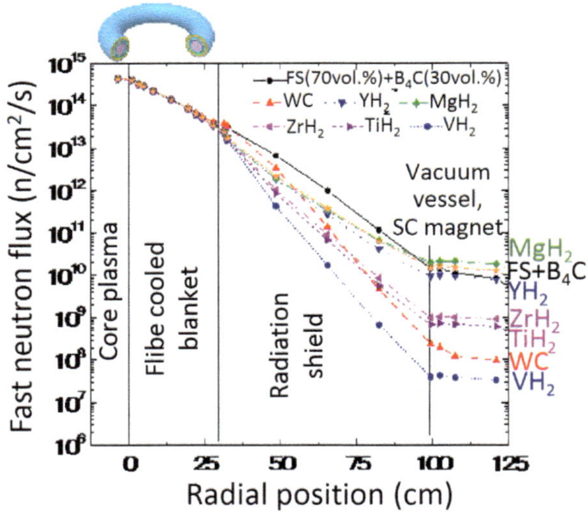

Figure 3.10 A comparison of different basic one-layer material structure radiation shielding for fast neutrons (E > 0.1 MeV) in a helical fusion reactor design. FS is ferritic steel and SC is superconducting magnetic material. These results were calculated with MCNP-5 code with idealized materials parameters.
Reproduced with permission from Ref. [219].

3.5.2.2 Yttrium hydride

The use of yttrium hydride as a nuclear material started with an investigation for the ANP reactor's moderating material [174,220]. YH_2 presented superior decomposition kinetics compared with ZrH_2 [221]. However, the use of YH_2 was paused due to unavailability of highly-pure Y metal and high cost per kg compared with Zr in the 1950s (Zr was 30 times less [222]). The use of YH_2 has been proposed for radiation shielding in fusion reactors [219] and under development as the moderator for the Transformational Challenge Reactor at Oak Ridge National Laboratory [209,223] due to both reduction in Y metal cost per kg (~\$30 per kg) and superior thermal stability properties compared with ZrH_2.

3.5.2.3 Titanium hydride

Titanium hydride (TiH_2) has been proposed as a moderating material for nuclear reactors and radiation shielding for fusion reactors due to it possessing one of the highest hydrogen densities and its low activation characteristics [224,225]. The mechanical properties of TiH_2 have been investigated [226,227]. However, as discussed in the next sections, TiH_2 suffers from a low dissociation temperature that could eliminate the use of this material in a fusion shield due to the safety considerations with unwanted hydrogen release.

3.5.2.4 Hafnium hydride

The use of hafnium hydride for any fusion application had not been proposed until the Japanese program titled "Development study of fast reactor core with hydride neutron absorber" (led by Tohoku University from 2008 to 2014), mainly due to the lack of development studies and high raw Hf metal cost. The purpose of this program was to investigate the possible use of HfH_2 as a control rod material for fast breeder fission reactors as a replacement of the traditional boron-10 enriched boron carbide (B_4C) control rod material. HfH_2 provides a high neutron absorption element (Hf), which improves its neutronics performance over other hydrides such as Y, Ti, and Zr. The main advantage over B_4C is due to the lack of helium production during neutron irradiation with the hydride, which increases the control rod lifetime (since B_4C is highly susceptible to helium damage, which leads to cracking and failure [164]). Since that study, there have been a few subsequent publications on the hydride's chemistry [228–230], physical properties [231], thermal properties [231,232], and superconducting properties [233]. Additional studies have addressed other aspects of the use of HfH_2, including as a control rod material [212,234,235], neutronics studies [212,235–237], physical properties (density, hardness, elastic modulus) [234,236], and thermal properties [234,237]. An experiment in the Russian test reactor BOR-60 studied neutron irradiation damage effects (a dose of 2.5×10^{22} n/cm^2 ($E > 0.1$ MeV) produced a 2.2% increase in linear dimension [238]), and coating compatibility has been investigated to reduce the buildup of hydrogen release under accident conditions [239]. The program led by Tohoku University was thorough in characterizing the material in both unirradiated and irradiated conditions, which provides a good knowledge base for using HfH_2 as a radiation shielding material for fusion reactors.

3.5.3 Properties of metal hydrides

The main properties, such as crystal structure, density, and hydrogen density, for the above hydrides are found in Table 3.7. YH_2 shows the best thermal stability, but suffers from a relatively low hydrogen density, and thus has inferior shielding capability, as evidence in shielding simulations [174]. ZrH_2 and HfH_2 show intermediate properties. While HfH_2 also carries with it the advantage of the high neutron cross section of Hf, which will enhance its shielding capability, although Hf will lead to the production of Hf-178m under neutron irradiation, which has a half-life of 31 years [174], and therefore could be undesirable from a waste management perspective.

3.5.4 Dissociation characteristics of metal hydrides

The hydrogen-metal bond is rather weak at high temperatures and breaks down, releasing the hydrogen in gaseous form. The explosive nature of hydrogen gas is of concern for high temperature operation (either under normal or accident conditions). Thus, the dissociation characteristics are a primary materials selection criterion. The equilibrium partial pressure of hydrogen varies with composition and temperature. Fig. 3.11 provides the dissociation characteristics of the aforementioned hydrides as a function of hydrogen-to-metal ratio at 600 and 800°C.

3.6 Summary

For the realization of using ceramics and metal hydrides for radiation shielding, greater development studies that focus on radiation damage of carbides/borides and the upscaling in manufacturing of metal-hydrides would increase the likelihood of use in fusion reactors. The main unknown for carbides and borides is related to the

Table 3.7 Structure, mass density, hydrogen density, and dissociation pressures of various candidate group III and IV hydrides.

Hydride	Structure	Density	N_H 10^{22} H	σth (natural)	Dissociation pressure
		(g/cc)	(at /$cm^3_{hydride}$)	Barns	at 900K[a] (torr)
YH_2	FCC	4.29	5.68	1.28	0.0102
TiH_2	FCC	3.75	9.05	6.09	693[b]
ZrH_2	FCT	5.61	7.25	0.185	1.14
HfH_2	FCT	11.36	7.58	104.1 373 (^{177}Hf, 18.6% natural abundance)	3.61

The neutron cross section data can be found in Ref. [158] and is readily available in the ENDF/B-VIII.0 nuclear database [159].

[a]Dissociation pressures taken from the "plateau region", i.e., the two-condensed phase region where the metal and hydride coexist.

[b]Data for TiH_2 is an extrapolation from data from 400 to 650 K [174].

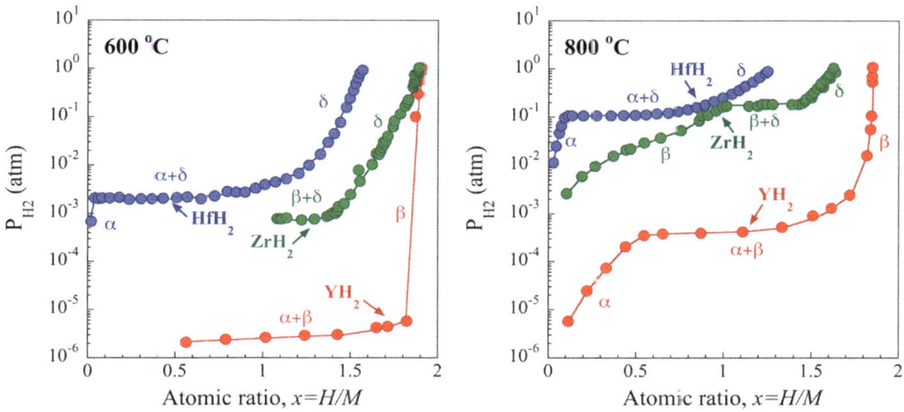

Figure 3.11 The dissociation pressure (H_2) as a function of atomic ratio of hydrogen to metal atoms at 600 and 800°C. α, β, and δ represent crystallographic structure type for each hydride system.
The data for ZrH_2, YH_2, and HfH_2 were taken from sources [240,241], and [226], respectively.

helium production from neutron irradiation under fusion (and fission) conditions, which significantly embrittle the material during service.

4. Plasma-facing components

Due to their favorable material properties (thermo-mechanical strength, high melting point, neutron resistance, lightweight) refractory ceramic materials have long been recognized to be a promising material for fusion armor [242]. Various fusion power plant design studies have called for ceramic cladding of at least some of the plasma-facing surfaces [78]. As most ceramics tend to have only low to moderate thermal conductivity, they are most strongly considered as plasma-facing materials away from the highest heat flux regions of the plasma. Thermal conductivity values of some of the more commonly considered ceramics are shown in Fig. 3.12.

Low-Z ceramics may have the added benefit as acting as a "wall conditioning" material as they erode from the wall, acting as a sacrificial layer preventing the erosion of heavy impurities from the high thermal conductivity refractory metals (e.g., tungsten) located at the highest heat flux regions of the reactor wall [243]. Finally, ceramics tend to simply ablate and reshape themselves under strong plasma bombardment, in contrast to metals, which tend to melt, resolidify, and form sharp leading edges, potentially resulting in metallic droplet contamination of the fusion core. One significant concern for the use of ceramics as plasma facing components (PFC) in DT fusion reactors, however, is their tendency to retain significant amounts of tritium through implantation and codeposition [244]. If such issues can be overcome, ceramics could pose significant advantages over the use of metals as a first-wall armor material. Candidate ceramic materials for plasma-facing components are discussed below.

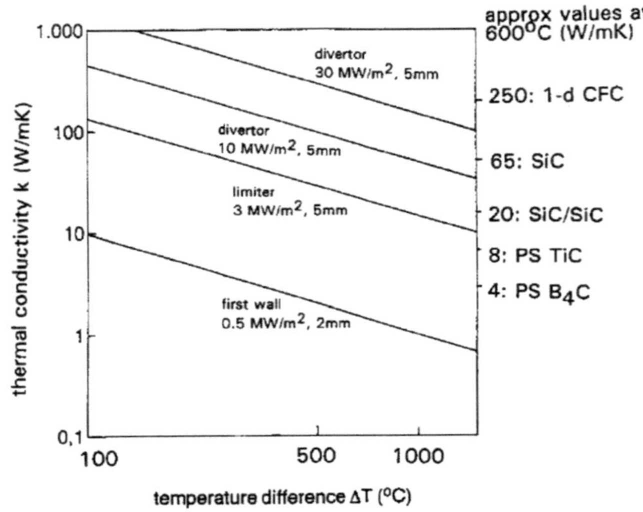

Figure 3.12 The required value of thermal conductivity, as a function of the temperature difference across the material, for different plasma-facing components in a fusion reactor system. Thermal conductivity values in the unirradiated condition at 600°C for some of the most promising ceramics for PFC application are provided.
Figure reproduced with permission from Ref. [242].

4.1 Established ceramics

Two ceramics can be considered at the highest (although still relatively low) TRL for fusion applications: graphite (TRL 3–4) and silicon carbide (TRL 2–3). Commercial manufacturing is available for both the pure form of each material as well as their respective composite forms: carbon fiber composite (CFC) and silicon fiber composite (SiC_f/SiC), respectively. The properties of these materials as PFCs in fusion devices are briefly reviewed.

4.1.1 Graphite and CFCs

Historically graphite and CFCs have been the most commonly used, and thus most studied, plasma-facing material for fusion applications. Armor-grade graphite is readily available in large quantities for scientific use and tend to be very forgiving to strong heat and plasma loading. Nearly every major fusion experiment over the last 40 years has utilized graphite armor in high heat flux regions for at least one significant operational phase. Several literature reviews are already available on the subject [245,246].

CFC monoblocks originally were planned for installation in much of the high heat flux regions of ITER divertor [32]. However, increasing understanding of the physics of fuel retention through codeposition in the years after the ITER Physics Basis was published led to a concern that graphite PFCs in ITER would trap unacceptable levels of tritium, resulting in ITER reaching its T safety limit and thus become unable to achieve its physics missions. This concern, in addition to the desire to reduce the overall expense of the ITER divertor project, led to the decision to remove graphite PFCs entirely from the ITER design and proceed directly to a full-tungsten divertor [43]. Tritium retention may be of significantly less issue for a commercial fusion reactor because all plasma-facing surfaces will be maintained at high temperature (>600°C) in the interest of thermal efficiency. When graphite PFCs are maintained at such high temperatures the tritium inventory can be reduced significantly [247]. A recent synthesis of the current understanding of implantation and codeposition on tritium breeding ratios (TBRs) for various plasma-facing materials in reactor-level devices can be found in Ref. [244].

Aside from tritium-related limitations the extreme degradation of the thermal conductivity of graphite under any significant irradiation damage [248] implies that bulk graphite tiles are very unlikely to ever be of relevance for a fusion reactor. It is conceivable that thin graphite coatings on top of more neutron-resilient substrates, such as SiC or tungsten, could be used. Such systems have been proposed [249] but remain in the early stages of serious evaluation.

4.1.2 Silicon carbide

SiC, while nowhere near as widely explored as graphite for fusion applications, was used as a PFC material on a number of fusion experiments in the 1980s and 1990s [250–252]. Technological limitations regarding hermeticity and joining, however, prevented serious consideration as a plasma-facing material for ITER [242]. In the subsequent years few, if any, tests were conducted in fusion experiments. As discussed in Section 2.2 and in other chapters of this book, in the intervening years progressive advances in the performance of nuclear-grade SiC fiber composites have occurred primarily for Generation-IV fission reactors [253]. Modern grades of SiC_f/SiC share many of the attractive properties of CFCs (thermo-mechanical strength, high melting point, commercially available manufacturing) and may have additional benefits:

1. Minimal degradation of thermomechanical properties of SiC even under high levels of neutron damage, particularly at high irradiation temperature [83].
2. Significantly lower C chemical sputtering yields relative to graphite and negligible levels of Si chemical sputtering [254,255], potentially implying lower wall erosion rates and reduced levels of codeposited fuel retention.
3. Lower tritium diffusivity relative to tungsten, which can be further decreased by a factor of $\sim 10^6$ by depositing a thin coating of monolithic SiC on top of a SiC/SiC_f substrate—as in the ARIES-AT reactor design study (1 mm SiC on 4 mm SiC_f/SiC [50]). This makes SiC very attractive as a tritium permeation barrier [256].

Points 2 and 3 above, in particular, have stimulated renewed interest in silicon carbide as a plasma-facing component for fusion applications in recent years [257–259], particularly away from the highest heat flux regions of the plasma. To date, however, no large-scale tests of SiC as a primary wall armor material have been performed in magnetic confinement experiments. SiC is considered favorable as a plasma-facing

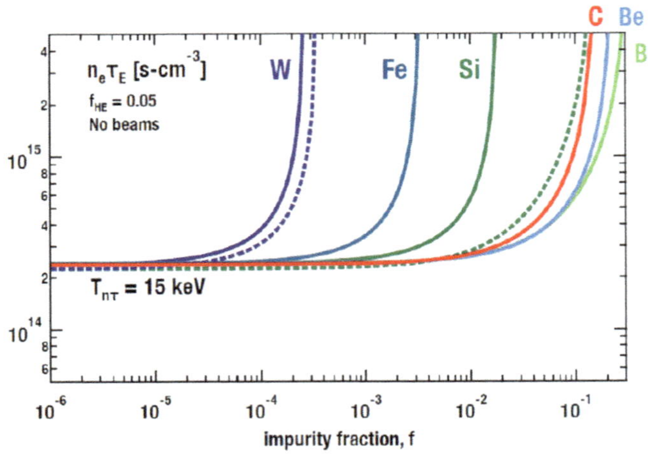

Figure 3.13 Required value of the fusion Lawson criterion, $\boldsymbol{nT\tau}$ (as discussed in Section 1.1) as a function of the core plasma impurity fraction for selected elements proposed for use as plasma-facing materials in future fusion devices.
Reprinted with permission from Ref. [.249].

material relative to refractory metals such as tungsten due to the higher acceptable impurity content in the core plasma, as shown in Fig. 3.13.

4.2 Emerging ceramics

4.2.1 Ultrahigh-temperature ceramics

Of all the UHTCs, TiC is the most widely studied for fusion first-wall applications. It was extensively studied in the 1980–90s due to its low atomic number (Z = 22) and high thermal stability. However, it was abandoned as chemical sputtering of carbon was identified as a problem (although this is about 1 order of magnitude slower than graphite [260]). Another disadvantage of the carbides is their relatively high vapor pressures, e.g., 2–7 Pa for group IV and V carbides at 3000 K [107] as compared with ~10^{-4} Pa for pure W.

Nuclear fusion tokamak experience with TiC has mainly been focused in the US and Japan. Princeton Plasma Physics Laboratory applied 20-micron thick TiC coatings to their graphite limiter in TFTR using chemical vapor deposition (CVD). However, the project was abandoned due to excessive Ti impurities in the plasma due to problems with adhesion [261]. TiC coatings were also studied at the ISX-B tokamak at Oak Ridge National Lab and GA DIII-D [262]. Preferential (i.e., chemical) sputtering was not reported to be prevalent under low energy deuterium bombardment [263]. TiC coatings were also used in the JT-60 reactor first wall [264]. Coatings of 20 microns thick were applied to Mo tile substrate using CVD. An extensive development program was carried out, which included the development of intermediate TiN coatings for improved adhesion.

Thermo-mechanical behavior of TiC coatings has been studied extensively on Mo and Inconel 625 substrates [265]. Thermal fatigue between 200 and 1400°C showed no cracking in Inconel-625 up to 50 cycles, while the coatings on Mo cracked after a single cycle [266]. The low cycle fatigue life of TiC-coated Mo was studied [267]. At 500°C, the coating decreased the fatigue life. However, at 800°C, the TiC coating improved lifetime, apparently due to its restriction of slip deformation of the underlying substrate.

In all of the above studies, the coatings were relatively thin (~20 microns) and vapor deposited. However, thicker (up to 120 micron) plasma sprayed coatings have been successful applied to Cu, Al and steel substrates and studied under disruption simulations at heat fluxes of 10^8 W/m^2. The porosity was 6%. In all samples, some coating melting occurred when irradiated for longer than 10 ms. The coating adherence was not good for aluminum [268]. Groot et al. studied thicker coatings (approx. 0.3–1.3 mm) in various atmospheres with 10%–26% porosity. The thermal shock resistance was improved with increasing thickness and lower porosity (10%–18%) [269]. Other carbides may have improved plasma facing performance. For example, dense spark-plasma sintered tungsten carbide tiles showed improved thermal shock resistance compared with rolled tungsten [270].

4.2.2 MAX phase ceramics

MAX phase ceramics are a recently emerging class of materials that may have promise as plasma-facing components due to their relatively high thermal stability, electrical and thermal conductivity, fracture toughness, machinability, and thermal shock resistance [271,272]. These materials are layered, hexagonal compounds with chemical formula $M_{n+1}AX_n$, where $n = 1$ to 4, consisting of an early transition metal M, and A-Group element A, and carbon and/or nitrogen, X. The operating temperature window for MAX phase materials tends to lie between that of UHTCs and metals, making them good candidates for functionally graded high-heat flux composite materials. Few irradiation experiments have been performed on MAX phase ceramics. Initial results indicate good irradiation stability, including saturation of radiation damage mechanisms at fluences <10 dpa and radiation defect annealing at temperatures >700°C [273–275]. Tallman et al. found that certain titanium-based MAX phase ceramics disassociated to form TiC and/or TiN phases under exposure to 0.1 dpa neutron irradiation. Si-based MAX phases generally appear more robust to radiation damage than their Al-based counterparts [276]. The first plasma loading experiments performed on Ti_3SiC_2 in the DIII-D tokamak indicated erosion rates of 0–9 nm/s, relative to 27–73 nm/s for SiC under equivalent conditions [258].

4.2.3 Diamond-coated PFC

Graphite has been eliminated for use in a fusion reactor due to the high retention in tritium/hydrogen [277]. Diamond-like carbons (DLCs), however, pose a possible candidate material for the PFC due to the material's high thermal conductivity, low thermal expansion, high resistance to thermal shock [278], high atomic density, chemical stability and high tolerance to radiation damage [278–283]. In particular for fusion, carbon intrinsically does not possess a liquid phase at expected conditions (i.e., it sublimates upon extreme temperatures), which enable this type of material class to be forgiving under tokamak disruptions and edge local modes [284].

CVD diamond has been shown to not retain hydrogen well [285–287] and the diamond surface remains crystalline after hydrogen and deuterium plasma exposure [285,287,288]. The principle behind the superior resistance to etching and retention of hydrogen species is likely due to the sp^3 crystal structure of carbon is known to be slower rate of degradation than that of the sp^2 crystal structure of carbon (such as graphite) [289–291]. The implication of this fact should improve the erosion performance of diamond-based materials under hydrogen plasmas. Diamond is not immune to etching from plasmas [279,292,293]. Thus, DLCs have become an emerging technology as coating on PFC tiles that could be promising for the long-term projection of fusion power stations.

4.2.4 SiC–W composites

Silicon Carbide-Tungsten composites have been proposed for plasma facing surface applications to leverage the performance benefits of both SiC and tungsten. The GAMBL (GA Modular BLanket) concept is a dual-cooled fusion blanket design that uses a range of SiC components and is described by Tillack et al. [294]. Specifically, in addition to a SiC-based PbLi breeder containment system, and SiC-composite structural components, this design also uses a novel SiC-tungsten gradient composite monoblock design for a first wall plasma facing component with integrated channels for helium coolant. This utilizes the favorable properties of tungsten for plasma interaction resistance and heat transfer and incorporates a compositional gradient transitioning from tungsten to SiC away from the plasma and in regions where the heat flux is lower. This reduces the total tungsten content in the design, reducing the impact of tungsten on tritium breeding. This gradient monoblock component incorporates an integral helium coolant channel, enabled by recent progress in making hermetic SiC-composite tube components for nuclear fission applications (Fig. 3.14). Some fabrication considerations for SiC–W alloys have been previously described in Section 2.3.

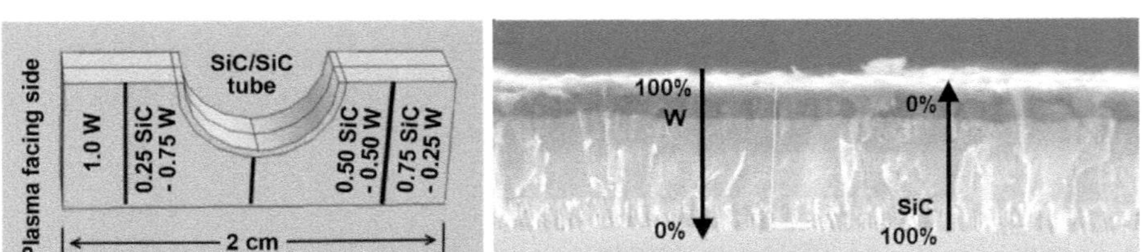

Figure 3.14 (Left) schematic cross section of SiC–W gradient monoblock plasma facing component design, (right) scanning electron micrograph of a gradient W–SiC layer cross section.
Adapted from Ref. [294].

4.3 Summary

Refractory ceramic materials have long been recognized to be a promising material for PFCs and first-wall armor due to their high operational temperatures, low sputtering yields, and high thermal shock resistance. There is promise for these materials to be used as the PFC material, however, further development must be conducted to understand the degradation under fusion relevant plasma exposure, potential tritium retention effects through erosion and codeposition, neutron irradiation damage, transmutation effects, and thermal shock resistance under accident conditions.

References

[1] M. Keilhacker, A. Gibson, C. Gormezano, P.H. Rebut, The scientific success of JET, Nucl. Fusion 41 (2001) 1925–1966, https://doi.org/10.1088/0029-5515/41/12/217.

[2] A.A. Harms, D.R. Kingdon, K.F. Schoepf, G.H. Miley, Principles of Fusion Energy: An Introduction to Fusion Energy for Students of Science and Engineering, World Scientific Publishing Company, 2000. https://play.google.com/store/books/details?id=c_Y7DQAAQBAJ.

[3] T.E.G. Nicholas, et al., Re-examining the role of nuclear fusion in a renewables-based energy mix, Energy Policy 149 (2021) 112043. https://dx.doi.org/10.1016/j.enpol.2020.112043.

[4] K.S. Krane, Introductory Nuclear Physics, Wiley, 1987. https://play.google.com/store/books/details?id=ConwAAAAMAAJ.

[5] A.A. Harms, D.R. Kingdon, K.F. Schoepf, G.H. Miley, Principles of fusion energy (2000). https://doi.org/10.1142/4447.

[6] Nuclear fusion: Status report and future prospects, Energy Policy 96 (2016) 770–778. https://doi.org/10.1016/j.enpol.2016.05.037.

[7] G. Federici, W. Biel, M.R. Gilbert, R. Kemp, N. Taylor, R. Wenninger, European DEMO design strategy and consequences for materials, Nucl. Fusion 57 (2017) 092002, https://doi.org/10.1088/1741-4326/57/9/092002.

[8] J.G. van der Laan, J.G. van der Laan, J. Reimann, A.V. Fedorov, Ceramic breeder materials, comprehensive nuclear, Materials (2016) 114–175, https://doi.org/10.1016/b978-0-12-803581-8.00733-5.

[9] T. Chikada, Ceramic coatings for fusion reactors, Comprehensive Nucl. Mater. (2020) 274–283, https://doi.org/10.1016/b978-0-12-803581-8.11674-1.

[10] D. Clery, Dark horse scores a fusion coup, Science 349 (2015) 912–913, https://doi.org/10.1126/science.349.6251.912.

[11] J. Lyons, E. Love, K. Burns, TEACUP: A tritium management and supplemental core follow program, Fusion Sci. Technol. 71 (2017) 616–621, https://doi.org/10.1080/15361055.2017.1290944.

[12] https://www.energy.gov/sites/prod/files/2015/08/f26/Senor%20-%20TMIST-3%20Irradiation%20Experiment.pdf (accessed May 18, 2020).

[13] Y. Koga, H. Matsuura, Y. Ida, R. Okamoto, K. Katayama, T. Otsuka, M. Goto, S. Nakagawa, S. Nagasumi, E. Ishitsuka, Y. Shimazaki, Study on lithium rod test module and irradiation method for tritium production using high temperature gas-cooled reactor, Fusion Eng. Des. 136 (2018) 587–591, https://doi.org/10.1016/j.fusengdes.2018.03.029.

[14] Fuelling the Fusion Reaction, ITER (n.d.). http://www.iter.org/sci/fusionfuels (accessed May 18, 2020).

[15] R.J. Pearson, A.B. Antoniazzi, W.J. Nuttall, Tritium supply and use: A key issue for the development of nuclear fusion energy, Fusion Eng. Des. 136 (2018) 1140–1148, https://doi.org/10.1016/j.fusengdes.2018.04.090.

[16] J.D. Lawson, Some criteria for a power producing thermonuclear reactor, Proc. Phys. Soc. B 70 (1957) 6–10, https://doi.org/10.1088/0370-1301/70/1/303.

[17] M. Turnyanskiy, R. Neu, R. Albanese, R. Ambrosino, C. Bachmann, S. Brezinsek, T. Donne, T. Eich, G. Falchetto, G. Federici, D. Kalupin, X. Litaudon, M.L. Mayoral, D.C. McDonald, H. Reimerdes, F. Romanelli, R. Wenninger, J.-H. You, European roadmap to the realization of fusion energy: Mission for solution on heat-exhaust systems, Fusion Eng. Des. 96–97 (2015) 361–364, https://doi.org/10.1016/j.fusengdes.2015.04.041.

[18] National Academies of Sciences, Engineering, and Medicine, Division on Engineering and Physical Sciences, Board on Physics and Astronomy, Committee on a Strategic Plan for U.S. Burning Plasma Research, Final Report of the Committee on a Strategic Plan for U.S. Burning Plasma Research, National Academies Press, 2019. https://play.google.com/store/books/details?id=5DebDwAAQBAJ.

[19] T. Muroga, S. Fukada, T. Hayashi, Overview of fusion engineering research in Japan focusing on activities in NIFS and universities, Fusion Sci. Technol. 75 (2019) 559–574, https://doi.org/10.1080/15361055.2019.1603499.

[20] A preliminary conceptual design study for Korean fusion DEMO reactor, Fusion Eng. Des. 88 (2013) 488–491. https://doi.org/10.1016/j.fusengdes.2013.02.123.

[21] Y. Wu, Conceptual design of the China fusion power plant FDS-II, Fusion Eng. Des. 83 (2008) 1683–1689, https://doi.org/10.1016/j.fusengdes.2008.06.048.

[22] S. Li, H. Jiang, Z. Ren, C. Xu, Optimal tracking for a divergent-type parabolic PDE system in current profile control, Abstr. Appl. Anal. 2014 (2014), https://doi.org/10.1155/2014/940965.

[23] J. Wesson, D.J. Campbell, Tokamaks, OUP Oxford, 2011. https://play.google.com/store/books/details?id=XJssMXjHUr0C.

[24] J.P. Freidberg, Plasma Physics and Fusion Energy, Cambridge University Press, 2007. https://play.google.com/store/books/details?id=ZGU-ngEACAAJ.

[25] Final Operations of the Tokamak Fusion Test Reactor (TFTR)—IEEE Conference Publication (n.d.). https://doi.org/10.1109/FUSION.1997.685665 (accessed September 10, 2020).

[26] C.C. Petty, The DIII-D Team, DIII-D research towards establishing the scientific basis for future fusion reactors, Nucl. Fusion 59 (2019) 112002, https://doi.org/10.1088/1741-4326/ab024a.

[27] H. Meyer, , for the AUG Team, D. Aguiam, C. Angioni, C.G. Albert, N. Arden, R. Arredondo Parra, O. Asunta, M. de Baar, M. Balden, V. Bandaru, K. Behler, A. Bergmann, J. Bernardo, M. Bernert, A. Biancalani, R. Bilato, G. Birkenmeier, T. C. Blanken, V. Bobkov, A. Bock, T. Bolzonella, A. Bortolon, B. Böswirth, C. Bottereau, A. Bottino, H. van den Brand, S. Brezinsek, D. Brida, F. Brochard, C. Bruhn, J. Buchanan, A. Buhler, A. Burckhart, Y. Camenen, D. Carlton, M. Carr, D. Carralero, C. Castaldo, M. Cavedon, C. Cazzaniga, S. Ceccuzzi, C. Challis, A. Chankin, S. Chapman, C. Cianfarani, F. Clairet, S. Coda, R. Coelho, J.W. Coenen, L. Colas, G.D. Conway, S. Costea, D.P. Coster, T.B. Cote, A. Creely, G. Croci, G. Cseh, A. Czarnecka, I. Cziegler, O. D'Arcangelo, P. David, C. Day, R. Delogu, P. de Marné, S.S. Denk, P. Denner, M. Dibon, A. Di Siena, D. Douai, A. Drenik, R. Drube,

M. Dunne, B.P. Duval, R. Dux, T. Eich, S. Elgeti, K. Engelhardt, B. Erdös, I. Erofeev, B. Esposito, E. Fable, M. Faitsch, U. Fantz, H. Faugel, I. Faust, F. Felici, J. Ferreira, S. Fietz, A. Figuereido, R. Fischer, O. Ford, L. Frassinetti, S. Freethy, M. Fröschle, G. Fuchert, J.C. Fuchs, H. Fünfgelder, K. Galazka, J. Galdon-Quiroga, A. Gallo, Y. Gao, S. Garavaglia, A. Garcia-Carrasco, M. Garcia-Muñoz, B. Geiger, L. Giannone, L. Gil, E. Giovannozzi, C. Gleason-González, S. Glöggler, M. Gobbin, T. Görler, I. Gomez Ortiz, J. Gonzalez Martin, T. Goodman, G. Gorini, D. Gradic, A. Gräter, G. Granucci, H. Greuner, M. Griener, M. Groth, A. Gude, S. Günter, L. Guimarais, G. Haas, A.H. Hakola, C. Ham, T. Happel, N. den Harder, G.F. Harrer, J. Harrison, V. Hauer, T. Hayward-Schneider, C.C. Hegna, B. Heinemann, S. Heinzel, T. Hellsten, S. Henderson, P. Hennequin, A. Herrmann, M.F. Heyn, E. Heyn, F. Hitzler, J. Hobirk, K. Höfler, M. Hölzl, T. Höschen, J.H. Holm, C. Hopf, W.A. Hornsby, L. Horvath, A. Houben, A. Huber, V. Igochine, T. Ilkei, I. Ivanova-Stanik, W. Jacob, A.S. Jacobsen, F. Janky, A.J. van Vuuren, A. Jardin, F. Jaulmes, F. Jenko, T. Jensen, E. Joffrin, C.-P. Käsemann, A. Kallenbach, S. Kálvin, M. Kantor, A. Kappatou, O. Kardaun, J. Karhunen, S. Kasilov, Y. Kazakov, W. Kernbichler, A. Kirk, S. Kjer Hansen, V. Klevarova, G. Kocsis, A. Köhn, M. Koubiti, K. Krieger, A. Krivska, A. Krämer-Flecken, O. Kudlacek, T. Kurki-Suonio, B. Kurzan, B. Labit, K. Lackner, F. Laggner, P.T. Lang, P. Lauber, A. Lebschy, N. Leuthold, M. Li, O. Linder, B. Lipschultz, F. Liu, Y. Liu, A. Lohs, Z. Lu, T.L. di Cortemiglia, N.C. Luhmann, R. Lunsford, T. Lunt, A. Lyssoivan, T. Maceina, J. Madsen R. Maggiora, H. Maier, O. Maj, J. Mailloux, R. Maingi, E. Maljaars, P. Manas, A. Mancini, A. Manhard, M.-E. Manso, P. Mantica, M. Mantsinen, P. Manz, M. Maraschek, C. Martens, P. Martin, L. Marrelli, A. Martitsch, M. Mayer, D. Mazon, P.J. McCarthy, R. McDermott, H. Meister, A. Medvedeva, R. Merkel, A. Merle, V. Mertens, D. Meshcheriakov, O. Meyer, J. Miettunen, D. Milanesio, F. Mink, A. Mlynek, F. Monaco, C. Moon, F. Nabais, A. Nemes-Czopf, G. Neu, R. Neu, A.H. Nielsen, S.K. Nielsen, V. Nikolaeva, M. Nocente, J.-M. Noterdaeme, I. Novikau, S. Nowak, M. Oberkofler, M. Oberparleiter, R. Ochoukov, T. Odstrcil, J. Olsen, F. Orain, F. Palermo, O. Pan, G. Papp, I. Paradela Perez, A. Pau, G. Pautasso, F. Penzel, P. Petersson, J. Pinzón Acosta, P. Piovesan, C. Piron, R. Pitts, U. Plank, B. Plaum, B. Ploeckl, V. Plyusnin, G. Pokol, E. Poli, L. Porte, S. Potzel, D. Prisiazhniuk, T. Pütterich, M. Ramisch, J. Rasmussen, G.A. Rattá, S. Ratynskaia, G. Raupp, G.L. Ravera, D. Réfy, M. Reich, F. Reimold, D. Reiser, T. Ribeiro, J. Riesch, R. Riedl, D. Rittich, J.F. Rivero-Rodriguez, G. Rocchi, M. Rodriguez-Ramos, V. Rohde, A. Ross, M. Rott, M. Rubel, D. Ryan, F. Ryter, S. Saarelma, M. Salewski, A. Salmi, L. Sanchis-Sanchez, J. Santos, O. Sauter, A. Scarabosio, G. Schall, K. Schmid, O. Schmitz, P.A. Schneider, R. Schrittwieser, M. Schubert, T. Schwarz-Selinger, J. Schweinzer, B. Scott, T. Sehmer, E. Seliunin, M. Sertoli, A. Shabbir, A. Shalpegin, L. Shao, S. Sharapov, G. Sias, M. Siccinio, B. Sieglin, A. Sigalov, A. Silva, C. Silva, D. Silvagni, P. Simon, J. Simpson, E. Smigelskis, A. Snicker, C. Sommariva, C. Sozzi, M. Spolaore, A. Stegmeir, M. Stejner, J. Stober, U. Stroth, E. Strumberger, G. Suarez, H.-J. Sun, W. Suttrop, E. Sytova, T. Szepesi, B. Tál, T. Tala, G. Tardini, M. Tardocchi, M. Teschke, D. Terranova, W. Tierens, E. Thorén, D. Told, P. Tolias, O. Tudisco, W. Treutterer, E. Trier, M. Tripský, M. Valisa, M. Valovic, B. Vanovac, D. van Vugt, S. Varoutis, G. Verdoolaege, N. Vianello, J. Vicente, T. Vierle, E. Viezzer, U. von Toussaint, D. Wagner, N. Wang, X. Wang, M. Weiland, A.E. White, S. Wiesen, M. Willensdorfer, B. Wiringer, M. Wischmeier, R. Wolf, E. Wolfrum, L. Xiang, Q. Yang, Z. Yang, Q. Yu, R. Zagórski, I. Zammuto, W. Zhang, M. van Zeeland,

T. Zehetbauer, M. Zilker, S. Zoletnik, H. Zohm, the EUROfusion MST1 Team, Overview of physics studies on ASDEX upgrade, Nucl. Fusion 59 (2019) 112014, https://doi.org/10.1088/1741-4326/ab18b8.

[28] B.N. Wan, Y.F. Liang, X.Z. Gong, J.G. Li, N. Xiang, G.S. Xu, Y.W. Sun, L. Wang, J. P. Qian, H.Q. Liu, X.D. Zhang, L.Q. Hu, J.S. Hu, F.K. Liu, C.D. Hu, Y.P. Zhao, L. Zeng, M. Wang, H.D. Xu, G.N. Luo, A.M. Garofalo, A. Ekedahl, L. Zhang, X. J. Zhang, J. Huang, B.J. Ding, Q. Zang, M.H. Li, F. Ding, S.Y. Ding, B. Lyu, Y.W. Yu, T. Zhang, Y. Zhang, G.Q. Li, T.Y. Xia, the EAST team, Collaborators, Overview of EAST experiments on the development of high-performance steady-state scenario, Nucl. Fusion 57 (2017) 102019, https://doi.org/10.1088/1741-4326/aa7861.

[29] https://doi.org/10.1063/1.874017 (accessed September 10, 2020).

[30] B. Lehnert, Half a century of fusion research towards ITER, Physica Scripta 87 (2013) 018201, https://doi.org/10.1088/0031-8949/87/01/018201.

[31] P. Rincon, Iter: World's largest nuclear fusion project begins assembly, BBC News (2020). https://www.bbc.co.uk/news/science-environment-53573294. (Accessed 5 September 2020). accessed.

[32] I.P.B. Editors, ITER Physics Expert Group Chairs and Co-Chairs, ITER Joint Central Team and Physics Integration Unit, Chapter 1: Overview and summary, Nucl. Fusion 39 (2002) 2137, https://doi.org/10.1088/0029-5515/39/12/301.

[33] https://www.iter.org/doc/www/content/com/Lists/ITER%20Technical%20Reports/Attachments/9/ITER-Research-Plan_final_ITR_FINAL-Cover_High-Res.pdf (accessed September 5, 2020).

[34] K. Ioki, V. Barabash, J. Cordier, M. Enoeda, G. Federici, B.C. Kim, I. Mazul, M. Merola, M. Morimoto, M. Nakahira, M. Pick, V. Rozov, M. Shimada, S. Suzuki, M. Ulrickson, Y. Utin, X. Wang, S. Wu, J. Yu, ITER vacuum vessel, in-vessel components and plasma facing materials, Fusion Eng. Des. 83 (2008) 787–794, https://doi.org/10.1016/j.fusengdes.2008.05.027.

[35] L.M. Giancarli, M. Abdou, D.J. Campbell, V.A. Chuyanov, M.Y. Ahn, M. Enoeda, C. Pan, Y. Poitevin, E. Rajendra Kumar, I. Ricapito, Y. Strebkov, S. Suzuki, P. C. Wong, M. Zmitko, Overview of the ITER TBM program, Fusion Eng. Des. 87 (2012) 395–402, https://doi.org/10.1016/j.fusengdes.2011.11.005.

[36] http://www.euro-fusion.org/fileadmin/user_upload/EUROfusion/Documents/2018_Research_roadmap_long_version_01.pdf (accessed September 5, 2020).

[37] Overview of EU DEMO design and R&D activities, Fusion Eng. Des. 89 (2014) 882–889. https://doi.org/10.1016/j.fusengdes.2014.01.070.

[38] Developing structural, high-heat flux and plasma facing materials for a near-term DEMO fusion power plant: The EU assessment, J. Nucl. Mater. 455 (2014) 277–291. https://doi.org/10.1016/j.jnucmat.2014.06.014.

[39] N. Soneda, Irradiation Embrittlement of Reactor Pressure Vessels (RPVs) in Nuclear Power Plants, Elsevier, 2014. https://play.google.com/store/books/details?id=A_TCAgAAQBAJ.

[40] Handbook of Generation IV Nuclear Reactors (2016). https://doi.org/10.1016/c2014-0-01699-1.

[41] Y. Katoh, L.L. Snead, I. Szlufarska, W.J. Weber, Radiation effects in SiC for nuclear structural applications, Curr. Opin. Solid State Mater. Sci. 16 (3) (2012) 143–152, https://doi.org/10.1016/j.cossms.2012.03.005.

[42] T. Hirai, H. Maier, M. Rubel, P. Mertens, R. Neu, E. Gauthier, J. Likonen, C. Lungu, G. Maddaluno, G.F. Matthews, R. Mitteau, O. Neubauer, G. Piazza, V. Philipps, B. Riccardi, C. Ruset, I. Uytdenhouwen, R&D on full tungsten divertor and beryllium

wall for JET ITER-like wall project, Fusion Eng. Des. 82 (2007) 1839–1845, https://doi.org/10.1016/j.fusengdes.2007.02.024.

[43] R.A. Pitts, S. Carpentier, F. Escourbiac, T. Hirai, V. Komarov, S. Lisgo, A. S. Kukushkin, A. Loarte, M. Merola, A. Sashala Naik, R. Mitteau, M. Sugihara, B. Bazylev, P.C. Stangeby, A full tungsten divertor for ITER: Physics issues and design status, J. Nucl. Mater. 438 (2013) S48–S56, https://doi.org/10.1016/j.jnucmat.2013.01.008.

[44] R. Villari, V. Barabash, F. Escourbiac, L. Ferrand, T. Hirai, V. Komarov, M. Loughlin, M. Merola, F. Moro, L. Petrizzi, S. Podda, E. Polunovsky, G. Brolatti, Nuclear analysis of the ITER full-tungsten divertor, Fusion Eng. Des. 88 (2013) 2006–2010, https://doi.org/10.1016/j.fusengdes.2013.02.156.

[45] L.A. El-Guebaly, Nuclear performance assessment of ARIES-AT, Fusion Eng. Des. 80 (2006) 99–110, https://doi.org/10.1016/j.fusengdes.2005.06.355.

[46] T. Maruyama, S.-I. Kitazawa, H. Ogawa, K. Itami, Preliminary thermal analysis of critical components in ITER lower port #02, Plasma Fusion Res. 11 (2016) 2405083, https://doi.org/10.1585/pfr.11.2405083.

[47] M. Merola, D. Loesser, A. Martin, P. Chappuis, R. Mitteau, V. Komarov, R.A. Pitts, S. Gicquel, V. Barabash, L. Giancarli, J. Palmer, M. Nakahira, A. Loarte, D. Campbell, R. Eaton, A. Kukushkin, M. Sugihara, F. Zhang, C.S. Kim, R. Raffray, L. Ferrand, D. Yao, S. Sadakov, A. Furmanek, V. Rozov, T. Hirai, F Escourbiac, T. Jokinen, B. Calcagno, S. Mori, ITER plasma-facing components, Fusion Eng. Des. 85 (2010) 2312–2322, https://doi.org/10.1016/j.fusengdes.2010.09.013.

[48] EU contribution to the procurement of the ITER blanket first wall, Fusion Eng. Des. 109–111 (2016) 661–665. https://doi.org/10.1016/j.fusengdes.2016.02.024.

[49] J.H. You, E. Visca, T. Barrett, B. Böswirth, F. Crescenzi, F. Domptail, M. Fursdon, F. Gallay, B.-E. Ghidersa, H. Greuner, M. Li, A.v. Müller, J. Reiser, M. Richou, S. Roccella, C. Vorpahl, European divertor target concepts for DEMO: Design rationales and high heat flux performance, Nucl. Eng. Des. Fusion 16 (2018) 1–11, https://doi.org/10.1016/j.nme.2018.05.012.

[50] F. Najmabadi, A. Abdou, L. Bromberg, T. Brown, V.C. Chan, M.C. Chu, F. Dahlgren, L. El-Guebaly, P. Heitzenroeder, D. Henderson, H.E.S. John, C.E. Kessel, L.L. Lao, G. R. Longhurst, S. Malang, T.K. Mau, B.J. Merrill, R.L. Miller, E. Mogahed, R.L. Moore, T. Petrie, D.A. Petti, P. Politzer, A.R. Raffray, D. Steiner, I. Sviatoslavsky, P. Synder, G.M. Syaebler, A.D. Turnbull, M.S. Tillack, L.M. Waganer, X. Wang, P. West, P. Wilson, The ARIES-AT advanced tokamak, Advanced technology fusion power plant, Fusion Eng. Des. 80 (2006) 3–23, https://doi.org/10.1016/j.fusengdes.2005.11.003.

[51] M. Abdou, N.B. Morley, S. Smolentsev, A. Ying, S. Malang, A. Rowcliffe, M. Ulrickson, Blanket/first wall challenges and required R&D on the pathway to DEMO, Fusion Eng. Des. 100 (2015) 2–43, https://doi.org/10.1016/j.fusengdes.2015.07.021.

[52] American Society of Mechanical Engineers. Boiler and Pressure Vessel Committee, BPVC Code Cases: Boilers and Pressure Vessels (1998). https://books.google.com/books/about/BPVC_Code_Cases.html?hl=&id=61_sPQAACAAJ.

[53] I. Saito, T. Shimakawa, Outline of the JSME (Japan Society of Mechanical Engineers) Rules on Design and Construction for Nuclear Power Plants, American Society of Mechanical Engineers Digital Collection, 2008, pp. 197–204, https://doi.org/10.1115/PVP2004-2690.

[54] RCC-MRx Code – Design and Construction Rules for Mechanical Components in High-Temperature Structures, Experimental Reactors and Fusion Reactors, (n.d.). https://www.afcen.com/en/106-rcc-mrx (accessed October 1, 2020).
[55] W.K. Sowder, ASME Section III Division 4 Fusion Energy Devices Code Rules, Online Companion Guide to the ASME Boiler and Pressure Vessel Codes (2020). https://doi.org/10.1115/1.861981_ch39.
[56] W. Sowder, R.W. Barnes, ASME Division IV Magnetic Confinement Fusion Energy Devices, ASME 2010 Pressure Vessels and Piping Conference: Volume 1 (2010). https://doi.org/10.1115/pvp2010-25128.
[57] M.J. Gorley, Critical assessment 12: Prospects for reduced activation steel for fusion plant, Mater. Sci. Technol. 31 (2015) 975–980, https://doi.org/10.1179/1743284714y.0000000732.
[58] R.W. Conn, Report of the DOE Panel on Low Activation Materials for Fusion Applications, 1983, https://doi.org/10.2172/7092294.
[59] H. Tanigawa, E. Gaganidze, T. Hirose, M. Ando, S.J. Zinkle, R. Lindau, E. Diegele, Development of benchmark reduced activation ferritic/martensitic steels for fusion energy applications, Nucl. Fusion 57 (2017) 092004, https://doi.org/10.1088/1741-4326/57/9/092004.
[60] E. Gaganidze, F. Gillemot, I. Szenthe, M. Gorley, M. Rieth, E. Diegele, Development of EUROFER97 database and material property handbook, Fusion Eng. Des. 135 (2018) 9–14, https://doi.org/10.1016/j.fusengdes.2018.06.027.
[61] A. Kohyama, A. Hishinuma, D.S. Gelles, R.L. Klueh, W. Dietz, K. Ehrlich, Low-activation ferritic and martensitic steels for fusion application, J. Nucl. Mater. 233–237 (1996) 138–147, https://doi.org/10.1016/s0022-3115(96)00327-3.
[62] L. El-Guebaly, M. Zucchetti, Progress and challenges of handling fusion radioactive materials, Fusion Sci. Technol. 68 (2015) 484–491, https://doi.org/10.13182/fst14-952.
[63] M.R. Gilbert, T. Eade, C. Bachmann, U. Fischer, N.P. Taylor, Waste assessment of European DEMO fusion reactor designs, Fusion Eng. Des. 136 (2018) 42–48, https://doi.org/10.1016/j.fusengdes.2017.12.019.
[64] E.E. Bloom, S.J. Zinkle, F.W. Wiffen, Materials to deliver the promise of fusion power – Progress and challenges, J. Nucl. Mater. 329–333 (2004) 12–19, https://doi.org/10.1016/j.jnucmat.2004.04.141.
[65] M. Sawan, Y. Katoh, L.L. Snead, Transmutation of silicon carbide in fusion nuclear environment, J. Nucl. Mater. 442 (1–3) (2013) S370.
[66] L.L. Snead, The Effect of Neutron Irradiation on PFC's at DEMO-Relevant Conditions, 2010.
[67] S.J. Zinkle, N.M. Ghoniem, Operating temperature windows for fusion reactor structural materials, Fusion Eng. Des. 51–52 (2000) 55–71, https://doi.org/10.1016/S0920-3796(00)00320-3.
[68] Y. Katoh, K. Ozawa, C. Shih, T. Nozawa, R.J. Shinavski, A. Hasegawa, L.L. Snead, Continuous SiC fiber, CVI SiC matrix composites for nuclear applications: Properties and irradiation effects, J. Nucl. Mater. 448 (2014) 448–476, https://doi.org/10.1016/j.jnucmat.2013.06.040.
[69] R.J. Price, Effects of fast-neutron irradiation on pyrolytic silicon carbide, J. Nucl. Mater. 33 (1969) 17–22, https://doi.org/10.1016/0022-3115(69)90003-8.
[70] C.H. Carter Jr., R.F. Davis, J. Bentley, Kinetics and mechanisms of high-temperature creep in silicon carbide: II, chemically vapor deposited, J. Am. Ceram. Soc. 67 (1984) 732–740. https://ceramics.onlinelibrary.wiley.com/doi/abs/10.1111/j.1151-2916.1984.tb19510.x?casa_token=cdXZx35UcJsAAAAA:T-Nk61x7vB5F1Srdwe4sc7lu3JxhwM5e2fMdxiE3FRALR_PeOhAgV-sNX54uhNAL5OElWhv2ZH36SWY.

[71] E.A. Gulbransen, S.A. Jansson, The high-temperature oxidation, reduction, and volatilization reactions of silicon and silicon carbide, Oxid. Met. 4 (1972) 181–201. https://idp.springer.com/authorize/casa?redirect_uri=https://link.springer.com/content/pdf/10.1007/bf00613092.pdf&casa_token=Zu04D49heKMAAAAA:HCV6uR7VoIyrOqZKyDhIVy3bJWZYDJOJbQZyXt2oZtAF7GFrhfHbzNEuDb_hUC7sN525jZCAYEjeUSPw.

[72] W. Zhuo, F. Niu, Y. Zhao, J. Liu, Hermeticity and tensile experiment of small plate-type SiC ceramic composites for advanced reactor applications, Ann. Nucl. Energy 110 (2017) 1098–1106, https://doi.org/10.1016/j.anucene.2017.08.036.

[73] T. Noda, M. Fujita, H. Araki, A. Kohyama, Effect of nuclear data and impurities on the evaluation of induced activity of CVI SiCf/SiC composites, Fusion Eng. Des. 61–62 (2002) 711–716, https://doi.org/10.1016/S0920-3796(02)00152-7.

[74] Y. Katoh, L.L. Snead, Silicon carbide and its composites for nuclear applications – Historical overview, J. Nucl. Mater. 526 (2019) 151849, https://doi.org/10.1016/j.jnucmat.2019.151849.

[75] L.L. Snead, T. Nozawa, Y. Katoh, T.-S. Byun, S. Kondo, D.A. Petti, Handbook of SiC properties for fuel performance modeling, J. Nucl. Mater. 371 (2007) 329–377, https://doi.org/10.1016/j.jnucmat.2007.05.016.

[76] S. Yajima, J. Hayashi, M. Omori, K. Okamura, Development of a silicon carbide fibre with high tensile strength, Nature 261 (1976) 683–685, https://doi.org/10.1038/261683a0.

[77] C.P. Deck, G.M. Jacobsen, J. Sheeder, O. Gutierrez, J. Zhang, J. Stone, H.E. Khalifa, C. A. Back, Characterization of SiC–SiC composites for accident tolerant fuel cladding, J. Nucl. Mater. 466 (2015) 667–681, https://doi.org/10.1016/j.jnucmat.2015.08.020.

[78] C.E. Kessel, M.S. Tillack, F. Najmabadi, F.M. Poli, K. Ghantous, N. Gorelenkov, X. R. Wang, D. Navaei, H.H. Toudeshki, C. Koehly, L. EL-Guebaly, J.P. Blanchard, C. J. Martin, L. Mynsburge, P. Humrickhouse, M.E. Rensink, T.D. Rognlien, M. Yoda, S. I. Abdel-Khalik, M.D. Hageman, B.H. Mills, J.D. Rader, D.L. Sadowski, P.B. Snyder, H.S.T. John, A.D. Turnbull, L.M. Waganer, S. Malang, A.F. Rowcliffe, The ARIES advanced and conservative tokamak power plant study, Fusion Sci. Technol. 67 (2015) 1–21, https://doi.org/10.13182/FST14-794.

[79] M.S. Tillack, X.R. Wang, D. Navaei, H.H. Toudeshki, A.F. Rowcliffe, F. Najmabadi, A. Team, Design and analysis of the ARIES-ACT1 fusion power core, Fusion Sci. Technol. 67 (2015) 49–74, https://doi.org/10.13182/FST14-790.

[80] A.R. Raffray, R. Jones, G. Aiello, M. Billone, L. Giancarli, H. Golfier, A. Hasegawa, Y. Katoh, A. Kohyama, S. Nishio, B. Riccardi, M.S. Tillack, Design and material issues for high performance SiCf/SiC-based fusion power cores, Fusion Eng. Des. 55 (2001) 55–95, https://doi.org/10.1016/S0920-3796(01)00181-8.

[81] L.L. Snead, T. Nozawa, M. Ferraris, Y. Katoh, R. Shinavski, M. Sawan, Silicon carbide composites as fusion power reactor structural materials, J. Nucl. Mater. 417 (2011) 330–339, https://doi.org/10.1016/j.jnucmat.2011.03.005.

[82] T. Nozawa, T. Hinoki, A. Hasegawa, A. Kohyama, Y. Katoh, L.L. Snead, C. H. Henager, J.B.J. Hegeman, Recent advances and issues in development of silicon carbide composites for fusion applications, J. Nucl. Mater. 386–388 (2009) 622–627, https://doi.org/10.1016/j.jnucmat.2008.12.305.

[83] Y. Katoh, L.L. Snead, C.H. Henager Jr., T. Nozawa, T. Hinoki, A. Iveković, S. Novak, S.M. Gonzalez de Vicente, Current status and recent research achievements in SiC/SiC composites, J. Nucl. Mater. 455 (2014) 387–397, https://doi.org/10.1016/j.jnucmat.2014.06.003.

[84] Y. Katoh, L.L. Snead, T. Cheng, C. Shih, W. Daniel Lewis, T. Koyanagi, T. Hinoki, C. H. Henager, M. Ferraris, Radiation-tolerant joining technologies for silicon carbide ceramics and composites, J. Nucl. Mater. 448 (2014) 497–511, https://doi.org/10.1016/j.jnucmat.2013.10.002.

[85] T. Koyanagi, Y. Katoh, T. Hinoki, C. Henager, M. Ferraris, S. Grasso, Progress in development of SiC-based joints resistant to neutron irradiation, J. Eur. Ceram. Soc. 40 (2020) 1023–1034, https://doi.org/10.1016/j.jeurceramsoc.2019.10.055.

[86] J.W. Cree, M.F. Amateau, Degradation of silicon carbide by molten lithium, J. Am. Ceram. Soc. 70 (1987), https://doi.org/10.1111/j.1151-2916.1987.tb05646.x. C-318–C-321.

[87] C. Park, T. Nozawa, R. Kasada, S. Tosti, S. Konishi, H. Tanigawa, The effect of wall flow velocity on compatibility of high-purity SiC materials with liquid Pb-Li alloy by rotating disc testing for 3000 h up to 900°C, Fusion Eng. Des. 136 (2018) 623–627, https://doi.org/10.1016/j.fusengdes.2018.03.042.

[88] C. Park, K. Noborio, R. Kasada, Y. Yamamoto, S. Konishi, Compatibility of SiCf/SiC composite exposed to liquid Pb–Li flow, J. Nucl. Mater. 417 (2011) 1218–1220, https://doi.org/10.1016/j.jnucmat.2011.06.016.

[89] J. Li, H. Huang, Q. Huang, M. Tang, B. Zhao, G. Ji, W. Zhang, R. Xie, L. Yan, Effect of irradiation damage on corrosion of 4H-SiC in FLiNaK molten salt, Corros. Sci. 125 (2017) 194–197, https://doi.org/10.1016/j.corsci.2017.05.028.

[90] Y. Katoh, T. Nozawa, C. Shih, K. Ozawa, T. Koyanagi, W. Porter, L.L. Snead, High-dose neutron irradiation of Hi-Nicalon Type S silicon carbide composites. Part 2: Mechanical and physical properties, J. Nucl. Mater. 462 (2015) 450–457, https://doi.org/10.1016/j.jnucmat.2014.12.121.

[91] A.G. Perez-Bergquist, T. Nozawa, C. Shih, K.J. Leonard, L.L. Snead, Y. Katoh, High dose neutron irradiation of Hi-Nicalon Type S silicon carbide composites, part 1: Microstructural evaluations, J. Nucl. Mater. 462 (2015) 443–449, https://doi.org/10.1016/j.jnucmat.2014.06.038.

[92] Y. Yamamoto, Y. Murakami, H. Yamaguchi, T. Yamamoto, D. Yonetsu, K. Noborio, S. Konishi, Re-evaluation of SiC permeation coefficients at high temperatures, Fusion Eng. Des. 109–111 (2016) 1286–1290, https://doi.org/10.1016/j.fusengdes.2015.12.041.

[93] T. Koyanagi, Y. Katoh, T. Nozawa, L.L. Snead, S. Kondo, C.H. Henager, M. Ferraris, T. Hinoki, Q. Huang, Recent progress in the development of SiC composites for nuclear fusion applications, J. Nucl. Mater. 511 (2018) 544–555, https://doi.org/10.1016/j.jnucmat.2018.06.017.

[94] Y. Zayachuk, P. Karamched, C. Deck, P. Hosemann, D.E.J. Armstrong, Linking microstructure and local mechanical properties in SiC-SiC fiber composite using micromechanical testing, Acta Mater. 168 (2019) 178–189, https://doi.org/10.1016/j.actamat.2019.02.001.

[95] F. Goesmann, R. Schmid-Fetzer, Stability of W as electrical contact on 6H SiC: Phase relations and interface reactions in the ternary system W Si C, Mater. Sci. Eng. B. 34 (1995) 224–231, https://doi.org/10.1016/0921-5107(95)01311-3.

[96] H.-K. Kang, Microstructures of high volume SiC reinforced tungsten composites produced by plasma spray, Scr. Mater. 51 (2004) 1051–1055, https://doi.org/10.1016/j.scriptamat.2004.08.008.

[97] S.J. Son, K.H. Park, Y. Katoh, A. Kohyama, Interfacial reactions and mechanical properties of W–SiC in-situ joints for plasma facing components, J. Nucl. Mater. 329–333 (2004) 1549–1552, https://doi.org/10.1016/j.jnucmat.2004.04.285.

[98] H. Kishimoto, T. Shibayama, T. Abe, K. Shimoda, S. Kawamura, A. Kohyama, Diffusion bonding technology of tungsten and SiC/SiC composites for nuclear applications, IOP Conf. Ser. Mater. Sci. Eng. 18 (2011) 162015, https://doi.org/10.1088/1757-899x/18/16/162015.

[99] H. Kishimoto, T. Shibayama, K. Shimoda, T. Kobayashi, A. Kohyama, Microstructural and mechanical characterization of W/SiC bonding for structural material in fusion, J. Nucl. Mater. 417 (2011) 387–390, https://doi.org/10.1016/j.jnucmat.2010.12.079.

[100] Y. Asakura, H. Kishimoto, J.-S. Park, N. Nakazato, T. Shibayama, A. Kohyama, Thermal property and microstructural characterization of W/SiC clad plates, Fusion Eng. Des. 125 (2017) 484–489, https://doi.org/10.1016/j.fusengdes.2017.05.105.

[101] Website (n.d.). https://burningplasma.org/activities/uploads_tec/FESAC_TEC_White_Paper_EHTM%20Rev20170614-2p.pdf (accessed September 24, 2020).

[102] W.G. Fahrenholtz, E.J. Wuchina, W.E. Lee, Y. Zhou (Eds.), Ultra-High Temperature Ceramics, John Wiley & Sons, Inc, Hoboken, NJ, 2014, https://doi.org/10.1002/9781118700853.

[103] W.G. Fahrenholtz, G.E. Hilmas, I.G. Talmy, J.A. Zaykoski Refractory diborides of zirconium and hafnium, J. Am. Ceram. Soc. 90 (2007) 1347–1364, https://doi.org/10.1111/j.1551-2916.2007.01583.x.

[104] V.I. Khripunov, C-14 Production in CTR Materials and Blankets (2006). http://inis.iaea.org/Search/search.aspx?orig_q=RN:37110183 (accessed September 24, 2020).

[105] Y. Katoh, Emerging High Temperature Materials for Potential Application to Fusion (June 20–22, 2017). https://www.burningplasma.org/activities/uploads_tec/170621_Katoh_FESAC-TEC_Emerging%20Materials.pdf.

[106] H.O. Pierson, Handbook of Refractory Carbides & Nitrides: Properties, Characteristics, Processing and Apps, William Andrew, 1996. https://play.google.com/store/books/details?id=K_K7q3jaqXEC.

[107] I.L. Shabalin, Ultra-High Temperature Materials II (2019). https://doi.org/10.1007/978-94-024-1302-1.

[108] I.L. Shabalin, Ultra-High Temperature Materials III: Refractory Carbides II (Ti and V Carbides), Springer, 2020. https://books.google.com/books/about/Ultra_High_Temperature_Materials_III.html?hl=&id=A-2NzQEACAAJ.

[109] B.R. Golla, A. Mukhopadhyay, B. Basu, S.K. Thimmappa, Review on ultra-high temperature boride ceramics, Prog. Mater. Sci. 111 (2020) 100651, https://doi.org/10.1016/j.pmatsci.2020.100651.

[110] J.M. Lonergan, W.G. Fahrenholtz, G.E. Hilmas, Zirconium diboride with high thermal conductivity, J. Am. Ceram. Soc. 97 (2014) 1689–1691, https://doi.org/10.1111/jace.12950.

[111] Large Tungsten Carbide Blank Manufactured by Spark Plasma Sintering (2015). https://www.pm-review.com/large-tungsten-carbide-blank-manufactured-by-spark-plasma-sintering/ (accessed September 24, 2020).

[111a] https://inl.gov/integrated-energy/idaho-lab-unveils-industrial-scale-advancedmanufacturing-technology/.

[112] M. Singh, R. Asthana, Joining of ZrB_2-based ultra-high-temperature ceramic composites to Cu-Clad-Molybdenum for advanced aerospace applications, Int. J. Appl. Ceram. Technol. 6 (2009) 113–133, https://doi.org/10.1111/j.1744-7402.2008.02291.x.

[113] D.S. King, G.E. Hilmas, W.G. Fahrenholtz, Plasma arc welding of ZrB2–20vol% ZrC ceramics, J. Eur. Ceram. Soc. 34 (2014) 3549–3557, https://doi.org/10.1016/j.jeurceramsoc.2014.05.011.

[114] D.S. King, G. Hilmas, W. Fahrenholtz, Mechanical behavior and applications of plasma arc welded ceramics, Int. J. Appl. Ceram. Technol. 13 (2016) 41–49, https://doi.org/10.1111/ijac.12402.
[115] B.-R. Kim, M.-S. Moon, K.-D. Woo, I.-J. Shon, Properties and rapid consolidation of binderless nanostuctured NbC by pulsed current activated sintering, Electron. Mater. Lett. 5 (2009) 119–122, https://doi.org/10.3365/eml.2009.09.119.
[116] K. Hackett, S. Verhoef, R.A. Cutler, D.K. Shetty, Phase constitution and mechanical properties of carbides in the Ta-C system, J. Am. Ceram. Soc. 92 (2009) 2404–2407, https://doi.org/10.1111/j.1551-2916.2009.03201.x.
[117] A. Kelly, D.J. Rowcliffe, Deformation of poly crystalline transition metal carbides, J. Am. Ceram. Soc. 50 (1967) 253–256, https://doi.org/10.1111/j.1151-2916.1967.tb15098.x.
[118] S. Ukai, Oxide dispersion strengthened steels, Comprehensive Nucl. Mater. (2012) 241–271, https://doi.org/10.1016/b978-0-08-056033-5.00069-0.
[119] J. Wang, L.J. Vandeperre, Deformation and Hardness of UHTCs as a Function of Temperature, Ultra-High Temperature Ceramics, 2014, pp. 236–266, https://doi.org/10.1002/9781118700853.ch10.
[120] A.G. Atkins, Deformation-mechanism maps (the plasticity and creep of metals and ceramics), J. Mech. Work. Technol. 9 (1984) 224–225, https://doi.org/10.1016/0378-3804(84)90015-9.
[121] K. Minato, T. Ogawa, Advanced concepts in TRISO fuel, Comprehensive Nucl. Mater. (2012) 215–236, https://doi.org/10.1016/b978-0-08-056033-5.00056-2.
[122] Y. Katoh, G. Vasudevamurthy, T. Nozawa, L.L. Snead, Properties of zirconium carbide for nuclear fuel applications, J. Nucl. Mater. 441 (2013) 718–742, https://doi.org/10.1016/j.jnucmat.2013.05.037.
[123] L.L. Snead, Y. Katoh, S. Kondo, Effects of fast neutron irradiation on zirconium carbide, J. Nucl. Mater. 399 (2010) 200–207, https://doi.org/10.1016/j.jnucmat.2010.01.020.
[124] S. Agarwal, T. Koyanagi, A. Bhattacharya, L. Wang, Y. Katoh, X. Hu, M. Pagan, S. J. Zinkle, Neutron irradiation-induced microstructure damage in ultra-high temperature ceramic TiC, Acta Mater. 186 (2020) 1–10, https://doi.org/10.1016/j.actamat.2019.12.022.
[125] E.U. Grinik, V.V. Ogorodnikov, A.G. Krainy, Fracturing of borides under neutron irradiation, J. Nucl. Mater. 233–237 (1996) 1349–1354, https://doi.org/10.1016/s0022-3115(96)00293-0.
[126] S.C. Middleburgh, D.C. Parfitt, P.R. Blair, R.W. Grimes, Atomic scale modeling of point defects in zirconium diboride, J. Am. Ceram. Soc. 94 (2011) 2225–2229, https://doi.org/10.1111/j.1551-2916.2010.04360.x.
[127] D.J. Hamman, P. Schall, Radiation Effects on Boron-Containing Compounds, 1960, https://doi.org/10.2172/4172091.
[128] L.M. Garrison, G.L. Kulcinski, G. Hilmas, W. Fahrenholtz, H.M. Meyer, The response of ZrB2 to simulated plasma-facing material conditions of He irradiation at high temperature, J. Nucl. Mater. 507 (2018) 112–125, https://doi.org/10.1016/j.jnucmat.2018.04.016.
[129] T. Koyanagi, Y. Katoh, C. Ang, D. King, G.E. Hilmas, W.G. Fahrenholtz, Response of isotopically tailored titanium diboride to neutron irradiation, J. Am. Ceram. Soc. 102 (2019) 85–89, https://doi.org/10.1111/jace.16036.
[130] A. Bhattacharya, C.M. Parish, T. Koyanagi, C.M. Petrie, D. King, G. Hilmas, W. G. Fahrenholtz, S.J. Zinkle, Y. Katoh, Nano-scale microstructure damage by neutron

irradiations in a novel Boron-11 enriched TiB2 ultra-high temperature ceramic, Acta Mater. 165 (2019) 26–39, https://doi.org/10.1016/j.actamat.2018.11.030.
[131] R.A. Causey, R.A. Karnesky, C. San Marchi, Tritium barriers and tritium diffusion in fusion reactors, Comprehensive Nucl. Mater. (2012) 511–549, https://doi.org/10.1016/b978-0-08-056033-5.00116-6.
[132] C. Shan, The behaviour of diffusion and permeation of tritium through 316L stainless steel with coating of TiC and TiN TiC, J. Nucl. Mater. 191–194 (1992) 221–225, https://doi.org/10.1016/0022-3115(92)90758-d.
[133] K.S. Forcey, A. Perujo, F. Reiter, F.L. Lolli-Ceroni, The formation of tritium permeation barriers by CVD, J. Nucl. Mater. 200 (1993) 417–420, https://doi.org/10.1016/0022-3115(93)90319-t.
[134] A.E. Pontau, K.L. Wilson, VB2: A thin film coating on vanadium for reduced tritium inventory in controlled fusion devices, Thin Solid Films 73 (1980) 109–114, https://doi.org/10.1016/0040-6090(80)90336-3.
[135] R.A. Kerst, W.A. Swansiger, Comparison of plasma and gas driven permeation of deuterium through titanium diboride, J. Nucl. Mater. 135 (1985) 77–81, https://doi.org/10.1016/0022-3115(85)90439-8.
[136] K.L. Wilson, A.E. Pontau, Deuterium trapping and release in titanium-based coatings for TFTR, J. Nucl. Mater. 93–94 (1980) 569–574, https://doi.org/10.1016/0022-3115(80)90176-2.
[137] F. Wang, X. Yan, T. Wang, Y. Wu, L. Shao, M. Nastasi, Y. Lu, B. Cui, Irradiation damage in (Zr0.25Ta0.25Nb0.25Ti0.25)C high-entropy carbide ceramics, Acta Mater. 195 (2020) 739–749, https://doi.org/10.1016/j.actamat.2020.06.011.
[138] R.-Z. Zhang, M.J. Reece, Review of high entropy ceramics: Design, synthesis, structure and properties, J. Mater. Chem. A 7 (2019) 22148–22162, https://doi.org/10.1039/c9ta05698j.
[139] J. Gild, Y. Zhang, T. Harrington, S. Jiang, T. Hu, M.C. Quinn, W.M. Mellor, N. Zhou, K. Vecchio, J. Luo, High-entropy metal diborides: A new class of high-entropy materials and a new type of ultrahigh temperature ceramics, Sci. Rep. 6 (2016), https://doi.org/10.1038/srep37946.
[140] J. Gu, J. Zou, S.-K. Sun, H. Wang, S.-Y. Yu, J. Zhang, W. Wang, Z. Fu, Dense and pure high-entropy metal diboride ceramics sintered from self-synthesized powders via boro/carbothermal reduction approach, Sci. China Mater. 62 (2019) 1898–1909, https://doi.org/10.1007/s40843-019-9469-4.
[141] Y. Zhang, Z.-B. Jiang, S.-K. Sun, W.-M. Guo, Q.-S. Chen, J.-X. Qiu, K. Plucknett, H.-T. Lin, Microstructure and mechanical properties of high-entropy borides derived from boro/carbothermal reduction, J. Eur. Ceram. Soc. 39 (2019) 3920–3924, https://doi.org/10.1016/j.jeurceramsoc.2019.05.017.
[142] E. Castle, T. Csanádi, S. Grasso, J. Dusza, M. Reece, Processing and properties of high-entropy ultra-high temperature carbides, Sci. Rep. 8 (2018) 8609, https://doi.org/10.1038/s41598-018-26827-1.
[143] X. Han, V. Girman, R. Sedlak, J. Dusza, E.G. Castle, Y. Wang, M. Reece, C. Zhang, Improved creep resistance of high entropy transition metal carbides, J. Eur. Ceram. Soc. 40 (2020) 2709–2715, https://doi.org/10.1016/j.jeurceramsoc.2019.12.036.
[144] B. Ye, T. Wen, D. Liu, Y. Chu, Oxidation behavior of (Hf0.2Zr0.2Ta0.2Nb0.2Ti0.2)C high-entropy ceramics at 1073–1473 K in air, Corros. Sci. 153 (2019) 327–332, https://doi.org/10.1016/j.corsci.2019.04.001.
[145] J. Gild, A. Wright, K. Quiambao-Tomko, M. Qin, J.A. Tomko, M.S.B. Hoque, J. L. Braun, B. Bloomfield, D. Martinez, T. Harrington, K. Vecchio, P.E. Hopkins, J. Luo,

Thermal conductivity and hardness of three single-phase high-entropy metal diborides fabricated by borocarbothermal reduction and spark plasma sintering, Ceram. Int. 46 (2020) 6906–6913, https://doi.org/10.1016/j.ceramint.2019.11.186.
[146] R. Prokopec, D.X. Fischer, H.W. Weber, M. Eisterer, Suitability of coated conductors for fusion magnets in view of their radiation response, Supercond. Sci. Technol. 28 (2015) 014005, https://doi.org/10.1088/0953-2048/28/1/014005.
[147] H.W. Weber, H. Böck, E. Unfried, L.R. Greenwood, Neutron dosimetry and damage calculations for the TRIGA MARK-II reactor in Vienna, J. Nucl. Mater. 137 (1986) 236–240, https://doi.org/10.1016/0022-3115(86)90225-4.
[148] A. Sykes, A.E. Costley, C.G. Windsor, O. Asunta, G. Brittles, P. Buxton, V. Chuyanov, J.W. Connor, M.P. Gryaznevich, B. Huang, J. Hugill, A. Kukushkin, D. Kingham, A. V. Langtry, S. McNamara, J.G. Morgan, P. Noonan, J.S.H. Ross, V. Shevchenko, R. Slade, G. Smith, Compact fusion energy based on the spherical tokamak, Nucl. Fusion 58 (2018) 016039, https://doi.org/10.1088/1741-4326/aa8c8d.
[149] C.G. Windsor, J.G. Morgan, P.F. Buxton, A.E. Costley, G.D.W. Smith, A. Sykes, Modelling the power deposition into a spherical tokamak fusion power plant, Nucl. Fusion 57 (2017) 036001, https://doi.org/10.1088/1741-4326/57/3/036001.
[150] C.G. Windsor, J.G. Morgan, P.F. Buxton, Heat deposition into the superconducting central column of a spherical tokamak fusion plant, Nucl. Fusion 55 (2015) 023014, https://doi.org/10.1088/0029-5515/55/2/023014.
[151] C.G. Windsor, J.M. Marshall, J.G. Morgan, J. Fair, G.D.W. Smith, A. Rajczyk-Wryk, J. M. Tarragó, Design of cemented tungsten carbide and boride-containing shields for a fusion power plant, Nucl. Fusion 58 (2018) 076014, https://doi.org/10.1088/1741-4326/aabdb0.
[152] J. Kenneth Shultis, R.E. Faw, Radiation Shielding, Amer Nuclear Society, 2000. https://books.google.com/books/about/Radiation_Shielding.html?hl=&id=HiN3QgAACAAJ.
[153] G.F. Knoll, Radiation Detection and Measurement, John Wiley & Sons, 2010. https://books.google.com/books/about/Radiation_Detection_and_Measurement.html?hl=&id=4vTJ7UDel5IC.
[154] W.J. Veigele, Photon cross sections from 0.1 keV to 1 MeV for elements Z = 1 to Z = 94, Atomic Data Nucl. Data Tables 5 (1973) 51–111, https://doi.org/10.1016/s0092-640x(73)80015-4.
[155] W.J. Veigele, Mass Attenuation and Absorption Cross Sections for 94 Elements: 0.1 keV to 1 MeV, in: Practical Handbook of Spectroscopy, Routledge, 2017, pp. 28–154, https://doi.org/10.1201/9780203742433-3.
[156] E. Storm, J.H. Hancock, Electron Elastic Scattering Cross Sections from 1 keV to 100 MeV for Elements Z = 1 to 100 (1975). https://doi.org/10.2172/4067467.
[157] X-Ray Mass Attenuation Coefficients, NIST (n.d.). https://www.nist.gov/pml/x-ray-mass-attenuation-coefficients (accessed July 19, 2020).
[158] V.F. Sears, Neutron scattering lengths and cross sections, Neutron News 3 (1992) 26–37, https://doi.org/10.1080/10448639208218770.
[159] ENDF/B-VIII.0: The 8th major release of the nuclear reaction data library with CIELO-project cross sections, new standards and thermal scattering data, Nucl. Data Sheets 148 (2018) 1–142, https://doi.org/10.1016/j.nds.2018.02.001.
[160] D.G. Whyte, J. Minervini, B. LaBombard, E. Marmar, L. Bromberg, M. Greenwald, Smaller & sooner: Exploiting high magnetic fields from new superconductors for a more attractive fusion energy development path, J. Fusion Energy 35 (2016) 41–53, https://doi.org/10.1007/s10894-015-0050-1.

[161] ARC: A compact, high-field, fusion nuclear science facility and demonstration power plant with demountable magnets, Fusion Eng. Des. 100 (2015) 378–405. https://doi.org/10.1016/j.fusengdes.2015.07.008.
[162] A.Q. Kuang, N.M. Cao, A.J. Creely, C.A. Dennett, J. Hecla, B. LaBombard, R. A. Tinguely, E.A. Tolman, H. Hoffman, M. Major, J. Ruiz Ruiz, D. Brunner, P. Grover, C. Laughman, B.N. Sorbom, D.G. Whyte, Conceptual design study for heat exhaust management in the ARC fusion pilot plant, Fusion Eng. Des. 137 (2018) 221–242, https://doi.org/10.1016/j.fusengdes.2018.09.007.
[163] Radiation Damage in Austenitic Steels (2012) 33–95. https://doi.org/10.1016/B978-0-08-056033-5.00065-3.
[164] Absorber materials for generation IV reactors, in: Structural Materials for Generation IV Nuclear Reactors, Woodhead Publishing, 2017, pp. 533–567. https://doi.org/10.1016/B978-0-08-100906-2.00015-X.
[165] Nuclear properties of Tungsten under 14 MeV neutron irradiation for fusion-fission hybrid reactors, Ann. Nucl. Energy 147 (2020) 107739. https://doi.org/10.1016/j.anucene.2020.107739.
[166] R.G. Abernethy, Predicting the performance of tungsten in a fusion environment: A literature review, Mater. Sci. Technol. 33 (2017) 388–399, https://doi.org/10.1080/02670836.2016.1185260.
[167] M.R. Gilbert, J.-C. Sublet, Neutron-induced transmutation effects in W and W-alloys in a fusion environment, Nucl. Fusion 51 (2011) 043005, https://doi.org/10.1088/0029-5515/51/4/043005.
[168] Thermal Spectrum Control Rod Materials (2012) 485–507. https://doi.org/10.1016/B978-0-08-056033-5.00077-X.
[169] Fast Spectrum Control Rod Materials (2012) 509–534. https://doi.org/10.1016/B978-0-08-056033-5.00078-1.
[170] A. Gubernat, P. Rutkowski, G. Grabowski, D. Zientara, Hot pressing of tungsten carbide with and without sintering additives, Int. J. Refract. Metals Hard Mater. 43 (2014) 193–199, https://doi.org/10.1016/j.ijrmhm.2013.12.002.
[171] ASM International, Handbook Committee, Engineered Materials Handbook, Desk Edition, ASM International, 1995. https://books.google.com/books/about/Engineered_Materials_Handbook_Desk_Editi.html?hl=&id=Cbk5n7Ny4asC.
[172] M. Athanasakis, E. Ivanov, E. del Rio, S.A. Humphry-Baker, A high temperature W2B–W composite for fusion reactor shielding, J. Nucl. Mater. 532 (2020) 152062, https://doi.org/10.1016/j.jnucmat.2020.152062.
[173] T. Hayashi, K. Tobita, Y. Nakamori, S. Orimo, Advanced neutron shielding material using zirconium borohydride and zirconium hydride, J. Nucl. Mater. 386–388 (2009) 119–121, https://doi.org/10.1016/j.jnucmat.2008.12.073.
[174] J.B. Vetrano, Hydrides as neutron moderator and reflector materials, Nucl. Eng. Des. 14 (1971) 390–412, https://doi.org/10.1016/0029-5493(70)90159-7.
[175] S. Yamanaka, K. Kurosaki, Neutron reflector materials (Be, hydrides), Comprehensive Nucl. Mater. (2012) 307–321, https://doi.org/10.1016/b978-0-08-056033-5.00021-5.
[176] C.E. Curtis, A.G. Tharp, Ceramic properties of Europium oxide, J. Am. Ceram. Soc. 42 (1959) 151–156, https://doi.org/10.1111/j.1151-2916.1959.tb14084.x.
[177] Europium Oxide as a Potential LMFBR Control Material (1973). https://books.google.com/books/about/Europium_Oxide_as_a_Potential_LMFBR_Cont.html?hl=&id=kMXIwQEACAAJ.
[178] E.P. Klochkov, V.D. Risovanyi, Y.E. Vaneev, A.N. Dorofeev, Atom. Energy 93 (2002) 656–660, https://doi.org/10.1023/A:1021096715382.

[179] A. Sinha, B.P. Sharma, Development of dysprosium titanate based ceramics, J. Am. Ceram. Soc. 88 (2005) 1064–1066, https://doi.org/10.1111/j.1551-2916.2005.00211.x.
[180] Dysprosium titanate as an absorber material for control rods, J. Nucl. Mater. 281 (2000) 84–89. https://doi.org/10.1016/S0022-3115(00)00129-X.
[181] Dysprosium hafnate as absorbing material for control rods, J. Nucl. Mater. 355 (2006) 163–170. https://doi.org/10.1016/j.jnucmat.2006.05.029.
[182] P. Predeep, N.S. Saxena, Effective thermal conductivity and thermal diffusivity of some rare earth oxides, Phys. Scr. 55 (2006) 634, https://doi.org/10.1088/0031-8949/55/5/017.
[183] The use of gadolinium as a burnable poison within U3Si2 fuel pellets, J. Nucl. Mater. 509 (2018) 204–211, https://doi.org/10.1016/j.jnucmat.2018.06.044.
[184] Burnable Poison-Doped Fuel (2020) 106–124. https://doi.org/10.1016/B978-0-12-803581-8.11699-6.
[185] J.E. Menard, T. Brown, L. El-Guebaly, M. Boyer, J. Canik, B. Colling, R. Raman, Z. Wang, Y. Zhai, P. Buxton, B. Covele, C. D'Angelo, A. Davis, S. Gerhardt, M. Gryaznevich, M. Harb, T.C. Hender, S. Kaye, D. Kingham, M. Kotschenreuther, S. Mahajan, R. Maingi, E. Marriott, E.T. Meier, L. Mynsberge, C. Neumeyer, M. Ono, J.-K. Park, S.A. Sabbagh, V. Soukhanovskii, P. Valanju, R. Woolley, Fusion nuclear science facilities and pilot plants based on the spherical tokamak, Nucl. Fusion 56 (2016) 106023, https://doi.org/10.1088/0029-5515/56/10/106023.
[186] B.G. Hong, Y.S. Hwang, J.S. Kang, D.W. Lee, H.G. Joo, M. Ono, Conceptual design study of a superconducting spherical tokamak reactor with a self-consistent system analysis code, Nucl. Fusion 51 (2011) 113013, https://doi.org/10.1088/0029-5515/51/11/113013.
[187] J.M. Marshall, D. Walker, P.A. Thomas, HRXRD study of the theoretical densities of novel reactive sintered boride candidate neutron shielding materials, Nucl. Mater. Energy 22 (2020) 100732, https://doi.org/10.1016/j.nme.2020.100732.
[188] S.A. Humphry-Baker, R.W. Harrison, G. Greaves, A.J. Knowles, G.D.W. Smith, S. E. Donnelly, W.E. Lee, A candidate fusion engineering material, WC-FeCr, Scr. Mater. 155 (2018) 129–133, https://doi.org/10.1016/j.scriptamat.2018.06.027.
[189] A.G. Kvashnin, H.A. Zakaryan, C. Zhao, Y. Duan, Y.A. Kvashnina, C. Xie, H. Dong, A.R. Oganov, New tungsten borides, their stability and outstanding mechanical properties, J. Phys. Chem. Lett. 9 (2018) 3470–3477, https://doi.org/10.1021/acs.jpclett.8b01262.
[190] S.A. Humphry-Baker, G.D.W. Smith, Shielding materials in the compact spherical tokamak, Philos. Trans. A Math. Phys. Eng. Sci. 377 (2019) 20170443, https://doi.org/10.1098/rsta.2017.0443.
[191] https://core.ac.uk/download/pdf/132201627.pdf (accessed September 24, 2020).
[192] J. Marshall, A.E. Rajczyk-Wryk, A. Hirsch, Cemented Carbides Comprising an Fe-Cr Binder Based Metallic Binder, 3401414: A1 (2018). https://patentimages.storage.googleapis.com/75/64/3f/c21cff9aa15f7e/EP3401414A1.pdf (accessed September 25, 2020).
[193] J. Marshall, An Iron Tungsten Borocarbide Body for Nuclear Shielding Applications, 3401413: A1 (2018). https://patentimages.storage.googleapis.com/a2/1f/63/3c66dd32717c30/EP3401413A1.pdf (accessed September 25, 2020).
[194] D. Maisonnier, I. Cook, S. Pierre, B. Lorenzo, B. Edgar, B. Karin, D.P. Luigi, F. Robin, G. Luciano, H. Stephan, N. Claudio, N. Prachai, P. Aldo, T. Neill, W. David, The European power plant conceptual study, Fusion Eng. Des. 75–79 (2005) 1173–1179, https://doi.org/10.1016/j.fusengdes.2005.06.095.

[195] S.A. Humphry-Baker, W.E. Lee, Tungsten carbide is more oxidation resistant than tungsten when processed to full density, Scr. Mater. 116 (2016) 67–70, https://doi.org/10.1016/j.scriptamat.2016.01.007.
[196] S.A. Humphry-Baker, P. Ramanujam, G.D.W. Smith, J. Binner, W.E. Lee, Ablation resistance of tungsten carbide cermets under extreme conditions, Int. J. Refract. Metals Hard Mater. 93 (2020) 105356, https://doi.org/10.1016/j.ijrmhm.2020.105356.
[197] S.A. Humphry-Baker, K. Peng, W.E. Lee, Oxidation resistant tungsten carbide hardmetals, Int. J. Refract. Metals Hard Mater. 66 (2017) 135–143, https://doi.org/10.1016/j.ijrmhm.2017.03.009.
[198] I.N. Frantzevitch, V.A. Lavrenko, L.A. Glebov, High-temperature oxidation of tungsten boride in oxygen and the effect of scale evaporation, Oxid Metals 6 (1973) 1–19, https://doi.org/10.1007/bf00612044.
[199] G.W. Keilholtz, R.E. Moore, M.F. Osborne, Fast-neutron effects on the carbides of titanium, zirconium, tantalum, niobium, and tungsten, Nucl. Appl. 4 (1968) 330–336, https://doi.org/10.13182/nt68-a26398.
[200] J.S. Sun, P. Yan, X.B. Sun, G. Lu, F. Liu, W. Ye, J.Q. Yang, Tribological properties of nitrogen ion implanted WC-Co, Wear 213 (1997) 131–134, https://doi.org/10.1016/s0043-1648(97)00165-8.
[201] P.J. Burnett, T.F. Page, Criteria for mechanical property modifications of ceramic surfaces by ion implantation, Radiat. Eff. 97 (1986) 283–296, https://doi.org/10.1080/00337578608226019.
[202] C. Björkas, K. Vörtler, K. Nordlund, Major elemental asymmetry and recombination effects in irradiated WC, Phys. Rev. B 74 (2006), https://doi.org/10.1103/physrevb.74.140103.
[203] S.X. Oliver, M.L. Jackson, P.A. Burr, Radiation-induced evolution of tungsten carbide in fusion reactors: Accommodation of defect clusters and transmutation elements, ACS Appl. Energy Mater. 3 (2020) 868–878, https://doi.org/10.1021/acsaem.9b01990.
[204] Website (n.d.). https://www.researchgate.net/publication/324980715_Tungsten_Boride_A_New_Material_for_Fusion (accessed September 24, 2020).
[205] Metal Hydrides (n.d.). https://doi.org/10.1016/C2013-0-12462-2 (accessed September 15, 2020).
[206] G. Bambakidis (Ed.), Metal Hydrides, Springer US, Boston, MA, 1981, https://doi.org/10.1007/978-1-4757-5814-6.
[207] J.D. Gylfe, H. Rood, J. Greenleaf, K. Balkwill, L. Frem, L. Goldfisher, Evaluation of Zirconium Hydride as Moderator in Integral, Boiling Water-Superheat Reactors, North American Aviation, Atomics International Division, 1962. https://digital.library.unt.edu/ark:/67531/metadc957975/m2/1/high_res_d/metadc957975.pdf. (Accessed 15 September 2020). accessed.
[208] W.H. Jordan, S.J. Cramer, A.J. Miller, Aircraft Nuclear Propulsion Program: Quarterly Progress Report for Period Ending December 31, 1956, Part 1–5, Oak Ridge National Laboratory (ORNL), 1957, https://doi.org/10.2172/1373535.
[209] X. Hu, D. Schappel, C.M. Silva, K.A. Terrani, Fabrication of yttrium hydride for high-temperature moderator application, J. Nucl. Mater. 539 (2020) 152335, https://doi.org/10.1016/j.jnucmat.2020.152335.
[210] X. Hu, C. Silva, K. Terrani, Development of yttrium hydride moderator for the transformational challenge reactor, Trans. Am. Nucl. Soc. 122 (2020), https://doi.org/10.13182/t32423.

[211] K. Konashi, K. Itoh, T. Yokoyama, M. Yamawaki, Utilization Research and Development of Hydride Materials in Fast Reactors, Trans Tech Publications Ltd, 2014, pp. 23–31, https://doi.org/10.4028/www.scientific.net/AST.94.23.

[212] K. Ikeda, H. Moriwaki, Y. Ohkubo, T. Iwasaki, K. Konashi, Application of hafnium hydride control rod to large sodium cooled fast breeder reactor, Nucl. Eng. Des. 278 (2014) 97–107, https://doi.org/10.1016/j.nucengdes.2014.07.002.

[213] M.M.H. Bhuiya, A. Kumar, K.J. Kim, Metal hydrides in engineering systems, processes, and devices: A review of non-storage applications, Int. J. Hydrogen Energy 40 (2015) 2231–2247, https://doi.org/10.1016/j.ijhydene.2014.12.009.

[214] G.M. Begun, J.F. Land, J.T. Bell, High temperature equilibrium measurements of the yttrium–hydrogen isotope (H_2, D_2, T_2) systems, J. Chem. Phys. 72 (1980) 2959–2966, https://doi.org/10.1063/1.439496.

[215] W.M. Mueller, J.P. Blackledge, U.S. Atomic Energy Commission, Metal Hydrides. Editors and Major Contributors: W.M. Mueller, J.P. Blackledge, G.G. Libowitz (1968). https://books.google.com/books/about/Metal_Hydrides_Editors_and_Major_Contrib.html?hl=&id=O3YBcgAACAAJ.

[216] W.M. Mueller, J.P. Blackledge, G.G. Libowitz, Metal Hydrides, Elsevier, 2013. https://play.google.com/store/books/details?id=JgIcBQAAQBAJ.

[217] Mechanical properties of yttrium hydride, J. Alloys Compd. 394 (2005) 207–210. https://doi.org/10.1016/j.jallcom.2004.10.035.

[218] The U ZrHx alloy: Its properties and use in TRIGA fuel, Nucl. Eng. Des. 64 (1981) 403–422. https://doi.org/10.1016/0029-5493(81)90135-7.

[219] T. Tanaka, H. Muta, Y. Hishinuma, H. Tamura, T. Muroga, A. Sagara, Applicability of hydride materials for radiation shielding in helical reactor FFHR-d1, Fusion Sci. Technol. 68 (2015) 705–710, https://doi.org/10.13182/FST15-110.

[220] N.G. Author, Aircraft Nuclear Propulsion Project Semiannual Progress Report for Period Ending October 31, 1959, (1959). https://doi.org/10.2172/1232650.

[221] Properties of Hydrided Yttrium (n.d.). https://books.google.com/books/about/Properties_of_Hydrided_Yttrium.html?id=QcoGSuZ2opMC (accessed September 24, 2020).

[222] Selected engineering and fabrication aspects of nuclear metal hydrides (Li, Ti, Zr, and Y), Nucl. Eng. Des. 31 (1974) 434–448. https://doi.org/10.1016/0029-5493(75)90178-8.

[223] B.R. Betzler, B.J. Ade, A.J. Wysocki, P.K. Jain, P.C. Chesser, M.S. Greenwood, K. A. Terrani, Transformational challenge reactor preconceptual core design studies, Nucl. Eng. Des. 367 (2020) 110781, https://doi.org/10.1016/j.nucengdes.2020.110781.

[224] J.-C. Sublet, J.W. Eastwood, J.G. Morgan, M.R. Gilbert, M. Fleming, W. Arter, FISPACT-II: An advanced simulation system for activation, transmutation and material modelling, Nucl. Data Sheets 139 (2017) 77–137, https://doi.org/10.1016/j.nds.2017.01.002.

[225] M.R. Gilbert, J.-C. Sublet, Handbook of activation, transmutation, and radiation damage properties of the elements simulated using FISPACT-II and TENDL-2014, Nuclear Fission Plants: PWR Focus (2015). https://books.google.com/books/about/Handbook_of_Activation_Transmutation_and.html?hl=&id=lNVIzQEACAAJ.

[226] R.K. Edwards, E. Veleckis, Thermodynamic properties and phase relations in the system hydrogen-hafnium 1,2, J. Phys. Chem. 66 (1962) 1657–1661, https://doi.org/10.1021/j100815a022.

[227] D. Setoyama, J. Matsunaga, H. Muta, M. Uno, S. Yamanaka, Mechanical properties of titanium hydride, J. Alloys Compd. 381 (2004) 215–220, https://doi.org/10.1016/j.jallcom.2004.04.073.

[228] M. Tada, Phase diagram of Hf-H alloy, J. Mater. Sci. 25 (1990) 2934–2936, https://doi.org/10.1007/BF00584907.

[229] O. Madelung (Ed.), Ga-Gd – Hf-Zr, Springer-Verlag, Berlin/Heidelberg, 1996, https://doi.org/10.1007/b56869.

[230] K. Kurosaki, D. Araki, Y. Ohishi, H. Muta, K. Konashi, S. Yamanaka, The δ′/δ phase transition in hafnium hydride and deuteride, J. Nucl. Sci. Technol. (2014) 1–5, https://doi.org/10.1080/00223131.2014.962638.

[231] M. Ito, K. Kurosaki, H. Muta, M. Uno, K. Konashi, S. Yamanaka, Thermomechanical properties of hafnium hydride, J. Nucl. Sci. Technol. 47 (2010) 156–159, https://doi.org/10.1080/18811248.2010.9711940.

[232] M. Ito, K. Kurosaki, H. Muta, M. Uno, K. Konashi, S. Yamanaka, Thermal conductivity of hafnium hydride, J. Nucl. Sci. Technol. 46 (2009) 814–818, https://doi.org/10.1080/18811248.2007.9711590.

[233] A.M. Duda, K.A. Szewczyk, M.W. Jarosik, K.M. Szcześniak, M.A. Sowińska, D. Szcześniak, Characterization of the superconducting state in hafnium hydride under high pressure, Phys. B Condens. Matter 536 (2018) 275–279, https://doi.org/10.1016/j.physb.2017.10.107.

[234] K. Konashi, T. Iwasaki, K. Itoh, M. Hirai, J. Sato, K. Kurosaki, A. Suzuki, Y. Matsumura, S. Abe, Study on an Innovative Fast Reactor Utilizing Hydride Neutron Absorber—Final Report of Phase I Study, 2010, pp. 2098–2104. https://tohoku.pure.elsevier.com/en/publications/study-on-an-innovative-fast-reactor-utilizing-hydride-neutron-abs-3. (Accessed 18 September 2020). accessed.

[235] T. Iwasaki, K. Konashi, Development of hydride absorber for fast reactor—Application of hafnium hydride to control rod of large fast reactor, J. Nucl. Sci. Technol. 46 (2009) 874–882, https://doi.org/10.1080/18811248.2007.9711595.

[236] K. Konashi, M. Yamawaki, Utilization of Hydride Materials in Nuclear Reactors, Trans Tech Publications Ltd, 2010, pp. 51–58, https://doi.org/10.4028/www.scientific.net/AST.73.51.

[237] K. Konashi, M. Yamawaki, Environmentally friendly application of hydrides to nuclear reactor cores, Adv. Appl. Ceram. 111 (2012) 106–111, https://doi.org/10.1179/1743676111Y.0000000051.

[238] K. Konashi, K. Itoh, T. Kido, Y. Kosaka, S. Seino, DEVELopment of Hydride Neutron Absorber for Fast Reactor—Irradiation Experiment on Hydride Neutron Absorber in Bor-60—Irradiation e, Korean Nuclear Society, 2013, pp. 1034–1039. https://tohoku.pure.elsevier.com/en/publications/development-of-hydride-neutron-absorber-for-fast-reactor-irradiat. (Accessed 18 September 2020). accessed.

[239] T. Kido, K. Itoh, A. Suzuki, Y. Matsumura, K. Konashi, Study on an Innovative Fast Reactor Utilizing Hydride Neutron Absorber Development of Coating Technique on Cladding Inner Surface, Atomic Energy Society of Japan, 2009, pp. 814–820. https://tohoku.pure.elsevier.com/en/publications/study-on-an-innovative-fast-reactor-utilizing-hydride-neutron-abs-2. (Accessed 18 September 2020). accessed.

[240] E. Zuzek, J.P. Abriata, A. San-Martin, F.D. Manchester, The H-Zr (hydrogen-zirconium) system, Bull. Alloy Phase Diagr. 11 (1990) 385–395, https://doi.org/10.1007/BF02843318.

[241] L.N. Yannopoulos, R.K. Edwards, P.G. Wahlbeck, The thermodynamics of the yttrium-hydrogen system 1, J. Phys. Chem. 69 (1965) 2510–2515, https://doi.org/10.1021/j100892a004.
[242] H. Bolt, Nonmetallic materials for plasma facing and structural applications in fusion reactors, Fusion Eng. Des. 22 (1993) 85–98, https://doi.org/10.1016/S0920-3796(05)80010-9.
[243] P.C. Stangeby, A.W. Leonard, Obtaining reactor-relevant divertor conditions in tokamaks, Nucl. Fusion 51 (2011) 063001, https://doi.org/10.1088/0029-5515/51/6/063001.
[244] R.P. Doerner, G.R. Tynan, K. Schmid, Implications of PMI and wall material choice on fusion reactor tritium self-sufficiency, Nucl. Eng. Des. Fusion 18 (2019) 56–61, https://doi.org/10.1016/j.nme.2018.12.006.
[245] A. Miyahara, T. Tanabe, Graphite as plasma facing material, J. Nucl. Mater. 155–157 (1988) 49–57, https://doi.org/10.1016/0022-3115(88)90226-7.
[246] G. Federici, C.H. Skinner, J.N. Brooks, J.P. Coad, C. Grisolia, Plasma-material Interactions in Current Tokamaks and Their Implications for Next-step Fusion Reactors, 2001, https://doi.org/10.2172/773280.
[247] T. Tanabe, N. Bekris, P. Coad, C.H. Skinner, M. Glugla, N. Miya, Tritium retention of plasma facing components in tokamaks, J. Nucl. Mater. 313–316 (2003) 478–490, https://doi.org/10.1016/s0022-3115(02)01377-6.
[248] T. Maruyama, M. Harayama, Neutron irradiation effect on the thermal conductivity and dimensional change of graphite materials, J. Nucl. Mater. 195 (1992) 44–50, https://doi.org/10.1016/0022-3115(92)90362-o.
[249] http://burningplasma.org/web/ReNeW/whitepapers/3-26%20ReNeW%20C%20as%20flow-thru%20PFC%2020Feb09.pdf (accessed September 23, 2020).
[250] Y. Gomay, S. Konoshima, N. Fujisawa, S. Kasai, M. Maeno, N. Suzuki, T. Hirayama, M. Shimada, Molybdenum, carbon and silicon carbide limiter experiment in the JFT-2 tokamak, Jpn. J. Appl. Phys. 18 (1979) 1317–1324, https://doi.org/10.1143/jjap.18.1317.
[251] J. Megusar, T.F. Yang, J.R. Martin, A.S. Wan, Redeposition of SiC-coated graphite limiter in Alcator C, Nucl. Instrum. Methods Phys. Res. Sect. B Beam Interact. Mater. Atoms 23 (1987) 476–481, https://doi.org/10.1016/0168-583x(87)90473-3.
[252] J. Li, B.N. Wan, J.R. Luo, G.L. Kuang, Y.P. Zhao, J.Y. Zhao, X.D. Zhang, X.N. Liu, P. Fu, J.K. Xie, C. Zhang, X.M. Gu, J.S. Mao, J.F. Shan, H.Y. Bai, K. Gentle, B. Rowan, P. Philippe, H. Huang, L. Lao, V. Chan, T. Watari, T. Seki, N. Nakamura, HT-7 team, long pulse enhanced confinement discharges in the HT-7 superconducting tokamak by ion Bernstein wave heating and lower hybrid wave current drive, Phys. Plasmas 10 (2003) 1653–1658, https://doi.org/10.1063/1.1556297.
[253] H.E. Khalifa, T. Koyanagi, G.M. Jacobsen, C.P. Deck, C.A. Back, Radiation stable, hybrid, chemical vapor infiltration/preceramic polymer joining of silicon carbide components, J. Nucl. Mater. 487 (2017) 91–95, https://doi.org/10.1016/j.jnucmat.2017.02.012.
[254] J. Westerhout, D. Borodin, R.S. Al, S. Brezinsek, A. Kirschner, S. Lisgo, H.J. van der Meiden, V. Philipps, M.J. van de Pol, A.E. Shumack, G. De Temmerman, W.A. J. Vijvers, G.M. Wright, N.J. Lopes Cardozo, J. Rapp, G.J. van Rooij, Chemical erosion of different carbon composites under ITER-relevant plasma conditions, Physica Scripta T138 (2009) 014017, https://doi.org/10.1088/0031-8949/2009/t138/014017.
[255] M. Balden, S. Picarle, J. Roth, Mechanism of the chemical erosion of SiC under hydrogen irradiation, J. Nucl. Mater. 290–293 (2001) 47–51, https://doi.org/10.1016/s0022-3115(00)00505-5.

[256] G.M. Wright, M.G. Durrett, K.W. Hoover, L.A. Kesler, D.G. Whyte, Silicon Carbide as a tritium permeation barrier in tungsten plasma-facing components, J. Nucl. Mater. 458 (2015) 272–274, https://doi.org/10.1016/j.jnucmat.2014.12.074.

[257] M.T. Koller, J.W. Davis, M.E. Goodland, T. Abrams, S. Gonderman, G. Herdrich, M. Frieß, C. Zuber, Deuterium retention in silicon carbide, SiC ceramic matrix composites, and SiC coated graphite, Nucl. Mater. Energy 20 (2019) 100704, https://doi.org/10.1016/j.nme.2019.100704.

[258] J. Coburn, E. Unterberg, J. Barton, D. Rudakov, I. Bykov, C.M. Parish, R. Wilcox, C. Lasnier, T. Abrams, J. Watkins, D.L. Hillis, M. Bourham, Erosion characterization of SiC and Ti3SiC2 on DIII-D using focused ion beam micro-trenches, Nucl. Mater. Energy 19 (2019) 316–323, https://doi.org/10.1016/j.nme.2019.02.036.

[259] C.J. Beers, E.G. Lindquist, T.M. Biewer, J.F. Caneses, J.B.O. Caughman, R. H. Goulding, N. Kafle, H. Ray, M.A. Showers, S.J. Zinkle, J. Rapp, Characterization of the helicon plasma flux to the target of Proto-MPEX, Fusion Eng. Des. 138 (2019) 282–288, https://doi.org/10.1016/j.fusengdes.2018.11.056.

[260] R. Yamada, K. Nakamura, M. Saidoh, Chemical sputtering yields of titanium carbides, J. Nucl. Mater. 111–112 (1982) 744–749, https://doi.org/10.1016/0022-3115(82)90299-9.

[261] https://inis.iaea.org/collection/NCLCollectionStore/_Public/17/066/17066605.pdf (accessed September 24, 2020).

[262] D.M. Mattox, A.W. Mullendore, J.B. Whitley, H.O. Pierson, Thermal-shock- and fatigue-resistant coatings for magnetically confined fusion environments, Thin Solid Films 73 (1980) 101–107, https://doi.org/10.1016/0040-6090(80)90335-1.

[263] J.A. Borders, G.C. Nelson, Sputtering and surface modification of TiC coatings by low energy deuterium bombardment, J. Nucl. Mater. 103 (1981) 369–371, https://doi.org/10.1016/0022-3115(82)90625-0.

[264] T. Abe, Y. Murakami, K. Obara, S. Hiroki, K. Nakamura, T. Mizoguchi, A. Doi, K. Inagawa, Development of TiC coated wall materials for JT-60, J. Nucl. Mater. 133–134 (1985) 754–759, https://doi.org/10.1016/0022-3115(85)90251-x.

[265] K. Nakamura, R. Yamada, M. Saidoh, Y. Murakami, Thermal shock testing of low-Z coatings for JT-60, J. Nucl. Mater. 111–112 (1982) 852–855, https://doi.org/10.1016/0022-3115(82)90319-1.

[266] R. Yamada, K. Nakamura, M. Saidoh, Y. Murakami, Thermal fatigue testing of low-Z coatings on molybdenum and Inconel 625, J. Nucl. Mater. 111–112 (1982) 856–860, https://doi.org/10.1016/0022-3115(82)90320-8.

[267] H. Nishi, T. Oku, T. Kodaira, Low cycle fatigue behavior of titanium carbide coated molybdenum, J. Nucl. Mater. 141–143 (1986) 152–155, https://doi.org/10.1016/s0022-3115(86)80026-5.

[268] Brossa, et al., J. Nucl. Mater. 155–157 (Part 1) (1988) 267–272.

[269] Groot, et al., J. Nucl. Mater. 179–181 (Part 1) (1991) 370–374.

[270] S.A. Humphry-Baker, G.D.W. Smith, G. Pintsuk, Thermal shock of tungsten carbide in plasma-facing conditions, J. Nucl. Mater. 524 (2019) 239–246, https://doi.org/10.1016/j.jnucmat.2019.06.041.

[271] M.W. Barsoum, The MN 1AXN phases: A new class of solids, Prog. Solid State Chem. 28 (2000) 201–281, https://doi.org/10.1016/s0079-6786(00)00006-6.

[272] Y. Medkour, A. Roumili, D. Maouche, L. Louail, Electrical properties of MAX phases, in: Advances in Science and Technology of Mn 1AXn Phases, 2012, pp. 159–175, https://doi.org/10.1533/9780857096012.159.

[273] G.M. Song, Self-healing of MAX phase ceramics for high temperature applications: evidence from Ti 3AlC2, in: Advances in Science and Technology of Mn 1AXn Phases, 2012, pp. 271–288, https://doi.org/10.1533/9780857096012.271.
[274] Q. Huang, H. Han, R. Liu, G. Lei, L. Yan, J. Zhou, Q. Huang, Saturation of ion irradiation effects in MAX phase Cr2AlC, Acta Mater. 110 (2016) 1–7, https://doi.org/10.1016/j.actamat.2016.03.021.
[275] D.J. Tallman, E.N. Hoffman, E.N. Caspi, B.L. Garcia-Diaz, G. Kohse, R.L. Sindelar, M.W. Barsoum, Effect of neutron irradiation on select MAX phases, Acta Mater. 85 (2015) 132–143, https://doi.org/10.1016/j.actamat.2014.10.068.
[276] D.W. Clark, S.J. Zinkle, M.K. Patel, C.M. Parish, High temperature ion irradiation effects in MAX phase ceramics, Acta Mater. 105 (2016) 130–146, https://doi.org/10.1016/j.actamat.2015.11.055.
[277] Hydrogen retention in graphite and carbon materials under a fusion reactor environment, J. Nucl. Mater. 313–316 (2003) 543–547. https://doi.org/10.1016/S0022-3115(02)01464-2.
[278] G.D. Temmerman, G. De Temmerman, J. Dodson, J. Linke, S. Lisgo, G. Pintsuk, S. Porro, G. Scarsbrook, Thermal shock resistance of thick boron-doped diamond under extreme heat loads, Nucl. Fusion 51 (2011) 052001, https://doi.org/10.1088/0029-5515/51/5/052001.
[279] G.D. Temmerman, G. De Temmerman, R.P. Doerner, P. John, S. Lisgo, A. Litnovsky, L. Marot, S. Porro, P. Petersson, M. Rubel, D.L. Rudakov, G. Van Rooij, J. Westerhout, J.I.B. Wilson, Interactions of diamond surfaces with fusion relevant plasmas, Physica Scripta T138 (2009) 014013, https://doi.org/10.1088/0031-8949/2009/t138/014013.
[280] D. Matsuura, T. Kamiya, C.M. Fortmann, I. Simizu, Microstructure control of very thin polycrystalline silicon layers on glass substrate by plasma enhanced CVD, Sol. Energy Mater. Sol. Cell. 66 (2001) 305–311, https://doi.org/10.1016/s0927-0248(00)00188-4.
[281] S. Porro, G. De Temmerman, P. John, S. Lisgo, I. Villalpando, J.I.B. Wilson, Effects in CVD diamond exposed to fusion plasmas, Phys. Status Solidi 206 (2009) 2028–2032, https://doi.org/10.1002/pssa.200982201.
[282] A.M. Stoneham, J.R. Matthews, I.J. Ford, Innovative materials for fusion power plant structures: Separating functions, J. Phys. Condens. Matter 16 (2004) S2597–S2621, https://doi.org/10.1088/0953-8984/16/27/001.
[283] S. Porro, G. De Temmerman, S. Lisgo, P. John, I. Villalpando, J.W. Zimmer, B. Johnson, J.I.B. Wilson, Nanocrystalline diamond coating of fusion plasma facing components, Diam. Relat. Mater. 18 (2009) 740–744, https://doi.org/10.1016/j.diamond.2009.01.006.
[284] D. Hill, A review of ELMs in divertor tokamaks, J. Nucl. Mater. 241–243 (1997) 182–198, https://doi.org/10.1016/s0022-3115(96)00504-1.
[285] A. Deslandes, M.C. Guenette, C.M. Samuell, I. Karatchevtseva, M. Ionescu, D. D. Cohen, B. Blackwell, C. Corr, D.P. Riley, Initial damage processes for diamond film exposure to hydrogen plasma, Fusion Eng. Des. 88 (2013) 3101–3107, https://doi.org/10.1016/j.fusengdes.2013.08.010.
[286] H. Jiang, F. Liu, H. Yan, L. Si, Z. Dou, Etching effects of hydrogen plasma treatment on diamond surfaces, Surf. Coating. Technol. 363 (2019) 12–17, https://doi.org/10.1016/j.surfcoat.2019.02.007.
[287] M.C. Guenette, A. Deslandes, L. Thomsen, C.S. Corr, D.P. Riley, D and D/He plasma interactions with diamond: Surface modification and D retention, Diam. Relat. Mater. 49 (2014) 103–110, https://doi.org/10.1016/j.diamond.2014.08.008.

[288] A. Deslandes, M.C. Guenette, C.S. Corr, I. Karatchevtseva, L. Thomsen, G. R. Lumpkin, D.P. Riley, Deuterium retention and near-surface modification of ion-irradiated diamond exposed to fusion-relevant plasma, Nucl. Fusion 54 (2014) 073003, https://doi.org/10.1088/0029-5515/54/7/073003.
[289] P.W. May, Diamond thin films: A 21st-century material, Philos. Trans. R. Soc. London, Ser. A: Math. Phys. Eng. Sci. 358 (2000) 473–495, https://doi.org/10.1098/rsta.2000.0542.
[290] C.M. Donnelly, R.W. McCullough, J. Geddes, Etching of graphite and diamond by thermal energy hydrogen atoms, Diam. Relat. Mater. 5 (1997) 787–790, https://doi.org/10.1016/s0925-9635(96)00606-1.
[291] C. Kanai, K. Watanabe, Y. Takakuwa, Ab initiocalculations on etching of graphite and diamond surfaces by atomic hydrogen, Phys. Rev. B 63 (2001), https://doi.org/10.1103/physrevb.63.235311.
[292] S. Porro, G. De Temmerman, S. Lisgo, D.L. Rudakov, A. Litnovsky, P. Petersson, P. John, J.I.B. Wilson, Diamond coatings exposure to fusion-relevant plasma conditions, J. Nucl. Mater. 415 (2011) S161–S164, https://doi.org/10.1016/j.jnucmat.2010.10.005.
[293] S. Porro, G. De Temmerman, D.A. MacLaren, S. Lisgo, D.L. Rudakov, J. Westerhout, M. Wiora, P. John, I. Villalpando, J.I.B. Wilson, Surface analysis of CVD diamond exposed to fusion plasma, Diam. Relat. Mater. 19 (2010) 818–823, https://doi.org/10.1016/j.diamond.2010.01.051.
[294] M.S. Tillack, et al., Fusion Eng. Des. 180 (2022) 113155.

Design of irradiation-resistant SiC fiber–reinforced SiC matrix composites for nuclear applications

4

Takaaki Koyanagi and Yutai Katoh
Materials Science and Technology Division, Oak Ridge National Laboratory, Oak Ridge, TN, United States

1. Introduction

The SiC-based ceramic matrix composites, particularly continuous SiC fiber–reinforced SiC matrix (SiC/SiC) composites, are candidate structural materials for core components of nuclear fission and fusion energy systems [1]. Typical SiC/SiC composite materials consist of small-diameter SiC fibers coated with a fiber/matrix interphase surrounded by a SiC matrix (Fig. 4.1). The engineered composite

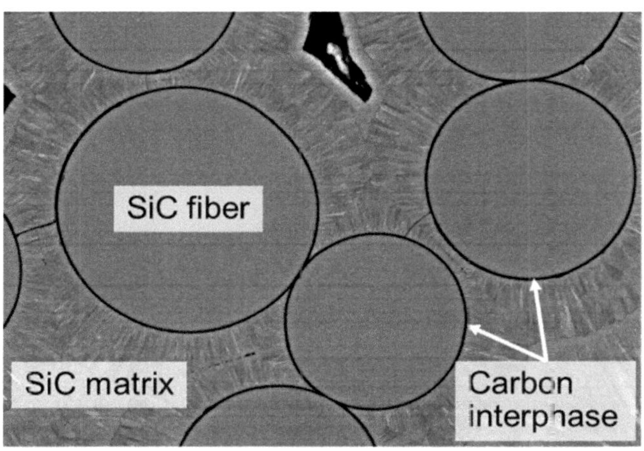

Figure 4.1 Constituents of continuous SiC fiber-reinforced SiC matrix composites: Chemical vapor infiltration of SiC matrix with Hi-Nicalon type S SiC fibers coated with pyrolytic carbon.
Figure reprinted from Koyanagi et al. [2].

Ceramics for Nuclear Applications. https://doi.org/10.1016/B978-0-12-823865-3.00005-5

microstructures are tailored to enhance the toughness of SiC/SiC composites because brittle monolithic SiC is unsuitable for structural applications.

These SiC/SiC composites exhibit quasi-ductile fracture behavior characterized by a brushlike fracture surface that results from fiber pullout (Fig. 4.2A). A properly engineered SiC/SiC composite material can maintain its component geometry after a burst test (Fig. 4.2B), whereas monolithic SiC fractures when subjected to such a mechanical test. The damage tolerance provided by fiber reinforcement makes SiC/SiC composites suitable for structural applications in nuclear systems.

In addition to the enhanced toughness provided by fiber reinforcement, SiC possesses intrinsic advantages for nuclear applications, including exceptional strength retention at elevated temperatures, low neutron absorption [4], low induced-radioactivity [5], and chemical inertness [6]. Furthermore, high-purity, crystalline SiC is radiation-resistant (i.e., remains stable) in a wide range of elevated temperatures in high-dose radiation environments [7]. These properties position SiC as a uniquely suitable material for nuclear reactor core applications.

Nuclear applications of SiC/SiC composites are often proposed for core components where alloys may be unsuitable due to their limitations, such as the restricted temperature design window. In fusion reactor applications, SiC/SiC composites were initially considered for the main blanket structure, with a goal of achieving high operating temperatures of up to about 1000°C [8–10]. More recently, using SiC/SiC composites for the flow channel inserts in dual-coolant helium/lead–lithium blankets [11–13] has emerged as a more viable application because the manufacturing hurdles for channel-shaped components (Fig. 4.3A) are lower than those for full blanket modules. The potential applications of SiC/SiC composites in fusion reactors are expanding to include coolant pipes in helium-cooled tungsten monoblock divertors [14] and structural components in molten salt breeding blankets [15].

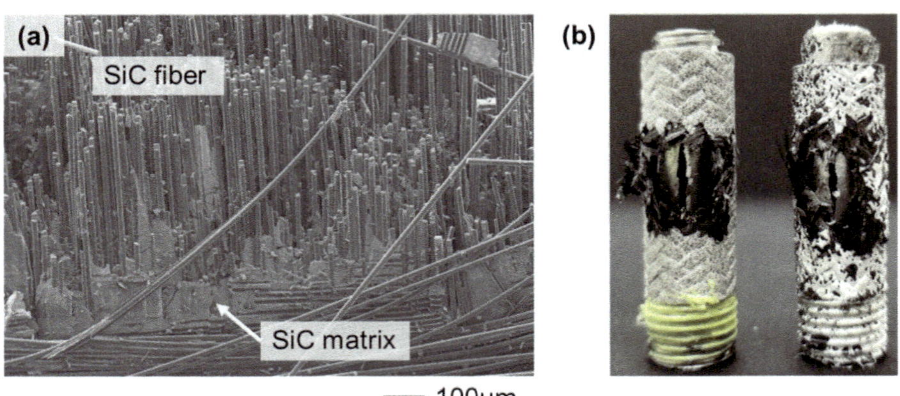

Figure 4.2 Fracture appearance of continuous SiC fiber-reinforced SiC matrix composites: (A) Fracture surface after flexural testing (reprinted from Koyanagi et al. [2]) and (B) SiC composite tubes after a burst test (reprinted from Cinbiz et al. [3]). The outer diameter of the tube specimens was approximately 10 mm.

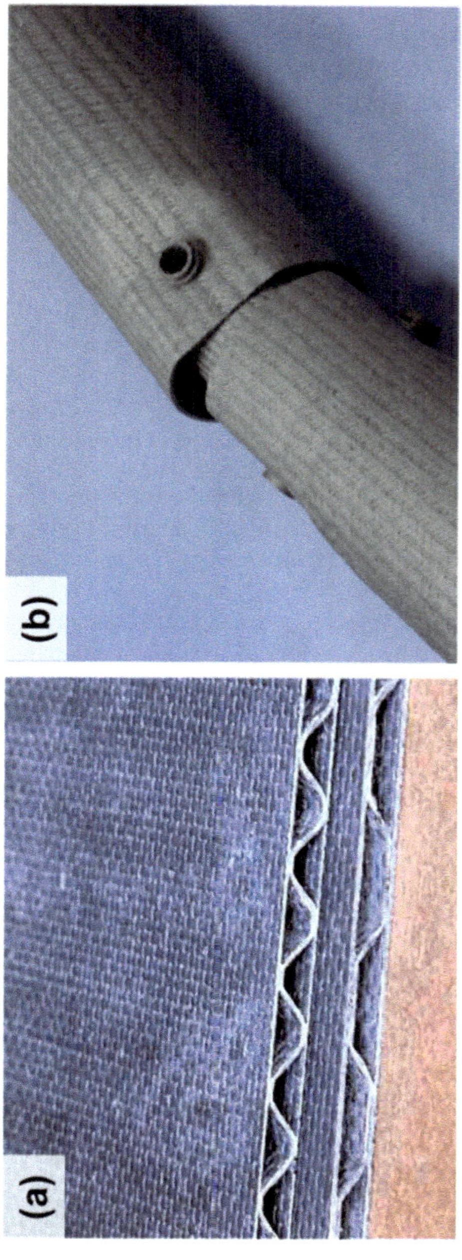

Figure 4.3 Examples of SiC composite components for structural nuclear applications: (A) Channel structure (reprinted from Rowcliffe et al. [13]) and (B) assembled thin-wall rods (reprinted from Snead et al. [16]).

Furthermore, SiC/SiC composite materials have been pursued for use in fission reactor components, including cladding, control rod sheaths (Fig. 4.3B), and other core structures in advanced fission reactors (e.g., gas-cooled fast reactors [17] and fluoride salt-cooled high-temperature reactors [18]). Among the various R&D efforts for nuclear fission reactors, accident-tolerant fuel cladding for light water reactors has recently garnered the most attention because of the urgent need for new materials prompted by the Fukushima Daiichi nuclear accident [19]. Research in this area has focused primarily on the processing, design, thermomechanical properties, and environmental resistance of tubular SiC/SiC composite geometries [19].

The foregoing applications of SiC/SiC composites require resistance to much higher doses of neutron irradiation than is required for out-of-core structures. Because neutron irradiation can significantly affect the mechanical and physical properties of SiC/SiC composites, the microstructural design of irradiation-resistant SiC/SiC composites is crucial to the successful deployment of these materials in nuclear core applications. For instance, the quasi-ductile fracture behavior of chemical vapor infiltration (CVI) SiC/SiC composites changes to brittle behavior after high-dose neutron irradiation at 320°C to 92 displacements per atom (dpa) (Fig. 4.4) [7]. In this chapter, 1×10^{25} n/m^2 [$E > 0.1$ MeV] is assumed to be equivalent to one dpa [20]. This change in behavior means that (1) the application temperature of ~300°C may need to be reconsidered in high-dose environments, (2) the cumulative neutron dose over a material's lifetime must be limited, or (3) materials with superior radiation resistance should be developed.

For SiC/SiC composites to meet the property requirements for specific nuclear applications, their constituents (e.g., fibers, matrix, and interphase at the fiber/matrix interface) must be selected based on a fundamental understanding of how SiC materials respond to nuclear reactor environments, including neutron irradiation.

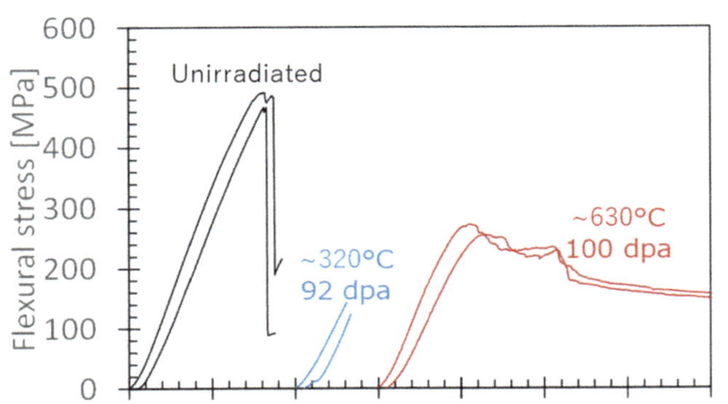

Figure 4.4 Flexural behavior of unirradiated and high-dose neutron-irradiated chemical vapor infiltration SiC fiber-reinforced SiC matrix composites [7]. Irradiation at approximately 320°C resulted in significant degradation in mechanical properties.
(Adapted from Koyanagi et al. [7].).

Variations in composite microstructure (e.g., variations associated with different fiber materials and matrix densification methods) can lead to notably different properties after irradiation, as explained in greater detail in Section 2.2. Understanding radiation damage processes in various SiC materials and their effects on macroscopic properties enables the prediction of the radiation responses of new classes of SiC/SiC composites and the identification of key technological challenges related to radiation resistance.

The primary objective of this chapter is to provide guidance on the microstructural design and selection of composite processing routes for the development of SiC/SiC composites tailored to applications in neutron irradiation environments. Example design approaches for SiC/SiC composites in nonnuclear applications, based on composite material theories, have been discussed elsewhere [21,22]. The Composite Materials Handbook Volume 5 [23] provides comprehensive design guidance on ceramic matrix composites for nonnuclear applications, covering material selection, processing, characterization, and quality control. For nuclear applications, Section III of the ASME Boiler and Pressure Vessel Code [24] explains the general requirements for nuclear ceramic matrix composites in the areas of design, manufacturing, and testing. To complement these documents, this capture discusses the composite processes and microstructures that are resistant to neutron irradiation environments. Section 2 of this chapter outlines the mechanisms of radiation damage in various grades of SiC, SiC/SiC composites, and their constituent materials. Section 3 presents a microstructural design approach for SiC/SiC composites. Next, Section 4 addresses the engineering design of SiC/SiC composites, considering essential property requirements such as corrosion resistance and fuel compatibility. Finally, Section 5 discusses remaining technological challenges and identifies future research directions.

2. Irradiation damage processes in SiC and SiC/SiC composites

This section discusses radiation damage processes in monolithic SiC ceramics and SiC/SiC composites fabricated using traditional manufacturing methods. The emphasis is on the effects of the initial microstructure on radiation-induced property changes. The goal is to apply this knowledge to the design of SiC/SiC composites for nuclear applications. The section highlights results of neutron irradiation experiments and describes irradiation experiments that mimic neutron irradiation, such as heavy and light ion irradiation, in a limited manner. General radiation damage is discussed at a high level. The effect of irradiation experiment variables, such as neutron spectrum and displacement damage rate [25] is beyond the scope of this chapter.

2.1 The effects of irradiation on monolithic SiC materials

Various monolithic SiC materials have been neutron irradiated in test reactors. The as-fabricated microstructure of a SiC ceramic is categorized based on feedstock material and densification method. The materials covered here include chemical vapor deposition (CVD) SiC, polymer-derived SiC, reaction-sintered SiC, and liquid phase sintered (LPS) SiC.

The properties of high-purity, polycrystalline β-phase CVD SiC after exposure to neutron irradiation have been extensively investigated. Comprehensive postirradiation examinations have established a foundational relationship between the radiation-induced defect microstructure and the resultant bulk property changes in SiC. Detailed findings can be found in the literature [26], but a summary is provided here. Radiation damage in SiC is highly temperature dependent. At moderate damage levels (e.g., several dpa), irradiation-induced amorphization occurs at temperatures below about 150°C. At intermediate temperatures (~150°C–~1000°C), a high density of point defects and defect clusters forms (Fig. 4.5); at higher temperatures, voids are observed [26].

In the intermediate temperature regime, which is most relevant to the proposed applications discussed in Section 1, CVD SiC exhibits irradiation-induced swelling that is temperature- and dose-dependent (Fig. 4.6). Swelling occurs rapidly after irradiation begins and approaches saturation values at approximately 1 dpa. The magnitude of swelling decreases as irradiation temperature increases (e.g., a saturation value of ~1% at 600°C vs. ~2% at 300°C) [28]. X-ray diffraction analysis has revealed that this swelling mechanism is due to lattice expansion resulting from the formation of interstitials and vacancy defects [29].

In SiC, swelling is accompanied by a reduction in thermal conductivity; the greater the swelling, the lower the thermal conductivity. The thermal conductivity of unirradiated CVD SiC (~400 W/m·K) at room temperature decreased to approximately 20 W/m·K after neutron irradiation at 600°C to 1 dpa [26]. By contrast, the effects of irradiation on elastic modulus and strength are insignificant [26]. In some CVD SiC materials, posttransient irradiation creep is much smaller than the magnitude of swelling. The creep coefficient, defined as the creep strain normalized by stress and neutron fluence, ranges from 1×10^{-7} to 2×10^{-7} $[\mathrm{MPa{\cdot}dpa}]^{-1}$ at damage levels between 1 and 30 dpa [27]. Retention of strength makes high-purity, crystalline

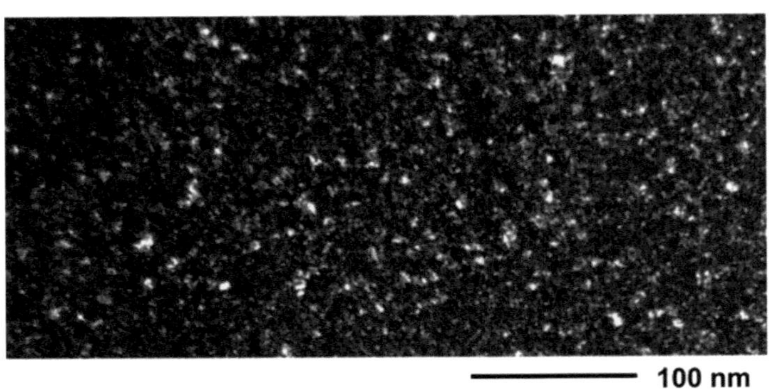

Figure 4.5 Irradiation defect clusters in chemical vapor deposition SiC as seen using weak-beam dark-field transmission electron microscopy imaging [27]. The specimen was neutron irradiated at 500°C to 12 dpa.

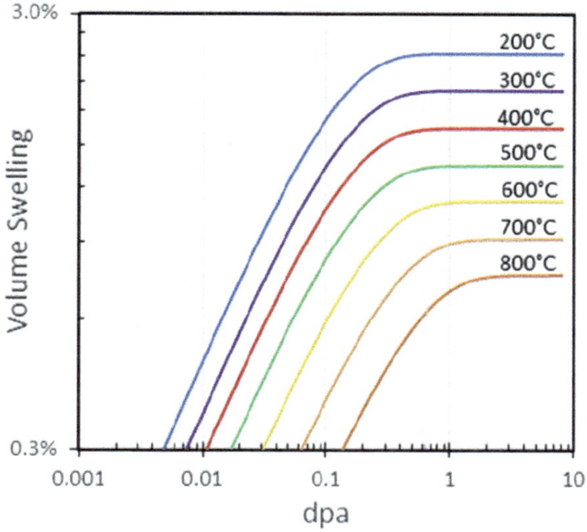

Figure 4.6 Irradiation temperature and dose dependence of swelling of chemical vapor deposition SiC [28].

β-SiC variants such as CVD SiC well-suited for use as a constituent phase in SiC/SiC composites.

The selection of cubic β-phase versus hexagonal α-phase is of interest in the design of SiC-based composites. CVD variants of α- and β-SiC exhibit identical irradiation-induced swelling and amorphization at approximately 60°C; α-SiC exhibits isotropic volume expansion under irradiation [30]. At an irradiation temperature of 1100°C, the general response of sintered α-SiC is similar to that of CVD β-SiC; however, some differences are evident in the magnitude of effects. Both materials experience swelling and a notable reduction in thermal diffusivity after irradiation [31]. Irradiation-induced formation of dislocation loops and voids in α-SiC changes to anisotropic behavior at extreme irradiation temperatures (~1450°C) (Fig. 4.7) [32]; this phenomenon is not evident in β-SiC [33]. Under such conditions, anisotropic thermophysical properties are expected in α-SiC. Anisotropy is also expected in α-SiC when helium transmutations occur under fusion reactor irradiation conditions; such behavior has been observed in microstructural studies of helium ion–irradiated α-SiC [34].

Crystallographic characteristics significantly influence the irradiation response of SiC materials. Amorphous like or nanocrystalline SiC, exemplified by polymer-derived SiC materials such as early generation SiC fibers (e.g., Nicalon and Hi-Nicalon fibers with grain sizes of approximately 2 and 5 nm, respectively [35]), exhibit radiation responses that are distinct from those of highly crystalline CVD SiC materials. Amorphous like SiC materials undergo substantial volumetric shrinkage (up to ~10%) upon irradiation at elevated temperatures from 650°C to 1000°C (Fig. 4.8) [36,37]. By contrast, crystalline SiC materials, including CVD SiC, demonstrate temperature-dependent swelling as shown in Fig. 4.6. The dimensional instability observed in amorphous like SiC materials is a significant concern;

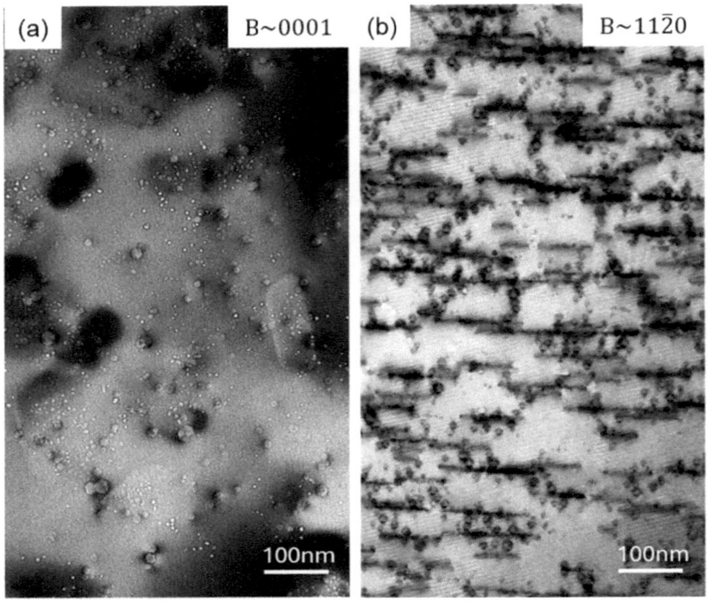

Figure 4.7 Transmission electron micrographs of SiC with a 6H crystal structure irradiated with neutrons at about 1450°C to six displacements per atom, shown from different crystallographic orientations: (A) [0001] and (B) [11 $\bar{2}$ 0].
(Adapted from Kondo et al. [32].).

consequently, this class of SiC materials has not been extensively studied for use in neutron irradiation environments.

Reaction-sintered SiC is a crystalline SiC-based material fabricated via the reaction of silicon and carbon to form SiC at or above the melting point of silicon (approximately 1400°C) [43]. The fabrication process typically involves the infiltration of molten silicon into a preformed carbon body or a mixture of carbon and SiC powders. This reaction typically yields a material consisting of SiC grains and residual silicon. The presence of residual silicon influences the radiation response of reaction-sintered SiC.

Irradiation-induced degradation of mechanical properties poses a significant challenge for silicon-rich reaction-sintered SiC. This effect has been consistently observed, even at the low damage levels (~0.1 dpa) shown in Fig. 4.9 [41,44,45]. The underlying mechanism has been identified via microstructural observations and X-ray diffraction analysis; the mechanical degradation is linked to spontaneous cracking caused by differential swelling between SiC and silicon grains [44]. Because SiC ceramics lack strain tolerance, small, irradiation-induced strains from differential swelling can initiate microcracking at phase boundaries. Differential swelling is a fundamental technical challenge to the use of reaction-sintered SiC under irradiation because the temperature and dose dependence of swelling differs in SiC versus silicon. Reaction-sintered SiC tends to experience volume expansion (Fig. 4.8) [41];

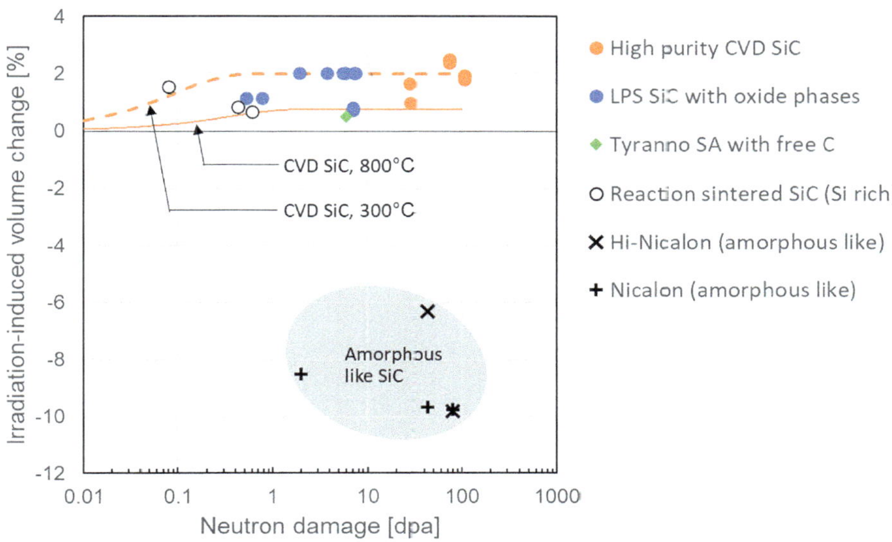

Figure 4.8 Irradiation-induced volume changes in various SiC ceramics. Data for chemical vapor deposition SiC [7,28,38], liquid phase sintered SiC [39,40], Tyranno SA SiC fibers [37], reaction-sintered SiC [41], and amorphous like SiC [36,37] obtained from previous studies. (Reproduced from Koyanagi et al. [42].).

this phenomenon can be explained by the swelling behavior of the crystalline SiC phase.

LPS SiC materials consist of crystalline SiC grains and secondary phases, which are attributed to sintering additives, typically oxides [51]. For the irradiation study, an Al_2O_3–Y_2O_3 additive system was employed to reduce the sintering temperature relative to solid-state sintering without additives [39,40]. When a minimal or moderate number of sintering additives is used, LPS SiC exhibits swelling behavior similar to that of CVD SiC [40]. Irradiation-induced degradation of strength is negligible up to approximately 10 dpa; however, irradiation-induced degradation, for reasons that remain unclear, has been reported under certain conditions [41,46,47]. Because the oxide secondary phases are finely distributed in LPS SiC (Fig. 4.10), this material does not exhibit the spontaneous cracking resulting from irradiation that is typically associated with differential swelling in SiC grains versus oxide phases.

Third-generation SiC fibers, including Hi-Nicalon Type S and Tyranno SA series fibers [35], are polymer-derived polycrystalline SiC containing free carbon [35] (Fig. 4.11). The crystalline structure of the SiC grains and the relatively small amount of free carbon result in a radiation response similar to that of CVD SiC. Third-generation fibers exhibit irradiation-induced swelling (Fig. 4.8) but do not exhibit significant radiation-induced degradation of tensile strength (Fig. 4.9). Radiation-induced dimensional changes of crystalline SiC fibers are markedly different from those of amorphous like SiC fibers (Fig. 4.8). The approximately

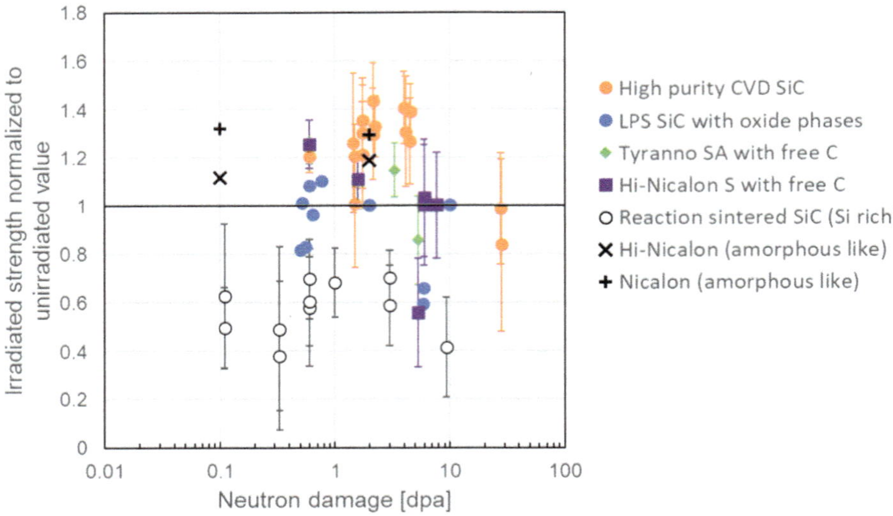

Figure 4.9 Effect of neutron irradiation on the strength of various SiC materials, highlighting the mechanical degradation of silicon-rich reaction-bonded SiC. Irradiation temperatures ranged from 280°C to 1300°C. Data for chemical vapor deposition SiC [20,38], liquid phase sintered SiC [41,46,47], Tyranno SA SiC fibers, Hi-Nicalon type S SiC fibers [48,49], reaction-sintered SiC [41,44,45], and amorphous like SiC [50] obtained from previous studies.
(Reproduced from Koyanagi et al. [42].).

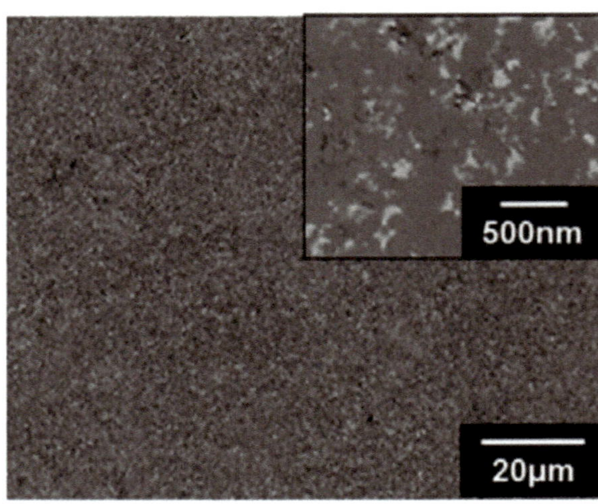

Figure 4.10 Backscattered electron micrograph of liquid phase sintered SiC fabricated by sintering SiC nanopowders with a total of 12 wt% Al_2O_3–Y_2O_3–SiO_2 additives. Bright contrast indicates oxide phases attributed to the sintering additives.
(Reproduced from Koyanagi et al. [46].).

Figure 4.11 High-angle annular dark-field scanning transmission electron micrograph of carbon-rich SiC materials: (A) Hi-Nicalon type S and (B) Tyranno SA3 SiC fibers. Dark spots indicate free carbon phases.

20 nm grain size in Hi-Nicalon Type S fibers [52] is sufficiently large to produce a radiation response distinct from that of amorphous like SiC fibers.

In as-fabricated conditions, SiC thermal conductivity typically decreases because of the presence of impurities and secondary phases. Of the materials discussed herein, dense, high-purity CVD SiC demonstrates the highest thermal conductivity, reaching about 400 W/m·K at ambient temperature [26]. However, the introduction of porosity, impurities, and secondary phases leads to a decrease in thermal conductivity [26,53]. Amorphous-like SiC materials, such as Nicalon and Hi-Nicalon SiC fibers, exhibit significantly lower room-temperature thermal conductivities, typically below 10 W/m·K [35].

Irradiation-induced defects further decrease the thermal conductivity of crystalline SiC. Defects resulting from irradiation, including point defects and their clusters, serve as phonon scattering centers [26,54]. The room-temperature thermal conductivity of CVD SiC decreases by an order of magnitude after irradiation at temperatures ranging from 200°C to 850°C to doses exceeding one dpa [26]. This decrease can be attributed to phonon scattering at the irradiation-induced defects [54]. Therefore, an irradiation-induced decrease in thermal conductivity is intrinsic to crystalline, high-purity SiC. Other SiC variants composed of crystalline SiC grains—including LPS SiC, reaction-sintered SiC, and crystalline SiC fibers such as Tyranno SA and Hi-Nicalon Type S—are expected to exhibit similar behavior; this result is supported by a limited experimental dataset for LPS SiC [47] and thermal property measurements of SiC/SiC composites containing Tyranno SA and Hi-Nicalon Type S fibers [55].

Table 4.1 summarizes the radiation responses of various grades of SiC. CVD SiC materials subjected to neutron irradiation have been studied extensively; thus, these materials can be considered a point of reference for studies on irradiation effects.

Table 4.1 Irradiation responses of SiC grades. The irradiation temperature and neutron damage considered are approximately 200–1000°C and up to 10 dpa, respectively.

Material class	Examples	Radiation responses
High-purity, crystalline SiC	CVD SiC	• Temperature- and dose-dependent swelling with saturation • Retention of strength • Reduction in thermal conductivity with increased swelling
Crystalline SiC with secondary phases	LPS SiC with oxides	• Swelling similar to that of CVD SiC • Retention of strength, in general; some degradation in specific cases (more sensitivity than CVD SiC) • Reduction in thermal conductivity with increased swelling
	Reaction-sintered SiC with free silicon	• Swelling (limited data) • Degradation of mechanical properties because of irradiation-induced cracking
	Tyranno SA3 and Hi-Nicalon type S SiC fibers with free carbon	• Swelling (limited data) • Retention of strength, in general; some degradation in specific cases (more sensitivity than CVD SiC)
Amorphous-like SiC	Nicalon and Hi-Nicalon SiC fibers	• Irradiation-induced shrinkage up to ~10%

Research highlights the saturation of swelling at elevated temperatures, the stability of mechanical properties after irradiation, and a significant decrease in thermal conductivity, particularly at lower irradiation temperatures. Reduced thermal conductivity is considered intrinsic to the SiC crystal structure. The incorporation of secondary phases into SiC materials increases the risk of irradiation-induced mechanical property degradation. LPS SiC is sensitive to irradiation-induced strength degradation, although strength is retained in some irradiation conditions. The presence of free silicon in reaction-sintered SiC leads to spontaneous cracking under irradiation, thereby reducing strength. Amorphous-like SiC materials, represented by preceramic polymer-derived SiC fibers such as Nicalon and Hi-Nicalon, exhibit behavior that differs from that of crystalline SiC materials, including irradiation-induced shrinkage up to about 10%, which poses significant challenges to using this class of materials in irradiation environments. Polymer-derived, crystalline SiC materials, such as Tyranno SA3 and Hi-Nicalon Type S fibers, exhibit swelling and retention of strength after irradiation that are qualitatively similar to those of other crystalline SiC materials. Correlations between SiC microstructure and radiation response provide a basis for selecting matrix and fiber combinations for targeted nuclear applications.

2.2 The effects of irradiation on SiC/SiC composites

Changes in the physical and mechanical properties of SiC/SiC composites under irradiation are highly dependent on the microstructures of their constituents. This section discusses the radiation resistance of SiC/SiC composite variants, emphasizing the stability or instability of interfaces within the composite microstructures. Special attention is given to SiC/SiC composites featuring a pyrolytic carbon (PyC) interphase at the fiber/matrix interface.

CVI SiC/SiC composites with crystalline SiC fibers, such as Hi-Nicalon Type S and Tyranno SA3, are exceptionally stable in terms of mechanical properties after irradiation at elevated temperatures (≥220°C) to approximately 10 dpa, as indicated by extensive data in the literature [55]. Figure 4.12 shows that irradiation has an insignificant effect on the stress–strain curves of CVI SiC/SiC composites with Hi-Nicalon Type S fibers under cyclic loading. A primary reason for the retention of mechanical properties after irradiation is the stability of the fiber/matrix microstructure, which showed no signs of debonding or cracking in the study highlighted in Fig. 4.12.

CVI SiC/SiC composites containing crystalline SiC fibers exhibit swelling induced by irradiation because the crystalline SiC matrix and fibers both undergo swelling [55]. Saturation swelling data for CVI SiC/SiC composites with Hi-Nicalon Type S and Tyranno SA3 SiC fibers fall within the data band for CVD SiC when irradiated at temperatures above 300°C [55]. Transient swelling behavior under irradiation to 1 dpa is also largely consistent between CVD SiC specimens and the in-plane directions for CVI SiC/SiC specimens (Fig. 4.13) [28].

The mechanical properties of CVI SiC/SiC composites with crystalline SiC fibers exhibit irradiation temperature–dependent degradation after exposure to high neutron doses. This degradation is a result of the instability at the fiber/matrix interface. After

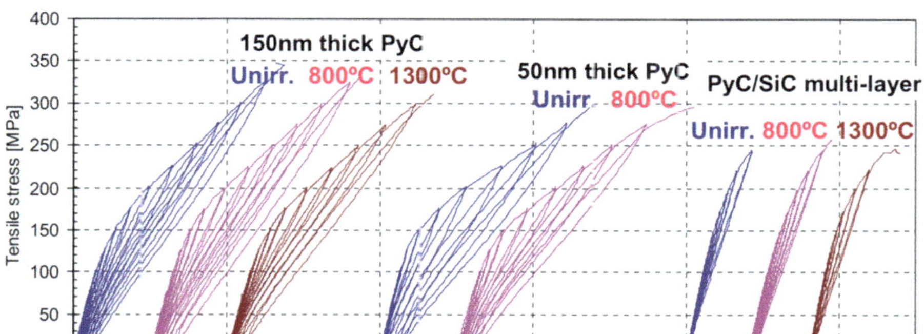

Figure 4.12 Stress–strain curves of chemical vapor infiltration SiC fiber-reinforced SiC matrix composite containing Hi-Nicalon type S SiC fibers with and without neutron irradiation at 800°C and 1300°C to about six displacements per atom. Three types of fiber/matrix interphase were examined via cyclic unloading–reloading tests at room temperature.
(Reproduced from Katoh et al. [56].).

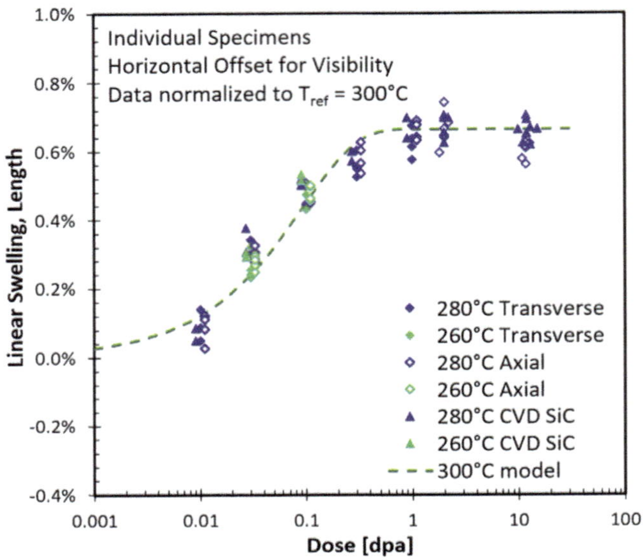

Figure 4.13 Neutron irradiation-induced length changes in Hi-Nicalon type S SiC fiber-reinforced SiC matrix composite specimens versus chemical vapor deposition (CVD) SiC specimens showing comparable dimensional stability between the in-plane directions (e.g., transverse vs. axial) and between SiC fiber-reinforced SiC matrix composites and CVD SiC irradiated at around 300°C.
(Reproduced from Katoh et al. [28].).

irradiation at approximately 320°C and 630°C to a high neutron dose (approaching 100 dpa), thin layers of the PyC phase in CVI SiC/SiC composites containing Hi-Nicalon Type S fibers exhibit interfacial debonding (Fig. 4.14) [7]. Additionally, interdiffusion is evident between the SiC and PyC phases after irradiation at about 320°C (Fig. 4.14B). Nozawa et al. conducted a critical investigation of interfacial instability in Hi-Nicalon Type S CVI SiC/SiC composites and confirmed, based on surface examinations, differences in swelling in SiC fibers versus CVI SiC matrix [57]. Figure 4.15 shows that the CVI SiC matrix swelled more than the fibers.

Differences in radiation response between the crystalline SiC fibers (e.g., Hi-Nicalon Type S vs. Tyranno SA3) are evident at high neutron irradiation doses. However, no clear distinctions are evident in terms of dimensional stability or mechanical property changes for these two types of fibers at low doses (less than 10 dpa) [55]. The surface morphology of a CVI SiC/SiC composite with Tyranno SA3 fibers did not change after neutron irradiation at about 600°C to 44 dpa (Fig. 4.16) [57], whereas the Hi-Nicalon Type S composite developed pits (Fig. 4.15). These results indicate that the swelling behavior of the Tyranno SA3 fibers and their CVI SiC matrix was similar, unlike what occurred in the Hi-Nicalon Type S composite. Although the microstructure appeared stable based on surface observations, high-magnification cross-sectional observations reveal irradiation-induced interfacial debonding at the fiber/matrix interface (Fig. 4.17).

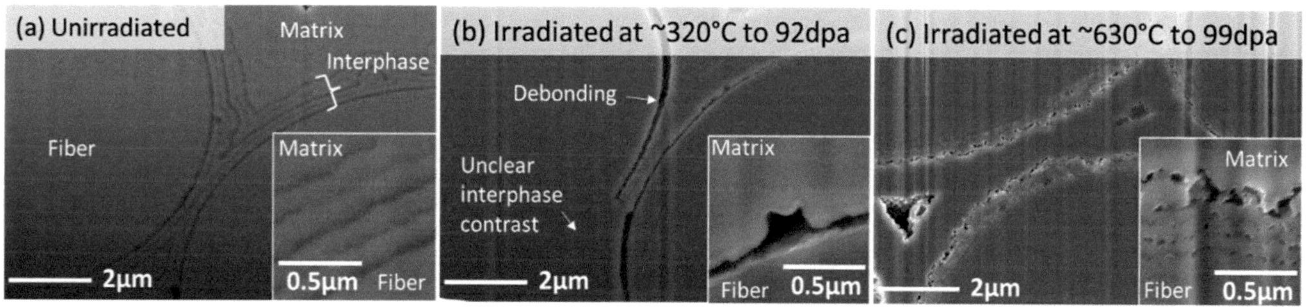

Figure 4.14 Secondary electron micrographs of polished crosssections of SiC fiber-reinforced SiC matrix composites with Hi-Nicalon type S fibers coated with multilayer SiC/pyrolytic carbon interphase: (A) Unirradiated, (B) neutron-irradiated at approximately 320°C to 92 displacements per atom (dpa), and (C) neutron-irradiated at approximately 630°C to 99 dpa. The absence of contrast in the interface region indicates interdiffusion between the SiC and pyrolytic carbon phases.
(Images reproduced from Koyanagi et al. [7].).

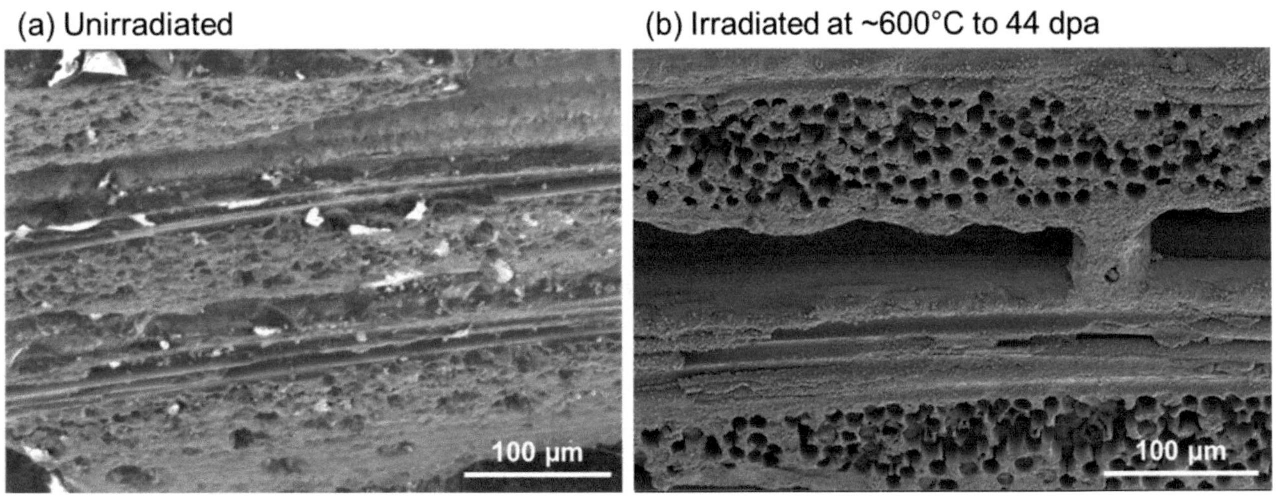

Figure 4.15 Surface morphology of chemical vapor infiltration SiC fiber-reinforced SiC matrix composite containing Hi-Nicalon type S SiC fibers coated with an approximately 100 nm thick pyrolytic carbon interphase: (A) Unirradiated and (B) neutron-irradiated at approximately 600°C to 44 displacements per atom. The pits on the irradiated surface indicate differential swelling between fiber and matrix.
(Images reproduced from Nozawa et al. [57].).

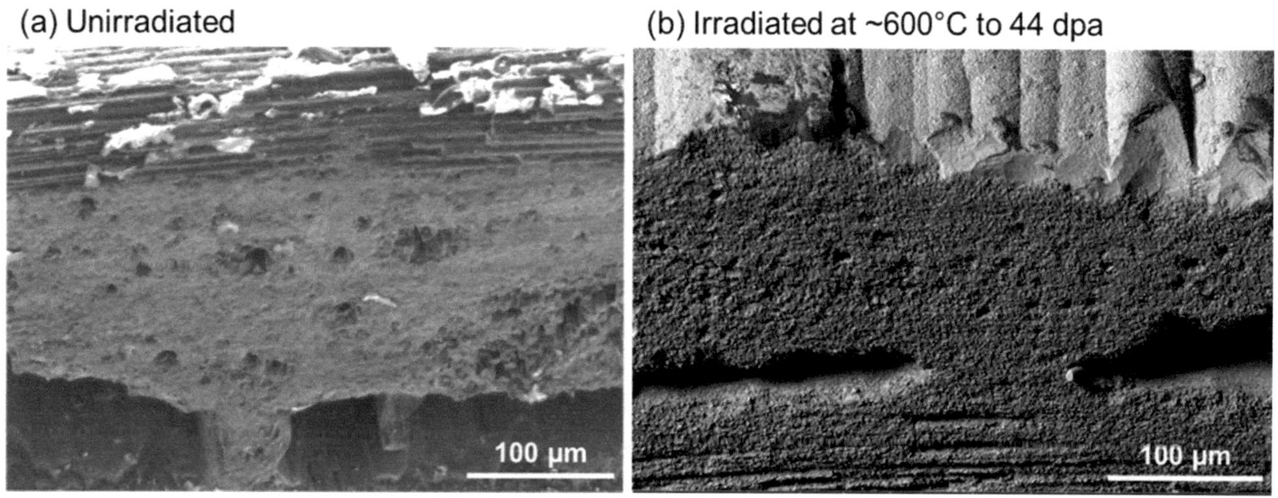

Figure 4.16 Surface morphology of chemical vapor infiltration SiC fiber-reinforced SiC matrix composite containing Tyranno SA3 SiC fibers coated with an approximately 100 nm thick pyrolytic carbon interphase: (A) Unirradiated and (B) neutron-irradiated at approximately 600°C to 44 displacements per atom. The images show the insignificant effects of irradiation on the surface.
(Images reproduced from Nozawa et al. [57].).

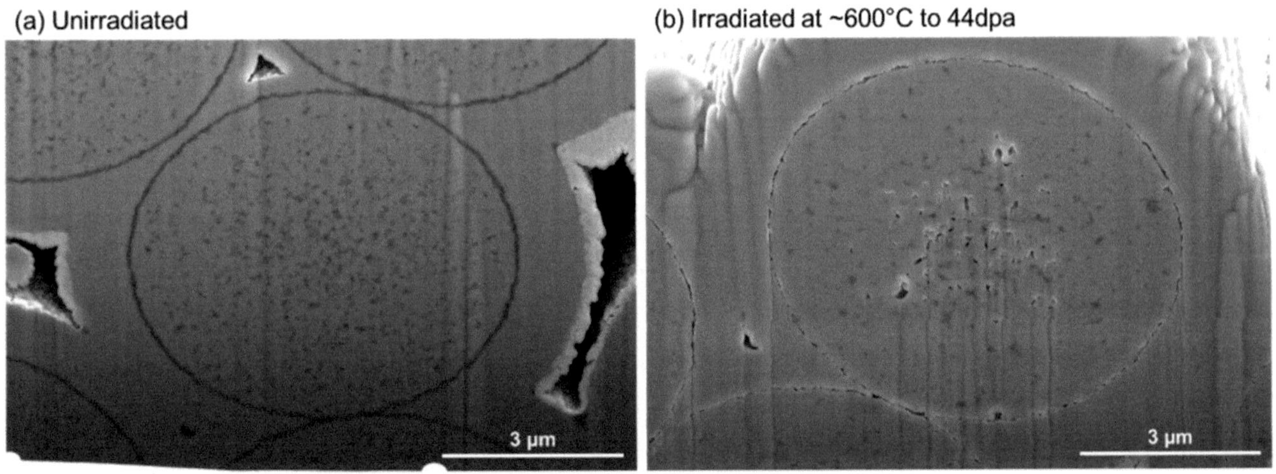

Figure 4.17 Cross-sectional observations of a chemical vapor infiltration SiC fiber-reinforced SiC matrix composite containing Tyranno SA3 SiC fibers coated with a <100 nm thick pyrolytic carbon interphase: (A) Unirradiated and (B) neutron irradiated at approximately 600°C to 44 displacements per atom. A comparison of the two micrographs reveals irradiation-induced interfacial debonding at the fiber/matrix interface. (Images reproduced from Nozawa et al. [57].).

After exhibiting interfacial debonding, irradiated CVI SiC/SiC composites show degradation in their mechanical properties. CVI SiC/SiC composites containing Hi-Nicalon Type S fibers coated with a multilayer PyC/SiC interphase exhibit an irradiation temperature–dependent degradation of ultimate flexural strength when evaluated at ambient temperature (Fig. 4.18) [7]. Although this specific grade of SiC/SiC composite maintains its strength after neutron exposure at 800°C up to 40 dpa [38], it exhibits a decrease in strength at lower irradiation temperatures and higher neutron doses [7,58]. These material property degradations are primarily due to interfacial debonding, which disrupts load transfer between fibers and matrix.

Mechanical property degradations associated with irradiation-induced debonding of the CVI SiC matrix and crystalline SiC fibers depend on fiber/matrix interface microstructure. For example, in one study, CVI SiC/SiC composites containing 50–160 nm thick PyC-coated Hi-Nicalon Type S fibers or 200–350 nm thick PyC-coated Tyranno SA3 fibers did not exhibit any notable degradation in mechanical properties, even when subjected to neutron irradiation at around 600°C to 30 dpa [59]. By contrast, in another study, CVI SiC/SiC composites containing approximately 100 nm thick PyC-coated Hi-Nicalon Type S or Tyranno SA3 fibers exhibited a decrease in ultimate flexural strength of up to 60% after irradiation at 500°C to only 11 dpa [57]. Although the sensitivity of the fiber/matrix interface microstructure to relatively high-dose neutron irradiation has not been fully identified, these results indicate a link between mechanical property degradation caused by irradiation and interfacial debonding.

Other nonCVI matrix types for SiC/SiC composites have been explored to a limited extent. A SiC/SiC composite densified by the nanoinfiltration and transient eutectic-phase (NITE) process—a densification method that involves liquid phase sintering of nanosized SiC powders and oxide additives—demonstrated tensile strength that was retained after high-temperature neutron irradiation. This class of SiC/SiC composites features a dense SiC matrix containing secondary phases resulting from sintering additives. The composite was fabricated using a 12 wt% Al_2O_3–Y_2O_3–SiO_2 additive system (Fig. 4.19A) and showed similar stress–strain curves under tensile loading–reloading tests before and after neutron irradiation at 600°C to 0.52 dpa, 830°C to 5.9 dpa, and 1270°C to 5.8 dpa (Fig. 4.19B) [46]. Additionally, SiC/SiC composites densified by liquid phase sintering with a 20 wt% Al_2O_3–Y_2O_3–$CaCO_3$ additive system exhibited swelling similar to that of CVD SiC under irradiation at 60°C–90°C to 0.20–0.25 dpa [60], the range in which irradiation-induced amorphization occurs [26]. Overall, the results indicate that LPS SiC/SiC composites with moderate amounts of oxide phases are promising; however, their high-dose neutron irradiation behavior remains largely unknown.

Polymer infiltration and pyrolysis (PIP) and reaction sintering of silicon and carbon are matrix densification techniques used to produce SiC/SiC composites. These techniques have been investigated predominantly for nonnuclear, high-temperature applications [61–63]. The radiation responses of these SiC/SiC composites—particularly those reinforced with third-generation crystalline SiC fibers such as Hi-Nicalon Type S and Tyranno SA3—have not been frequently studied. Nozawa et al. studied neutron-irradiated reaction-sintered SiC/SiC composites containing BN-coated

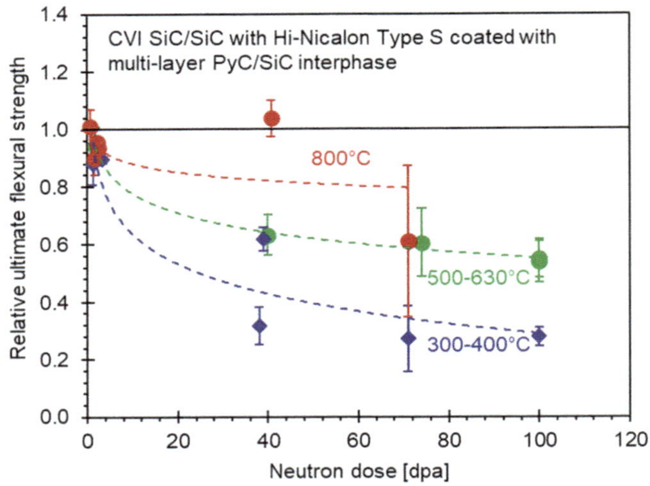

Figure 4.18 Neutron dose dependence of ultimate flexural strength normalized to the unirradiated value for chemical vapor infiltration SiC fiber-reinforced SiC matrix composite containing Hi-Nicalon type S SiC fibers coated with a multilayer pyrolytic carbon/SiC interphase. Irradiation temperatures are noted in the figure.
(Reproduced from Koyanagi et al. [7].).

Hi-Nicalon SiC fibers and PIP SiC/SiC composites containing carbon-coated Tyranno TE SiC fibers [64]. Specimens were irradiated at 800°C to 0.4 dpa. Hegeman et al. conducted a postirradiation characterization of three grades of PIP SiC/SiC composites containing Hi-Nicalon SiC fibers after neutron irradiation at 600°C and 900°C to 3.5 dpa [65]. Because these amorphous-like SiC fibers are unstable under neutron irradiation (Fig. 4.8), the results are not directly applicable to nuclear reactor core environments. Furthermore, these studies did not include postirradiation microstructural investigations of radiation damage mechanisms.

In summary, SiC/SiC composites consisting of a crystalline SiC matrix and fibers coated with a carbon interphase exhibit minimal degradation in their mechanical properties after neutron irradiation at elevated temperatures to approximately 10 dpa. However, high-dose neutron irradiation–induced, irradiation temperature–dependent microstructural instability at the fiber/matrix interface leads to a deterioration in mechanical properties (Fig. 4.18). Preserving the integrity of the fiber/matrix interface is crucial to the microstructural design of SiC/SiC composites intended for use in irradiation environments.

3. Microstructural design of SiC-based composite

Fundamental understanding of the responses of the various grades of monolithic SiC and SiC/SiC composites fabricated through traditional methods and discussed in Section 2 is crucial to informing the microstructural design of advanced SiC/SiC

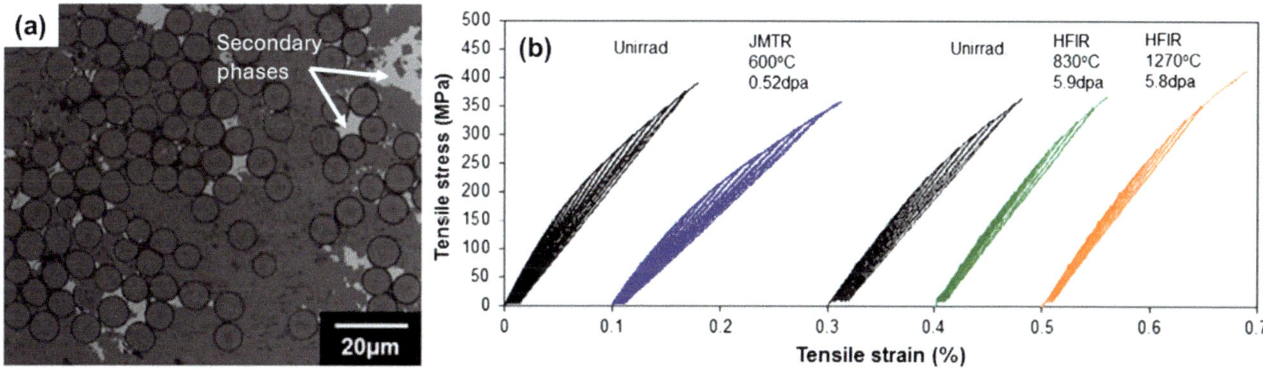

Figure 4.19 (A) Backscattered electron image of a SiC fiber-reinforced SiC matrix composite fabricated using the nano infiltration and transient eutectic-phase process and (B) the composite's stress–strain curves under tensile loading–reloading tests before and after neutron irradiation at 600°C to 0.52 displacements per atom (dpa), 830°C to 5.9 dpa, and 1270°C to 5.8 dpa. The micrograph shows the presence of secondary phases resulting from oxide sintering aids.
(Images reproduced from Koyanagi et al. [46].).

composites intended for nuclear applications that involve neutron irradiation. Section 3.1 presents a conceptual design approach for the microstructure of SiC/SiC composites by correlating the as-fabricated microstructures of the constituents of SiC/SiC composites with the resulting irradiation-induced effects. This approach will facilitate the identification of key technological challenges, thereby enabling the planning of targeted experiments to evaluate potential limitations. Section 3.2 reviews established and emerging options for SiC/SiC composite constituents as well as promising material combinations based on mechanisms of radiation damage. The intended applications are assumed to involve neutron irradiation at elevated temperatures (>200°C, beyond the threshold for radiation-induced amorphization [26]) and moderate to high neutron doses exceeding 1 dpa.

3.1 Design approach for radiation-tolerant SiC/SiC composites

Fiber-reinforced composites are designed to be damage-tolerant. Thus, the mechanical integrity of SiC/SiC composites under irradiation conditions must be prioritized in the design of composite microstructure. Two primary degradation mechanisms must be mitigated in the composite design process. The first is the degradation of the SiC matrix and fibers as a result of irradiation. The mechanical properties of the SiC fibers and matrix have a significant influence on the overall mechanical performance of a composite, as explained subsequently. Therefore, the selection of radiation-resistant matrix and fiber materials is crucial to ensuring the long-term structural integrity of a composite.

A matrix cracking stress σ_{mc}, which is nearly equal to the proportional limit stress (e.g., the elastic limit), can serve as a design parameter for structural components made from SiC/SiC composites [17]. For unidirectionally reinforced fiber composites, σ_{mc} is formulated for matrix cracking that involves interfacial debonding [21] —a case that applies to CVI SiC/SiC composites—as follows:

$$\sigma_{mc} = E_L \left[\frac{6\tau\Gamma_m f^2 E_f}{(1-f)E_m^2 R E_L} \right]^{\frac{1}{3}} - \sigma^T \tag{4.1}$$

where E_L is ply modulus in the longitudinal direction, E_f is Young's modulus of the fiber, E_m is Young's modulus of the matrix, τ is interfacial sliding stress, Γ_m is matrix fracture energy, R is fiber radius, and σ^T is misfit stress attributed to residual stress in the material. When the moduli of the matrix and fiber are equivalent and process-induced residual stresses are negligible (e.g., in specific CVI SiC/SiC composites that exhibit matrix porosity), matrix fracture energy influences matrix cracking stress (e.g., $\sigma_{mc} \propto \Gamma_m^{1/3}$). The formulation of matrix cracking in two-dimensionally reinforced composites is addressed in the literature [21].

Ultimate tensile strength σ_{UTS} is governed by the stochastic mechanical properties of SiC fibers. In a unidirectionally reinforced composite with multiple matrix cracks during failure, ultimate tensile strength is expressed as follows [66]:

$$\sigma_{UTS} = f\sigma_C \left[\frac{2(m+1)}{(m+2)m}\right]^{\frac{1}{m}+1} \left(\frac{m+1}{m+2}\right) \tag{4.2}$$

where σ_C is characteristic fiber strength, and m is the Weibull modulus.

The second degradation mechanism is irradiation-induced debonding at the fiber/matrix interface, which causes a loss of load transfer between the fiber and matrix that results in a low proportional limit stress and low ultimate strength. As Fig. 4.15 illustrates, a mismatch in swelling between the fiber and matrix can cause the debonding. The fiber/matrix interphase must remain adherent during irradiation.

Once a combination of materials is identified for the fiber, matrix, and interphase, the next step in composite design is to determine the fiber architecture, fiber volume fraction, and other microstructural parameters to tailor the anisotropic thermomechanical properties to meet the property requirements for a target application. The analytical tools for designing anisotropic composite properties are available elsewhere [67,68].

3.2 Available and emerging options for SiC/SiC composite constituents

3.2.1 SiC fibers

Table 4.2 summarizes the microstructure and mechanical properties of commercially available crystalline SiC fibers. Crystalline, near-stoichiometric SiC fibers such as Hi-Nicalon Type S and Tyranno SA3 are preferred for nuclear applications because of the challenges posed by irradiation-induced shrinkage in amorphous-like SiC fibers [55].

Tyranno SA4 fiber exhibits enhanced tensile strength relative to its predecessor, Tyranno SA3 [70]. SA4 fiber–reinforced CVI SiC/SiC composites perform better than SA3 fiber composites in terms of tensile properties [71]. SA4 fiber is expected to demonstrate radiation resistance comparable to that of SA3 fiber because they share a slightly carbon-rich chemistry, minimal presence of process-induced impurities, and a submicrometer grain size (Table 4.2).

Sylramic SiC fiber is a crystalline, nominally stoichiometric material with minimal free carbon content [35,72]. However, its use in nuclear applications is limited because it contains 2.3 wt% boron, which rapidly transmutes to helium upon exposure to thermal neutrons. Newsome et al. reported a significant reduction in the proportional limit stress and ultimate flexural strength of Sylramic fiber–reinforced CVI SiC composites after irradiation; values decreased to less than half those measured in the unirradiated state after neutron irradiation at temperatures from 300°C to 800°C to approximately five dpa [20]. This degradation was explained by helium transmutation. In one study, researchers synthesized and evaluated ^{11}B-enriched Sylramic SiC fibers, aiming to mitigate the effects of boron transmutation [27]. Irradiation-induced creep strains were comparable among Sylramic, Hi-Nicalon Type S, and Tyranno SA3 fibers. Thus, isotopically modified Sylramic fibers have potential for use in irradiation environments; however, the associated production costs would be considerable.

Table 4.2 Microstructural parameters and room-temperature strength of commercially available continuous SiC fibers.

Trade name	C/Si atomic ratio	Impurity elements (wt%)	Fiber diameter (μm)]	Grain size (nm)	Room temp. Strength (GPa)
Hi-Nicalon Type S [69]	1.07	0.3 O	13	10–50	2.5
Tyranno SA3 [69]	1.03 (edge), 1.19 (core)	0.1 O, 0.2–0.3 Al	7.2	50–400	2.8
Tyranno SA4 [70]	1.07	<1 O, <2 metal	8	150	4.0
Sylramic [35]	1.01	0.8 O, 2.3 B, 0.4 N, 2.1 Ti	10	100	3.2
Ultra-SCS [35]	>1 (SiC on carbon core)	Not reported	140	10–100	~6
SiC free form fibers [70]	0.96	Not reported	39	Amorphous (edge) 15–40 (core)	4.0

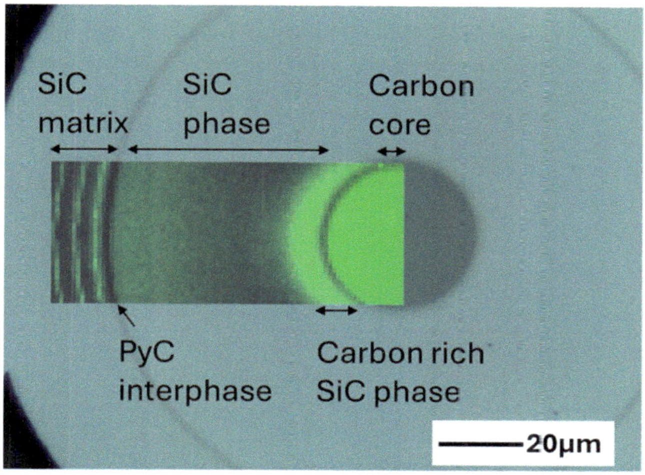

Figure 4.20 Optical micrograph of unirradiated Ultra-SCS SiC fiber surrounded by a chemical vapor infiltration SiC matrix. The overlaid Raman chemical map shows the relative intensity of the carbon D-band peak.
(Reproduced from Koyanagi and Katoh [74].).

Ultra-SCS SiC fiber is a unique material option for fiber reinforcement because of its large fiber diameter—approximately 140 μm—and carbon core [35,73]. The large diameter makes the fiber less flexible and limits the weaving of the fibers. The fiber is manufactured via CVD of SiC on the carbon core. The CVD SiC phase has a transitional microstructure, including a carbon-rich SiC phase near the carbon core and less free carbon near the fiber surface (Fig. 4.20) [73,74].

The Ultra-SCS SiC fibers did not exhibit stability of mechanical properties after neutron irradiation at elevated temperatures. CVI SiC/SiC composites incorporating Ultra-SCS SiC fibers exhibited a decrease in ultimate flexural strength of approximately 70% and loss of quasi-ductile fracture behavior after irradiation at about 300°C to 11.8 dpa (Fig. 4.21A) [74]. The degradation of the fibers was accompanied by irradiation-induced cracking in the SiC phase (Fig. 4.21B), which was due to strains resulting from differential swelling in the carbon core versus the SiC phase [74]. Similar degradation behavior has been observed after irradiation at approximately 500°C and 600°C [2].

Laser CVD SiC fibers are a relatively novel class of SiC fiber developed by Free Form Fibers. In contrast to the preceramic polymer conversion method used for Hi-Nicalon Type S and Tyranno series fibers, laser CVD offers an alternative approach for synthesizing high-strength SiC fibers with customizable small diameters and chemical compositions [75]. Laser CVD uses laser beams as an energy source to convert precursor gas species into SiC. This process does not require a carbon core; the absence of a carbon core helps overcome challenges associated with processing small-diameter fibers and the material degradation common to Ultra-SCS SiC fibers.

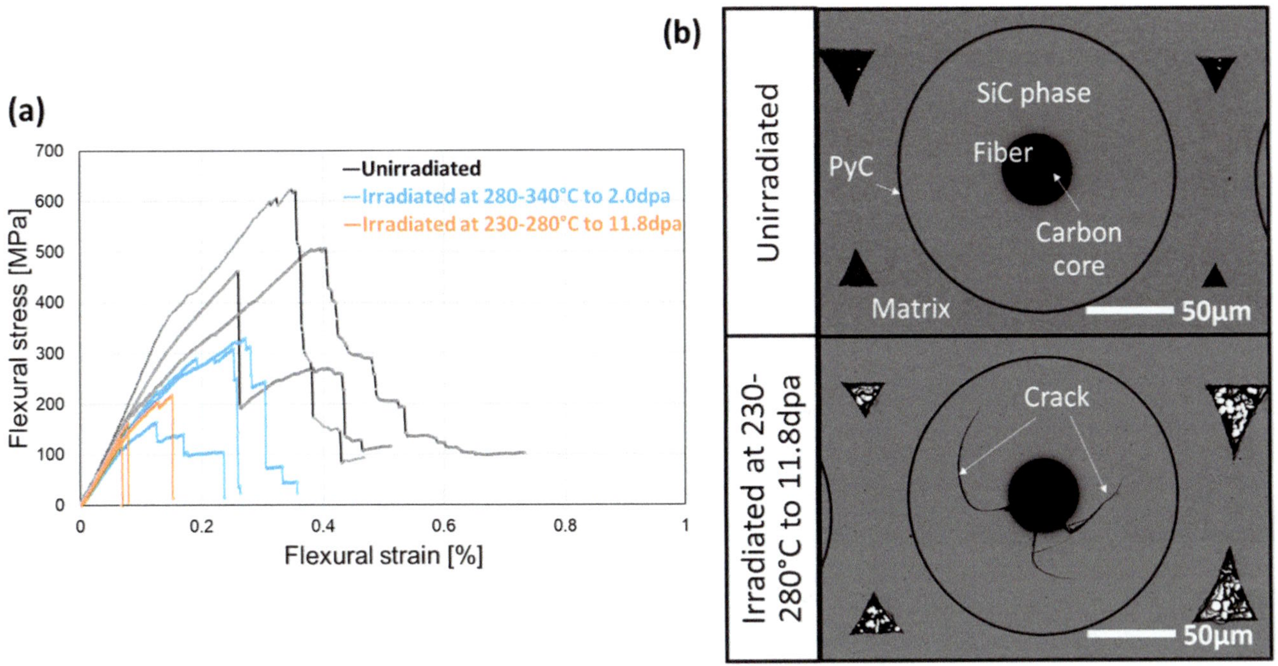

Figure 4.21 (a) Flexural stress–strain curves and (b) backscattered electron micrographs of chemical vapor infiltration SiC fiber-reinforced SiC matrix composites with Ultra-SCS SiC fibers unirradiated and neutron irradiated at approximately 300°C to 11.8 displacements per atom. (Reproduced from Koyanagi and Katoh [74].).

Figure 4.22 Laser-driven chemical vapor deposition SiC fibers fabricated by free form fibers.
(Reproduced from Koyanagi et al. [2].).

Laser CVD SiC fibers have a unique microstructure and chemistry (Table 4.2). These fibers typically have a relatively large diameter (approximately 40 μm) (Fig. 4.22) and a high room-temperature tensile strength (4 GPa) [70]. The cross-section of a laser CVD SiC fiber exhibits a graded microstructure that correlates with the powder distribution within the laser spot during deposition. Detailed analysis by Gietl et al. [70] revealed that the surface region of these fibers has an amorphous microstructure, and the microstructure is progressively more crystalline toward the core region (Fig. 4.23). Additionally, the laser CVD SiC fibers are slightly silicon-rich; free silicon phases (indicated by the bright contrasts in Fig. 4.23) have been identified using Raman spectroscopy.

The radiation stability of laser CVD SiC fibers is unknown. Their dimensional stability under irradiation is expected to differ from that of crystalline SiC fibers because amorphous regions are present near the surface.

Aside from the aforementioned commercially available SiC fibers, new SiC fibers are in development. Laboratory-grade KD-series SiC fibers are reported to be crystalline and carbon-rich, and researchers are actively evaluating and continuously improving the KD fibers [76]. The annealing of amorphous SiC fibers during matrix densification via hot-pressing at elevated temperatures has been found to be a cost-effective method for producing crystalline SiC fibers for composites [77].

3.2.2 SiC matrix

Densification methods for monolithic SiC ceramics may be suitable for applying a SiC matrix to a SiC fiber preform. As discussed in Section 2.1, CVI yields a high-purity, highly crystalline SiC matrix. The infiltration process inherently results in the formation of pores within and between fiber bundles (Fig. 4.24).

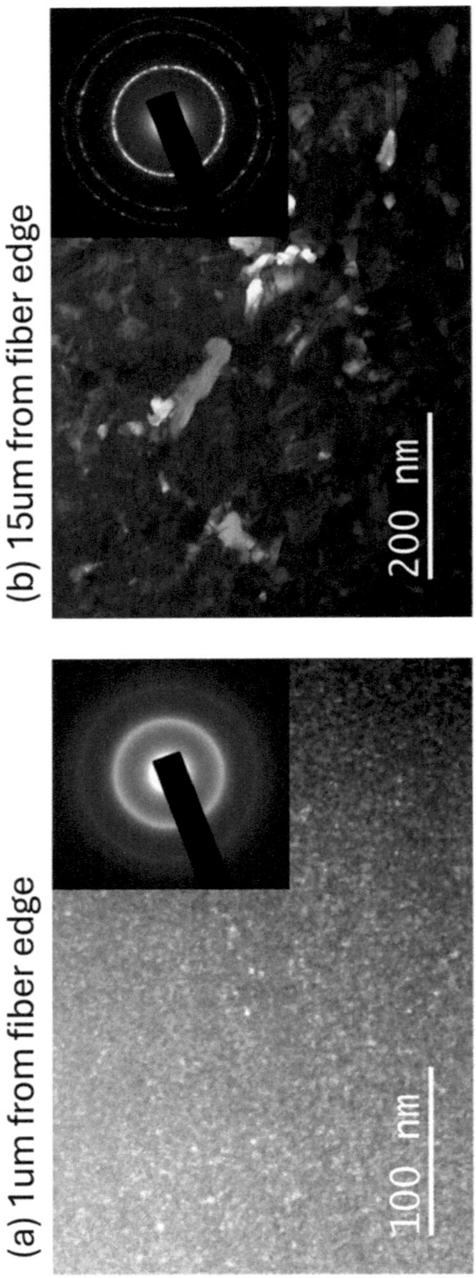

Figure 4.23 High-angle annular dark-field scanning transmission electron micrographs of a laser chemical vapor deposition SiC fiber: (a) Near-surface region and (b) core region. Representative selected area diffraction patterns are inserted. (Micrographs reproduced from Gietl et al. [70].).

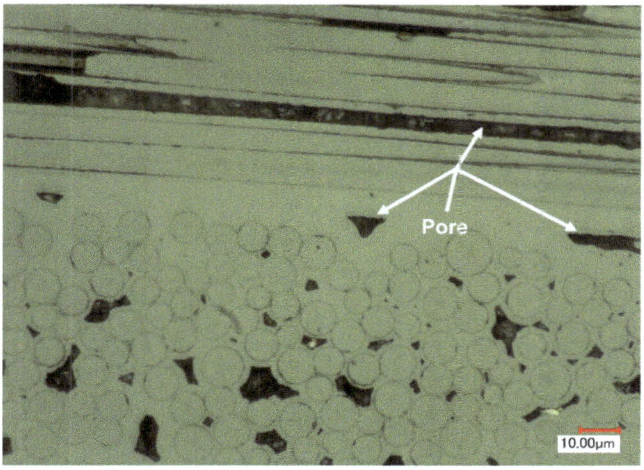

Figure 4.24 Cross-sectional optical micrograph of chemical vapor infiltration SiC composites with Tyranno SA3 SiC fibers showing pores that formed in the matrix phases during the densification process.

The SiC powder sintering–based processing techniques, particularly the LPS/NITE process, provide an alternative method for matrix densification. LPS of SiC powders has been achieved using various oxide additive systems to reduce the sintering temperature relative to solid-state sintering [51]. Oxide additive systems include Al_2O_3–Y_2O_3 and Al_2O_3–Y_2O_3–MgO, which do not react with SiC at the sintering temperature. The matrix of NITE SiC/SiC composites tends to exhibit segregation of secondary phases, which can be attributed to the presence of sintering additives within the matrix and among the fiber bundles. These additives facilitate the reduction of porosity (Fig. 4.19A). This class of SiC/SiC composites has been densified by hot-pressing. Field-assisted sintering (also known as spark plasma sintering) is an alternative densification method that has the advantage of reduced processing time [78].

The PIP process and associated methodologies have been the focus of continuous development for nuclear applications [79,80]. Researchers have successfully demonstrated the processing of thin-walled tubular geometries [80]. However, the resulting matrix microstructures are typically amorphous or nanostructured and may be unstable under irradiation conditions. A significant challenge inherent in the PIP process is the conversion of crystalline SiC under conditions (e.g., temperatures and durations) that prevent the degradation of SiC fibers. A breakthrough in the formation of a highly crystalline SiC phase at comparatively low conversion temperatures is needed before PIP technology can be widely used.

Reaction sintering silicon and carbon powders presents a notable risk of radiation-induced mechanical degradation primarily because of the spontaneous cracking associated with differential swelling in silicon-rich SiC [44]. The elimination of free silicon phases—a goal that remains unachieved [81]—must be addressed for applications in irradiated environments.

Some researchers have investigated a hybrid matrix densification approach for ceramic matrix composites [82]. This approach combines the foregoing matrix densification processes and can potentially be extended to the SiC/SiC composite system. The strategy involves using CVI to achieve partial SiC densification; CVI is applied for a brief period to safeguard the fiber/matrix interphase prior to subsequent densification steps [83]. The final stage of matrix densification is then carried out using more time- and cost-efficient methods than the CVI process [83]. The irradiation stability of such heterogeneous composites can be forecasted based on the performance of their individual matrix components.

3.2.3 Fiber/matrix interface

The toughening of SiC/SiC composites is enabled by the fiber/matrix interface, which provides adequate bonding and interfacial sliding strengths [21]. The fracture toughness of ceramic composites results primarily from the deflection of matrix cracks at the fiber/matrix interface, fibers bridging the cracks, and fiber pullout, which is associated with frictional energy dissipation [21]. The interphase is typically formed via CVI. Interphase thickness has been reported to have a slight effect on mechanical properties such as ultimate tensile strength, proportional limit stress, Young's modulus, and strain to failure [84].

Some grades of carbon-based interphase, such as monolayer PyC interphase (Fig. 4.1) and multilayer PyC/SiC interphase (Fig. 4.14A), retain the mechanical properties of the composite after neutron irradiation. In one study, a multilayer interphase retained thin PyC layers without noticeable debonding even after neutron irradiation at 800°C to 71 dpa (Fig. 4.25). However, accumulating irradiation test data

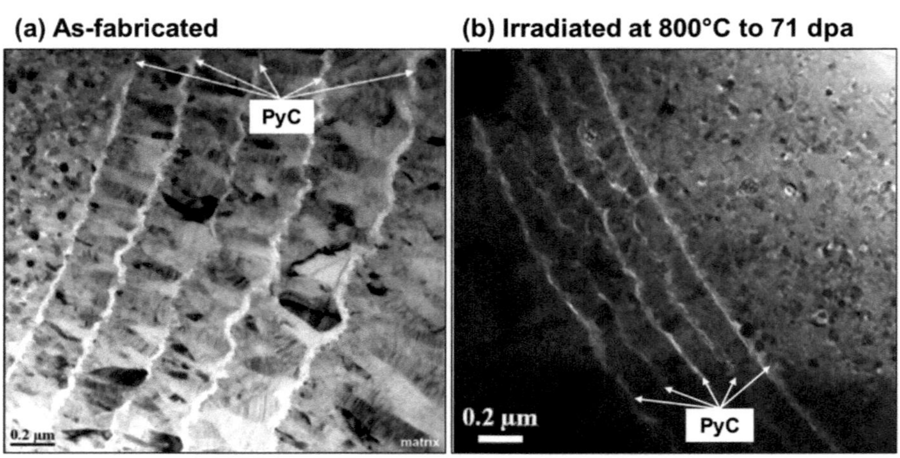

Figure 4.25 Bright-field transmission electron micrograph of a SiC fiber-reinforced SiC matrix composite with a multilayer pyrolytic carbon/SiC interphase: (a) As-fabricated and (b) neutron irradiated at 800°C to 71 displacements per atom.
(Reproduced from Perez-Bergquist et al. [85].).

increasingly indicate that different grades of PyC interphase result in distinct radiation responses and sometimes lead to mechanical degradation after irradiation to only 10 dpa. A significant knowledge gap exists with regard to the effects of interphase microstructure—specifically, interphase thickness and atomic-scale ordering—on the mechanical properties of irradiated SiC/SiC composites. Systematic experimental investigations are needed to address this gap.

If isotopically controlled ^{11}BN is used to eliminate ^{10}B content, then BN might serve as an alternative interphase material. This step would mitigate the risk of boron burnup and the generation of transmutant helium through ^{10}B (n, α) ^{7}Li reactions during irradiation. Although the radiation stability of ^{11}BN (e.g., swelling and changes in mechanical properties under irradiation) is largely unknown, the processing of interphases containing BN is well-established in aerospace applications [86,87].

The microstructural instability of PyC-based interphase irradiated at 300°C and 500°C to doses exceeding 90 dpa (Fig. 4.14) suggests the potential benefits of exploring alternative interphase materials to enhance the radiation resistance of SiC/SiC composites. A variety of nonPyC interphases, such as ZrO_2 [88,89], Ti_3SiC_2 [90], and TiC [91], have been deposited onto SiC fibers. Key requirements for an effective SiC/SiC composite interphase include (1) chemical compatibility with both the SiC fibers and the matrix during interphase coating and matrix densification, and (2) the ability to precisely control interphase thickness at the nano- or micrometer scale. Additionally, to facilitate composite toughening mechanisms, the interphase must exhibit specific mechanical and physical properties that enable effective deflection of matrix cracks at the fiber/matrix interface. Detailed descriptions of interphase processing and performance requirements are provided in the literature [2]. Although existing composite theory and processing technologies offer pathways for developing new SiC/SiC composite interphases, selecting appropriate interphase materials for nuclear applications remains challenging because high-dose neutron irradiation has a significant effect on dimensional stability and mechanical properties.

4. Engineering design of SiC-based composite

The microstructure of SiC/SiC composites can be engineered to meet specific property requirements for various applications. In certain cases, SiC/SiC composites are integrated with other materials. For example, environmental barrier coatings (EBCs) can be applied on SiC/SiC composite surfaces to mitigate corrosion [92]. Table 4.3 summarizes SiC/SiC composite configurations used in fuel cladding applications, in which external surfaces are exposed to corrosive coolants, and gas-tightness is essential to prevent the release of fission products.

The SiC/SiC composite layer may not be gas-tight because of process-induced pores; thus, a monolithic SiC or metallic layer is used as a hermetic seal (HS). Additionally, this layer is designed to function as an EBC, providing protection against coolant corrosion on the outer surface and mitigating fuel–cladding chemical interaction (FCCI) at the inner surface.

Table 4.3 Examples of engineered SiC/SiC composites designed for fuel cladding applications.

Class	Layer configuration (Inner to outer)	Configuration	Description	Reference
Full ceramics	Composite	Full composite	A monolithic surface layer may be present	[93]
	Composite–monolith	Duplex	Monolithic layer as an HS/EBC against hydrothermal corrosion	[94]
	Monolith–composite	Duplex	Monolithic layer as an HS/EBC against FCCI	[93,95]
	Monolith–composite–monolith	Triplex	Monolithic layer as an HS/EBC against hydrothermal corrosion and FCCI	[95,96]
Metal-assisted ceramics	Metal–composite	Duplex	Metallic layer as an HS/EBC against FCCI	[97]
	Composite–metal	Duplex	Metallic layer as an HS/EBC against hydrothermal corrosion	[98]
	Metal–composite–metal	Triplex	Metallic layer as an HS/EBC against hydrothermal corrosion and FCCI	Not proposed
	Composite–metal–composite	Triplex	Metallic layer as an HS	[17]

HS, = hermetic seal.; *and*, FCCI = fuel–cladding chemical interaction.

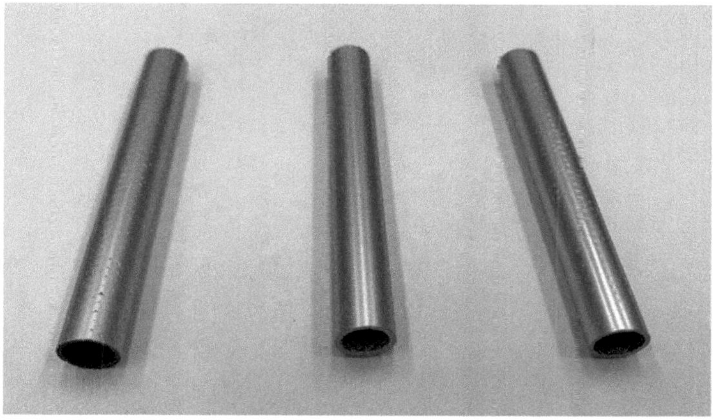

Figure 4.26 SiC fiber-reinforced SiC matrix composite tubes with a chromium coating. Each tube is 75 mm long. The processing method is reported in Mouche et al. [100].

The incorporation of a monolithic SiC layer influences the mechanical properties of the entire SiC-based tube [93]. Metallic layers (Fig. 4.26), in addition to affecting the mechanical characteristics, affect processing and neutronic performance. When a nonSiC phase is incorporated into the system, maintaining bond integrity at the phase boundary becomes challenging because of the differential swelling between the SiC and nonSiC phases [99]. This issue is similar to the challenges encountered in developing interfaces for SiC/SiC composites.

A duplex structure featuring a thick SiC monolith on its outer surface has been investigated as a SiC/SiC composite–based light-water reactor fuel cladding (Fig. 4.27) [101]. This design was developed with consideration of the in-pile stress state [102], in which the outer surface undergoes compression, thereby mitigating the risk of monolith failure, while the inner surface is under tension, which may induce cracking. The stress state is driven by the through-thickness temperature gradient and irradiation-induced property changes [102]. The composite configurations listed in Table 4.3 provide a basis for the design of SiC-based composites for other applications.

5. Future research opportunities

With regard to composite design and manufacturing, research is needed to address the irradiation-induced debonding observed after irradiation at low or intermediate temperatures below 800°C (Fig. 4.14 and Fig. 4.15). Key areas of focus include controlling the phase, thickness, bonding conditions, and microstructure of the fiber/matrix interface, all of which are crucial to the development process. Irradiation-induced debonding is caused by strain resulting from the differential swelling of the fiber, matrix, and interphase; thus, the dimensional stability of composite constituents under

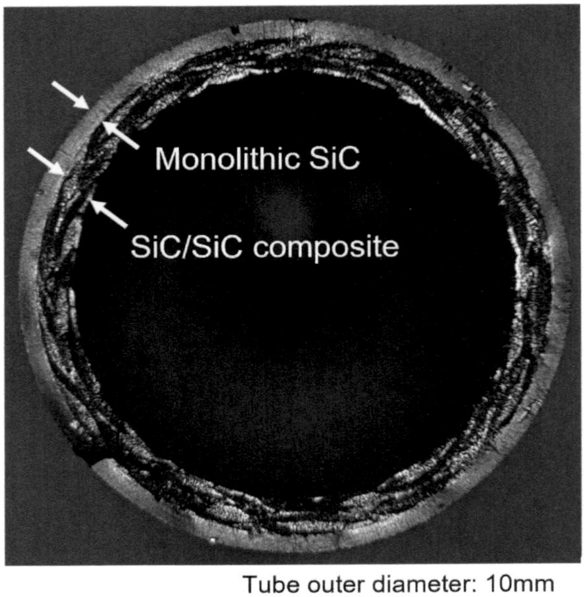

Figure 4.27 Cross-sectional X-ray radiograph of a duplex SiC fiber-reinforced SiC matrix composite tube with a monolithic SiC outer layer. (Reproduced from Koyanagi et al. [101].).

irradiation must be considered in composite design. Furthermore, research opportunities exist for the development of a new generation of SiC fibers that retain their strength after exposure to high-dose neutron irradiation at low temperatures. Fractographic analysis indicates that irradiation at 320°C to 92 dpa leads to a loss of strength in Hi-Nicalon Type S SiC fibers [7]. Quantitative analysis of strength loss remains challenging, but the complete disappearance of the fracture mirror pattern suggests that the reduction in fiber strength is significant [21]. Changes in the fracture surface have been linked to the irradiation-induced loss of free carbon phases [7]; this link may provide insight into the underlying degradation mechanisms and inform potential strategies for fiber enhancement.

Cost-effective manufacturing remains a significant challenge to using SiC/SiC composites in nuclear applications [35,103]. A viable solution is needed for cost-effectiveness and irradiation resistance, including dimensional stability and retention of strength under irradiation. Although reaction sintering of Si and C powders for SiC densification is more cost-effective than CVI SiC densification, this process has not been pursued extensively for nuclear applications because spontaneous cracking can occur under irradiation [44]. A new paradigm is needed in SiC material manufacturing to address low-cost production.

Advanced manufacturing techniques (including additive manufacturing) for SiC/SiC composites present significant opportunities for fabricating and testing prototypes to evaluate the feasibility of the material technology. Moreover, near-net shape processing may contribute to a reduction in manufacturing costs. The fiber-reinforced additive manufacturing process has led to considerable advancements in polymer-based composite technology [104]. However, the additive manufacturing of fiber-reinforced ceramic matrix composites is still in the developmental stage. The particular additive manufacturing process and feedstock/densification route must be carefully selected, considering manufacturability and material properties, including dimensional stability under irradiation and the effect of irradiation on mechanical properties. The literature provides recommendations regarding the additive manufacturing of radiation-stable monolithic SiC materials [42].

Incorporating dissimilar materials into the SiC/SiC composite system may facilitate the development of novel nuclear applications and provide mitigation technologies, such as surface coatings for enhanced corrosion resistance. This design approach is particularly relevant because SiC/SiC composites face inherent challenges associated with SiC chemistry and ceramics, including susceptibility to corrosion in specific chemical environments [105] and the loss of gas-tightness after microcracking [106]. Integrating alternative materials such as EBCs into the SiC/SiC composite system may address these limitations. The interface with a dissimilar material must exhibit tolerance to strain arising from the differential swelling of SiC and the secondary phase under irradiation conditions. Currently, no established methodology exists for developing a microstructurally stable dissimilar interface for SiC composites to be used in high-dose irradiation environments. One potential mitigation strategy is to engineer the residual stresses to accommodate irradiation-induced strain [100]. Investigations into the behavior of dissimilar interfaces under neutron irradiation are crucial to advancing the development of joining technologies for SiC/SiC composites [99].

6. Summary

The effect of neutron irradiation on the physical and mechanical properties of SiC/SiC composites must be thoroughly considered in their design and manufacture for nuclear core structures subjected to irradiated environments. These considerations are necessary because of the potential for significant strength degradation and undesirable volume changes under irradiation when the composite microstructure is not adequately engineered. Nuclear-grade SiC/SiC composites are characterized by crystalline, near-stoichiometric SiC phases with a PyC interphase. They exhibit saturation swelling when exposed to doses above approximately two dpa and show minimal degradation of mechanical properties when irradiated above the threshold temperature for irradiation-induced amorphization (~150°C) to doses below 10 dpa. CVI has been established as the reference method for producing this class of SiC/SiC composites. As neutron dose accumulates, SiC/SiC composites are increasingly susceptible to irradiation-induced degradation, depending on irradiation conditions and composite microstructure. Recent experimental data indicate that CVI SiC/SiC composites can be exposed to an irradiation temperature of 800°C, given that certain composites demonstrate no significant strength reduction with doses up to approximately 40 dpa.

A mechanistic understanding of the response of SiC-based materials with varying microstructures to neutron irradiation is the basis for developing novel SiC materials for use in irradiation environments. Phase-pure, highly crystalline SiC is generally considered stable under irradiation: reference CVD SiC ceramics exhibit temperature-dependent saturation swelling and negligible degradation in strength after irradiation. Based on this definition, nanocrystalline SiC, such as Hi-Nicalon Type S SiC fibers, can be regarded as stable under low-dose irradiation up to approximately 10 dpa. At higher doses, submicron-sized grain structures, such as those of Tyranno SA3 SiC fibers, exhibit dimensional stability comparable to that of CVD and CVI SiC. For future reference, more clarity is needed regarding the crystallinity required for irradiation stability (e.g., coherent crystal domain size, quantification of lattice disorder, and impurity levels). The introduction of secondary phases increases the risk of irradiation-induced strength degradation. The presence of free silicon in the SiC matrix and carbon cores in SiC fibers has been shown to promote spontaneous irradiation-induced cracking because of differential swelling. The microstructure of SiC/SiC composites must be engineered to minimize cracking and delamination at the constituent boundaries; in this process, one must ensure that fiber and matrix have similar dimensional stability under irradiation.

The SiC/SiC composites are engineered materials that can be tailored to specific applications. Duplex and layered structures that incorporate combinations of SiC/SiC composites, monolithic SiC, and dissimilar materials can potentially address the inherent challenges of SiC/SiC composites, including issues related to leak tightness and corrosion in specific chemical environments. Preventing irradiation-induced debonding and cracking because of differential swelling is crucial to maintaining the structural integrity of SiC/SiC composites under irradiation conditions.

Advancements in SiC/SiC composite technology necessitate material innovations. Key research opportunities include (1) developing a stable fiber/matrix interface under high-dose neutron irradiation at low to intermediate temperatures (e.g., 300°C–600°C) and (2) exploring advanced and additive manufacturing techniques for SiC/SiC composites that enable near-net shaping, cost-effectiveness, and the integration of novel designs, such as SiC/SiC composite–metal hybrid systems. As detailed in this chapter, the conceptual design of such new material systems should build on irradiation studies of conventional SiC and SiC/SiC composites to develop materials with predictable dimensional changes and minimal strength degradation under irradiation.

Acknowledgments

Preparation of this chapter was supported by the US Department of Energy Office of Fusion Energy Sciences and Office of Nuclear Energy, Advanced Fuels Campaign under contract DE-C05-00OR22725 with UT-Battelle LLC.

References

[1] Y. Katoh, L.L. Snead, Silicon carbide and its composites for nuclear applications – Historical overview, J. Nucl. Mater. 526 (2019), 151849-151849.

[2] T. Koyanagi, Y. Katoh, T. Nozawa, Design and strategy for next-generation silicon carbide composites for nuclear energy, J. Nucl. Mater. 540 (2020), 152375-152375.

[3] M.N. Cinbiz, T. Koyanagi, G. Singh, Y. Katoh, K.A. Terrani, N.R. Brown, Failure behavior of SiC/SiC composite tubes under strain rates similar to the pellet-cladding mechanical interaction phase of reactivity-initiated accidents, J. Nucl. Mater. 514 (2019) 66–73.

[4] N.M. George, K. Terrani, J. Powers, A. Worrall, I. Maldonado, Neutronic analysis of candidate accident-tolerant cladding concepts in pressurized water reactors, Ann. Nucl. Energy 75 (2015) 703–712.

[5] G.J. Butterworth, Low activation structural materials for fusion, Fusion Eng. Des. 11 (1–2) (1989) 231–244.

[6] K.A. Terrani, B.A. Pint, C.M. Parish, C.M. Silva, L.L. Snead, Y. Katoh, Silicon carbide oxidation in steam up to 2 MPa, J. Am. Ceram. Soc. 97 (8) (2014) 2331–2352.

[7] T. Koyanagi, T. Nozawa, Y. Katoh, L.L. Snead, Mechanical property degradation of high crystalline SiC fiber-reinforced SiC matrix composite neutron irradiated to similar to 100 displacements per atom, J. Eur. Ceram. Soc. 38 (4) (2018) 1087–1094.

[8] F. Najmabadi, R.W. Conn, C.G. Bathke, L. Bromberg, E.T. Cheng, D.R. Cohn, P.I. H. Cooke, R.L. Creedon, D.A. Ehst, K. Evans, N.M. Ghoniem, S.P. Grotz, M.Z. Hasan, J.T. Hogan, J.S. Herring, A.W. Hyatt, E. Ibrahim, S.A. Jardin, C. Kessel, M. Klasky, R. A. Krakowski, T. Kunugi, J.A. Leuer, J. Mandrekas, R.C. Martin, T.K. Mau, R.L. Miller, Y.K.M. Peng, R.L. Reid, J.F. Santarius, M.J. Schaffer, J. Schultz, K.R. Schultz, J. Schwartz, S. Sharafat, C.E. Singer, L. Snead, D. Steiner, D.J. Strickler, D.K. Sze, M. Valenti, D.J. Ward, J.E.C. Williams, L.J. Wittenberg, C.P.C. Wong, The ARIES-I tokamak reactor study, Fusion Technol. 19 (3P2A) (1991) 783–790.

[9] L. Giancarli, J.P. Bonal, a. Caso, G. Le Marois, N.B. Morley, J.F. Salavy, Design requirements for SiC/SiC composites structural material in fusion power reactor blankets, Fusion Eng. Des. 41 (1998) 165–171.

[10] S. Nishio, S. Ueda, I. Aoki, R. Kurihara, T. Kuroda, H. Miura, T. Kunugi, Y. Seki, T. Nagashima, M. Ohta, J. Adachi, S. Yamazak, I. Kawaguchi, T. Hashimoto, K. Shinya, Y. Murakami, H. Takase, T. Nakamura, Improved tokamak concept focusing on easy maintenance, Fusion Eng. Des. 41 (1) (1998) 357–364.

[11] P. Norajitra, L. Bühler, U. Fischer, K. Kleefeldt, S. Malang, G. Reimann, H. Schnauder, L. Giancarli, H. Golfier, Y. Poitevin, J.F. Salavy, The EU advanced lead lithium blanket concept using SiCf/SiC flow channel inserts as electrical and thermal insulators, Fusion Eng. Des. 58–59 (2001) 629–634.

[12] D.-K. Sze, M. Tillack, L. El-Guebaly, Blanket system selection for the ARIES-ST, Fusion Eng. Des. 48 (3–4) (2000) 371–378.

[13] A.F. Rowcliffe, L.M. Garrison, Y. Yamamoto, L. Tan, Y. Katoh, Materials challenges for the fusion nuclear science facility, Fusion Eng. Des. 135 (2018) 290–301.

[14] Y. Huang, M.S. Tillack, N.M. Ghoniem, Tungsten monoblock concepts for the fusion nuclear science facility (FNSF) first wall and divertor, Fusion Eng. Des. 135 (2017) 346–355

[15] R. Boullon, J.-C. Jaboulay, J. Aubert, Molten salt breeding blanket: Investigations and proposals of pre-conceptual design options for testing in DEMO, Fusion Eng. Des. 171 (2021), 112707-112707.
[16] L.L. Snead, T. Nozawa, M. Ferraris, Y. Katoh, R. Shinavski, M. Sawan, Silicon carbide composites as fusion power reactor structural materials, J. Nucl. Mater. (2011) 330–339.
[17] C. Sauder, Ceramic matrix composites: Nuclear applications, ceramic matrix composites, John Wiley & Sons, Inc., Hoboken, NJ, USA, 2014, pp. 609–646.
[18] J.J. Lee, S.S. Raiman, Y. Katoh, T. Koyanagi, C.I. Contescu, X. Hu, Y. Yang, Chemical compatibility of silicon carbide in molten fluoride salts for the fluoride salt-cooled high temperature reactor, J. Nucl. Mater. 524 (2019) 119–134.
[19] K.A. Terrani, Accident tolerant fuel cladding development: Promise, status, and challenges, J. Nucl. Mater. (2018) 13–30.
[20] G. Newsome, L.L. Snead, T. Hinoki, Y. Katoh, D. Peters, Evaluation of neutron irradiated silicon carbide and silicon carbide composites, J. Nucl. Mater. 371 (1–3) (2007) 76–89.
[21] A.G. Evans, F.W. Zok, The physics and mechanics of fibre-reinforced brittle matrix composites, J. Mater. Sci. 29 (15) (1994) 3857–3896.
[22] J. Lamon, *Chemical vapor infiltrated SiC/SiC composites (CVI SiC/SiC), handbook of ceramic composites*, Springer US2005, pp. 55-76.
[23] Composite materials handbook volume 5, SAE International, 2017.
[24] BPVC.III.NCA - *BPVC section III-rules for constructions of nuclear facility components-subsection NCA-General Requirements for Division 1 and Division* 2, ASME2025.
[25] H.L. Heinisch, L.R. Greenwood, W.J. Weber, R.E. Williford, Displacement damage in silicon carbide irradiated in fission reactors, J. Nucl. Mater. 327 (2–3) (2004) 175–181.
[26] L.L. Snead, T. Nozawa, Y. Katoh, T.S. Byun, S. Kondo, D.A. Petti, Handbook of SiC properties for fuel performance modeling, J. Nucl. Mater. 371 (1–3) (2007) 329–377.
[27] T. Koyanagi, Y. Katoh, K. Ozawa, K. Shimoda, T. Hinoki, L.L. Snead, Neutron-irradiation creep of silicon carbide materials beyond the initial transient, J. Nucl. Mater. 478 (2016) 97–111.
[28] Y. Katoh, T. Koyanagi, J.L. McDuffee, L.L. Snead, K. Yueh, Dimensional stability and anisotropy of SiC and SiC-based composites in transition swelling regime, J. Nucl. Mater. 499 (2018) 471–479.
[29] D.J. Sprouster, T. Koyanagi, E. Dooryhee, S.K. Ghose, Y. Katoh, L.E. Ecker, Microstructural evolution of neutron irradiated 3C-SiC, Scr. Mater. 137 (2017) 132–136.
[30] L.L. Snead, Y. Katoh, T. Koyanagi, K. Terrani, E.D. Specht, Dimensional isotropy of 6H and 3C SiC under neutron irradiation, J. Nucl. Mater. 471 (2016) 92–96.
[31] D.J. Senor, G.E. Youngblood, L.R. Greenwood, D.V. Archer, D.L. Alexander, M. C. Chen, G.A. Newsome, Defect structure and evolution in silicon carbide irradiated to 1 dpa-SiC at 1100 °C, J. Nucl. Mater. 317 (2–3) (2003) 145–159.
[32] S. Kondo, C.M. Parish, T. Koyanagi, Y. Katoh, Equilibrium shapes and surface selection of nanostructures in 6H-SiC, Appl. Phys. Lett. 110 (14) (2017).
[33] S. Kondo, Y. Katoh, L.L. Snead, Microstructural defects in SiC neutron irradiated at very high temperatures, J. Nucl. Mater. 382 (2–3) (2008) 160–169.
[34] J. Chen, P. Jung, H. Trinkaus, Evolution of helium platelets and associated dislocation loops in αSiC, Phys. Rev. Lett. 82 (13) (1999) 2709–2712.
[35] J.A. DiCarlo, H.-M. Yun, *Non-oxide (Silicon Carbide) Fibers, Handbook of Ceramic Composites*, Springer US2005, pp. 33-52.

[36] G.E. Youngblood, R.H. Jones, A. Kohyama, L.L. Snead, Radiation response of SiC-based fibers, J. Nucl. Mater. 258–263 (1998) 1551–1556.
[37] T. Hinoki, L.L. Snead, Y. Katoh, A. Hasegawa, T. Nozawa, A. Kohyama, The effect of high dose/high temperature irradiation on high purity fibers and their silicon carbide composites, J. Nucl. Mater. 307–311 (2002) 1157–1162.
[38] Y. Katoh, T. Nozawa, L.L. Snead, K. Ozawa, H. Tanigawa, Stability of SiC and its composites at high neutron fluence, J. Nucl. Mater. 417 (1–3) (2011) 400–405.
[39] K. Ozawa, Y. Katoh, L.L. Snead, T. Nozawa, T. Hinoki, A. Hasegawa, The effects of neutron irradiation on dimensional change in advanced silicon carbide composites, 2009, pp. 4–15.
[40] L.L. Snead, K.A. Terrani, Y. Katoh, C. Silva, K.J. Leonard, A.G. Perez-Bergquist, Stability of SiC-matrix microencapsulated fuel constituents at relevant LWR conditions, J. Nucl. Mater. 448 (1–3) (2014) 389–398.
[41] T. Iseki, T. Maruyama, T. Yano, T. Suzuki, T. Mori, Effects of neutron irradiation and subsequent annealing on strength and toughness of SiC ceramics, J. Nucl. Mater. 170 (1) (1990) 95–100.
[42] T. Koyanagi, K. Terrani, S. Harrison, J. Liu, Y. Katoh, Additive manufacturing of silicon carbide for nuclear applications, J. Nucl. Mater. 543 (2021).
[43] Y.-M. Chiang, R.P. Messner, C.D. Terwilliger, D.R. Behrendt, Reaction-formed silicon carbide, Mater. Sci. Eng. 144 (1–2) (1991) 63–74.
[44] R.B. Matthews, Irradiation damage in reaction-bonded silicon carbide, J. Nucl. Mater. 51 (2) (1974) 203–208.
[45] J.C. Corelli, J. Hoole, J. Lazzaro, C.W. Lee, Mechanical, thermal, and microstructural properties of neutron-irradiated SiC, J. Am. Ceram. Soc. 66 (7) (1983) 529–538.
[46] T. Koyanagi, K. Ozawa, T. Hinoki, K. Shimoda, Y. Katoh, Effects of neutron irradiation on mechanical properties of silicon carbide composites fabricated by nano-infiltration and transient eutectic-phase precess, J. Nucl. Mater. 448 (1–3) (2014) 478–486.
[47] K.A. Terrani, C. Ang, L.L. Snead, Y. Katoh, Irradiation stability and thermo-mechanical properties of NITE-SiC irradiated to 10 dpa, J. Nucl. Mater. 499 (2018) 242–247.
[48] Y. Katoh, L.L. Snead, T. Nozawa, S. Kondo, J.T. Busby, Thermophysical and mechanical properties of near-stoichiometric fiber CVI SiC/SiC composites after neutron irradiation at elevated temperatures, J. Nucl. Mater. 403 (1–3) (2010) 48–61.
[49] T. Nozawa, T. Hinoki, L.L. Snead, Y. Katoh, A. Kohyama, Neutron irradiation effects on high-crystallinity and near-stoichiometry SiC fibers and their composites, J. Nucl. Mater. 329–333 (2004) 544–548.
[50] M.C. Osborne, C.R. Hubbard, L.L. Snead, D. Steiner, Neutron irradiation effects on the density, tensile properties and microstructural changes in Hi-Nicalon™ and sylramic™ SiC fibers, J. Nucl. Mater. 253 (1–3) (1998) 67–77.
[51] A. Noviyanto, D.-H. Yoon, Metal oxide additives for the sintering of silicon carbide: Reactivity and densification, Curr. Appl. Phys. 13 (1) (2013) 287–292.
[52] C. Sauder, A. Brusson, J. Lamon, Influence of interface characteristics on the mechanical properties of Hi-Nicalon type-S or Tyranno-SA3 fiber-reinforced SiC/SiC mini-composites, Int. J. Appl. Ceram. Technol. 7 (3) (2010) 291–303.
[53] H.-S. Kim, Y.-W. Kim, Thermal conductivity of liquid-phase sintered silicon carbide ceramics: A review, J. Eur. Ceram. Soc. 43 (9) (2023) 3855–3874.
[54] L.L. Snead, S.J. Zinkle, D.P. White, Thermal conductivity degradation of ceramic materials due to low temperature, low dose neutron irradiation, J. Nucl. Mater. 340 (2–3) (2005) 187–202.

[55] Y. Katoh, K. Ozawa, C. Shih, T. Nozawa, R.J. Shinavski, A. Hasegawa, L.L. Snead, Continuous SiC fiber, CVI SiC matrix composites for nuclear applications: Properties and irradiation effects, J. Nucl. Mater. 448 (1–3) (2014) 448–476.
[56] Y. Katoh, K. Ozawa, T. Hinoki, Y. Choi, L.L. Snead, A. Hasegawa, Mechanical properties of advanced SiC fiber composites irradiated at very high temperatures, J. Nucl. Mater. 417 (1–3) (2011) 416–420.
[57] T. Nozawa, T. Koyanagi, Y. Katoh, H. Tanigawa, High-dose, intermediate-temperature neutron irradiation effects on silicon carbide composites with varied fiber/matrix interfaces, J. Eur. Ceram. Soc. 39 (8) (2019) 2634–2647.
[58] Y. Katoh, T. Nozawa, C. Shih, K. Ozawa, T. Koyanagi, W. Porter, L.L. Snead, High-dose neutron irradiation of Hi-Nicalon type S silicon carbide composites. Part 2: Mechanical and physical properties, J. Nucl. Mater. 462 (2015) 450–457.
[59] T. Nozawa, T. Koyanagi, Y. Katoh, H. Tanigawa, Failure evaluation of neutron-irradiated SiC/SiC composites by underwater acoustic emission, J. Nucl. Mater. 566 (2022), 153787-153787.
[60] M.I. Idris, H. Konishi, M. Imai, K. Yoshida, T. Yano, Neutron irradiation swelling of SiC and SiCf/SiC for advanced nuclear applications, Energy Proc. 71 (2015) 328–336.
[61] R.T. Bhatt, J.D. Kiser, Creep behavior and failure mechanisms of CVI and PIP SiC/SiC composites at temperatures to 1650 °C in air, J. Eur. Ceram. Soc. 41 (13) (2021) 6196–6206.
[62] Y. Wei, F. Ye, L. Cheng, L. Zhang, J. Yang, Effects of in-situ carbon layer and volume fraction of SiC fiber on the mechanical properties and fracture behaviors of SiCf/SiC composites fabricated by a modified PIP method, J. Eur. Ceram. Soc. 43 (4) (2023) 1397–1406.
[63] G.N. Morscher, V.V. Pujar, Design guidelines for in-plane mechanical properties of SiC fiber-reinforced melt-infiltrated SiC composites, Int. J. Appl. Ceram. Technol. 6 (2) (2009) 151–163.
[64] T. Nozawa, T. Hinoki, Y. Katoh, A. Kohyama, Effects of fibers and fabrication processes on mechanical properties of neutron irradiated SiC/SiC composites, J. Nucl. Mater. 307–311 (2002) 1173–1177.
[65] J.B.J. Hegeman, J.G. van der Laan, M. van Kranenburg, M. Jong, D. d'Hulst, P. ten Pierick, Mechanical and thermal properties of SiCf/SiC composites irradiated with neutrons at high temperatures, Fusion Eng. Des. 75–79 (2005) 789–793.
[66] W.A. Curtin, B.K. Ahn, N. Takeda, Modeling brittle and tough stress–strain behavior in unidirectional ceramic matrix composites, Acta Mater. 46 (10) (1998) 3409–3420.
[67] Y. Katoh, T. Nozawa, L.L. Snead, T. Hinoki, A. Kohyama, Property tailorability for advanced CVI silicon carbide composites for fusion, Fusion Eng. Des. 81 (8–14) (2006) 937–944.
[68] C.S. Lynch, A.G. Evans, Effects of off-axis loading on the tensile behavior of a ceramic-matrix composite, J. Am. Ceram. Soc. 79 (12) (2005) 3113–3123.
[69] C. Sauder, J. Lamon, Tensile creep behavior of SiC-Based fibers with a low oxygen content, J. Am. Ceram. Soc. 90 (4) (2007) 1146–1156.
[70] H. Gietl, O. Karakoc, T. Koyanagi, S. Harrison, Y. Katoh, Emerging continuous SiC fibers for high-temperature applications, Ceram. Int. 50 (18) (2024) 32893–32904.
[71] J. Braun, C. Sauder, Mechanical behavior of SiC/SiC composites reinforced with new tyranno SA4 fibers: Effect of interphase thickness and comparison with tyranno SA3 and Hi-Nicalon S reinforced composites, J. Nucl. Mater. 558 (2022).
[72] R.E. Jones, D. Petrak, J. Rabe, A. Szweda, SYLRAMIC™ SiC fibers for CMC reinforcement, J. Nucl. Mater. 283–287 (2000) 556–559.

[73] G. Chollon, R. Naslain, C. Prentice, R. Shatwell, P. May, High temperature properties of SiC and diamond CVD-Monofilaments, J. Eur. Ceram. Soc. 25 (11) (2005) 1929–1942.
[74] T. Koyanagi, Y. Katoh, Mechanical properties of SiC composites neutron irradiated under light water reactor relevant temperature and dose conditions, J. Nucl. Mater. 494 (2017) 46–54.
[75] F.T. Wallenberger, P.C. Nordine, M. Boman, Inorganic fibers and microstructures directly from the vapor phase, Compos. Sci. Technol. 51 (2) (1994) 193–212.
[76] S. Wu, Y.Z. Gou, Y. Xiang, W.F. Kang, Y.S. Wang, Q.Z. Song, N.N. Xu, Y.D. Wang, Effect of long-time annealing at high temperature on the microstructure and mechanical properties of different types of SiC fibers, Compos Part A-Appl S 185 (2024).
[77] K. Shimoda, H. Kakisawa, Preparation of enhanced erosive wear resistance SiC-fiber reinforced SiC ceramic matrix composites integrated with a knit fabric via high temperature crystallization of amorphous Si–C–O–Al fibers, Ceram. Int. 51 (2025) 6206–6219.
[78] U. Santoro, E. Novitskaya, K. Karandikar, H.E. Khalifa, O.A. Graeve, Phase stability of SiC/SiC fiber reinforced composites: The effect of processing on the formation of α and β phases, Mater. Lett. 241 (2019) 123–127.
[79] Q. Zhang, H. Liu, T. Qiao, P.J. Withers, P. Xiao, Simple and efficient densification of SiCf/SiC composites by gradated concentration polymer infiltration and pyrolysis, Mater. Sci. Eng. 874 (2023).
[80] L. Shen, W. Yan, T. Zhu, H. Li, C. Xiao, J. Xu, Analysis of multi-scale failure behavior of SiCf/SiC composite clad tube under flexible clamping damage, J. Eur. Ceram. Soc. 44 (11) (2024) 6269–6285.
[81] S. Suyama, T. Kameda, Y. Itoh, Development of high-strength reaction-sintered silicon carbide, Diam. Relat. Mater. 12 (3–7) (2003) 1201–1204.
[82] J. Magnant, R. Pailler, Y. Le Petitcorps, L. Maillé, A. Guette, J. Marthe, E. Philippe, Fiber-reinforced ceramic matrix composites processed by a hybrid technique based on chemical vapor infiltration, slurry impregnation and spark plasma sintering, J. Eur. Ceram. Soc. 33 (1) (2013) 181–190.
[83] T. Aoki, T. Ogasawara, Y. Okubo, K. Yoshida, T. Yano, Fabrication and properties of Si–Hf alloy melt-infiltrated tyranno ZMI fiber/SiC-based matrix composites, Compos. Appl. Sci. Manuf. 66 (2014) 155–162.
[84] Y. Katoh, T. Nozawa, L.L. Snead, Mechanical properties of thin pyrolitic carbon interphase SiC-matrix composites reinforced with near-stoichiometric SiC fibers, J. Am. Ceram. Soc. 88 (11) (2005) 3088–3095.
[85] A.G. Perez-Bergquist, T. Nozawa, C. Shih, K.J. Leonard, L.L. Snead, Y. Katoh, High dose neutron irradiation of Hi-Nicalon type S silicon carbide composites, part 1: Microstructural evaluations, J. Nucl. Mater. 462 (2015) 443–449.
[86] G.N. Morscher, H.M. Yun, J.A. DiCarlo, L. Thomas-Ogbuji, Effect of a boron nitride interphase that debonds between the interphase and the matrix in SiC/SiC composites, J. Am. Ceram. Soc. 87 (1) (2008) 104–112.
[87] S. Jacques, A. Lopez-Marure, C. Vincent, H. Vincent, J. Bouix, SiC/SiC Minicomposites with structure-graded BN interphases, J. Eur. Ceram. Soc. 20 (12) (2000) 1929–1938.
[88] H. Li, J. Lee, M.R. Libera, W.Y. Lee, A. Kebbede, M.J. Lance, H. Wang, G. N. Morscher, Morphological evolution and weak interface development within chemical-vapor-deposited zirconia coating deposited on Hi-Nicalon™ fiber, J. Am. Ceram. Soc. 85 (6) (2002) 1561–1568.
[89] A.V. Utkin, A.A. Matvienko, A.T. Titov, N.I. Baklanova, Multiple zirconia interphase for SiC/SiCf composites, Surf. Coating. Technol. 205 (8–9) (2011) 2724–2729.

[90] H.G. Lee, D. Kim, J.Y. Park, W.J. Kim, Formation of Ti 3 SiC 2 interphase coating on SiC f/SiC composite by electrophoretic deposition, Int. J. Appl. Ceram. Technol. 15 (3) (2018) 602–610.

[91] S. Jacques, I. Jouanny, O. Ledain, L. Maillé, P. Weisbecker, Nanoscale multilayered and porous carbide interphases prepared by pressure-pulsed reactive chemical vapor deposition for ceramic matrix composites, Appl. Surf. Sci. 275 (2013) 102–109.

[92] H.E. Eaton, G.D. Linsey, Accelerated oxidation of SiC CMC's by water vapor and protection via environmental barrier coating approach, J. Eur. Ceram. Soc. 22 (14–15) (2002) 2741–2747.

[93] C.P. Deck, G.M. Jacobsen, J. Sheeder, O. Gutierrez, J. Zhang, J. Stone, H.E. Khalifa, C. A. Back, Characterization of SiC-SiC composites for accident tolerant fuel cladding, J. Nucl. Mater. 466 (2015) 1–15.

[94] J.D. Arregui-Mena, T. Koyanagi, E. Cakmak, C.M. Petrie, W.-J. Kim, D. Kim, C. P. Deck, C. Sauder, J. Braun, Y. Katoh, Qualitative and quantitative analysis of neutron irradiation effects in SiC/SiC composites using X-ray computed tomography, Compos. B Eng. 238 (2022), 109896-109896.

[95] J.D. Stempien, D.M. Carpenter, G. Kohse, M.S. Kazimi, Characteristics of composite silicon carbide fuel cladding after irradiation under simulated PWR conditions, Nucl. Technol. 183 (1) (2013) 13–29.

[96] D. Kim, H.G. Lee, J.Y. Park, W.J. Kim, Fabrication and measurement of hoop strength of SiC triplex tube for nuclear fuel cladding applications, J. Nucl. Mater. 458 (2015) 29–36.

[97] Y.-H. Koo, J.-H. Yang, J.-Y. Park, K.-S. Kim, H.-G. Kim, D.-J. Kim, Y.-I. Jung, K.-W. Song, KAERI's development of LWR accident-tolerant fuel, Nucl. Technol. 186 (2) (2014) 295–304.

[98] C. Ang, C. Kemery, Y. Katoh, Electroplating chromium on CVD SiC and SiCf-SiC advanced cladding via PyC compatibility coating, J. Nucl. Mater. 503 (2018) 245–249.

[99] Y. Katoh, L.L. Snead, T. Cheng, C. Shih, W.D. Lewis, T. Koyanagi, T. Hinoki, C. H. Henager Jr., M. Ferraris, Radiation-tolerant joining technologies for silicon carbide ceramics and composites, J. Nucl. Mater. 448 (1–3) (2014) 497–511.

[100] P.A. Mouche, T. Koyanagi, D. Patel, Y. Katoh, Adhesion, structure, and mechanical properties of Cr HiPIMS and cathodic arc deposited coatings on SiC, Surf. Coating. Technol. 410 (2021), 126939-126939.

[101] T. Koyanagi, C. Hawkins, B. Lamm, E. Lara-Curzio, C. Deck, Y. Katoh, Mechanical degradation of duplex SiC-fiber reinforced SiC matrix composite tubes under a controlled high-temperature steam environment, Ceram. Int. 50 (2024) 25558–25567.

[102] M. Ben-Belgacem, V. Richet, K.A. Terrani, Y. Katoh, L.L. Snead, Thermo-mechanical analysis of LWR SiC/SiC composite cladding, J. Nucl. Mater. 447 (1–3) (2014) 125–142.

[103] S. Yilmaz, M. Theodore, S. Ozcan, Silicon carbide fiber manufacturing: Cost and technology, Compos. B Eng. 269 (2024).

[104] P. Parandoush, D. Lin, A review on additive manufacturing of polymer-fiber composites, Compos. Struct. 182 (2017) 36–53.

[105] K.A. Terrani, Y. Yang, Y.J. Kim, R. Rebak, H.M. Meyer, T.J. Gerczak, Hydrothermal corrosion of SiC in LWR coolant environments in the absence of irradiation, J. Nucl. Mater. 465 (2015) 488–498.

[106] T. Koyanagi, X. Hu, C.M. Petrie, G. Singh, C. Ang, C.P. Deck, W.-J. Kim, D. Kim, C. Sauder, J. Braun, Y. Katoh, Hermeticity of SiC/SiC composite and monolithic SiC tubes irradiated under radial high-heat flux, J. Nucl. Mater. 588 (2024).

Nuclear graphite

5

Dong Liu
Department of Engineering Science, University of Oxford, Oxford, United Kingdom

1. Introduction

Among the many nuclear materials, nuclear graphite is one of the most fascinating structural materials due to its complex nano-/micro-scale structures, the ability to withstand load at extremely high temperatures and high irradiation doses (fast neutrons) [1,2].

In particular, nuclear graphite is used to construct critical structural core components (e.g., neutron moderator, fuel element, reflectors, core support posts and so on [3]) in historical reactors (e.g., the very first artificial nuclear reactor Chicago Pile-1 and the following Clinton Pile), current operating fleets (e.g., the UK gas-cooled AGRs, Russia water-cooled RBMK reactors) and GenIV designs with more extra safety considerations (e.g., prismatic or pebble-bed HTRs). As a matter of fact, these high-temperature graphite-moderated, gas-cooled reactors (HTGR), either in the form of regular size or small modular reactor (SMR), have become one of the favorite designs in the largest number of countries in the EU and worldwide to develop and construct. The prototypes of HTGRs are already built in China (HTR-10 and HTR-PM) by licensing the German design (AVR) and in Japan (HTTR with an outlet temperature of 950°C). The high demand for HTRs in the EU is also reflected in the ongoing large programmes with the US, UK and Japan (e.g., GEMINI and GEMINI + on the design and regulatory framework of the first modern HTGR). In these HTRs, the volume of graphite to be used is considerable, for example, a typical graphite moderated regular size reactor may contain over 1000 t of graphite [2,4].

During operation, the graphite components will be subjected to irradiation damage in a gas (usually Helium) or molten salt environment. This will cause changes in its structure, dimensions, and physical and mechanical properties, which will be discussed in this chapter. Note the UK AGRs used CO_2 gas as a coolant, which introduced a unique working environment for graphite, causing radiolytic oxidation. A consequence of this is the weight loss of the graphite core component with service, introducing another variable in the prediction and understanding of graphite in-service performance. More reading about radiolytic oxidation can be found in Refs. [5–10]; it will not be discussed in detail in the current chapter.

In addition to the industrial imperative, nuclear graphite is a scientifically interesting material due to its complex nano- and microstructure and properties ranging over multiple length-scales changing with irradiation. In this chapter, three main aspects, including its nano-crystalline structure, irradiation-induced dimensional and

Ceramics for Nuclear Applications. https://doi.org/10.1016/B978-0-12-823865-3.00008-0

property changes, will be presented in terms of present understanding and existing challenges.

2. Manufacture and structure of nuclear graphite

Perfect graphite is a stack of graphene sheets, in which the carbon atoms in the basal planes are bound by sp^2 hybridization in a planar network to three equidistant nearest neighbors, 120 degrees apart in a given plane, to form the hexagonal graphene structure. Strong in-plane covalent double bonds of both σ-type and π-type are present, but the sheets are only weakly bound with van der Waals bonds with an interlayer spacing of 3.35 Å in an ABAB stacking sequence. This lattice structure results in highly anisotropic crystal properties [6,11], for example, the coefficient of thermal expansion (CTE) is ~$27.5 \times 10^{-6}\ K^{-1}$ (20–120°C) along the '*c*'-axis, but only $-1.5 \times 10^{-6}\ K^{-1}$ for the '*a*'-axis. The elastic moduli parallel and perpendicular to the basal planes are $C_{11} = 10^6 \times 10^{10}\ N\ m^{-2}$ and $C_{33} = 3.46 \times 10^{10}\ N\ m^{-2}$, respectively, with a basal plane shear modulus of $C_{44} = 0.45 \times 10^{10}\ N\ m^{-2}$. In nuclear-grade graphite, the layered structure exists at much shorter length scales, being surrounded by pores and cracks [12,13]. Nevertheless, the extreme properties parallel and perpendicular to the basal planes strongly influence the nano-/macro-scale properties of polycrystalline graphite composites, both before and after irradiation [14–16].

To manufacture nuclear-grade graphite, a coke material, typically petroleum or pitch-based, is crushed and graded into the required particle size, hot-mixed with a pitch binder, and formed into billets known as the 'green-article'. As the pitch-binder is mixed and cooled, it also passes through a mesophase forming graphitizable structures to bind the filler coke particles together [11]. Gas evolved during a subsequent baking stage (~800°C) leads to the generation of further porosity within the binder phase. To improve the final properties, there may be subsequent impregnations under vacuum using a lighter liquid pitch, followed by rebaking. The final step is graphitization, which is conducted at around 2700°C or above. The resulting product is a porous polycrystalline artifact, and the filler tends to graphitize easier than the binder, which is often a mixture of complex chemicals with different degrees of a graphitization at a certain temperature. A brief schematic is presented in Fig. 5.1A to illustrate the typical manufacturing procedure for nuclear graphite. More in-depth information and knowledge can be found in Refs. [3,17].

There are many different grades of graphite manufactured by varying the starting raw materials. One of the very well-known and well-studied graphite materials is the Gilsocarbon used in the UK advanced gas-cooled reactors (AGRs); and this material will be used as a typical example throughout this chapter for comparison with other grades. Gilsocarbon nuclear graphite is an IM1-24 (GCMB grade) polygranular nuclear graphite comprising spherical coke filler particles and a binder matrix. It has a complex structure with pores ranging from nano-size to macro-scale, Fig. 5.1 illustrates its typical structure at different length-scales. Gilsocarbon graphite was manufactured by GrafTech France S.N.C. (Rue des Garennes, France) and during

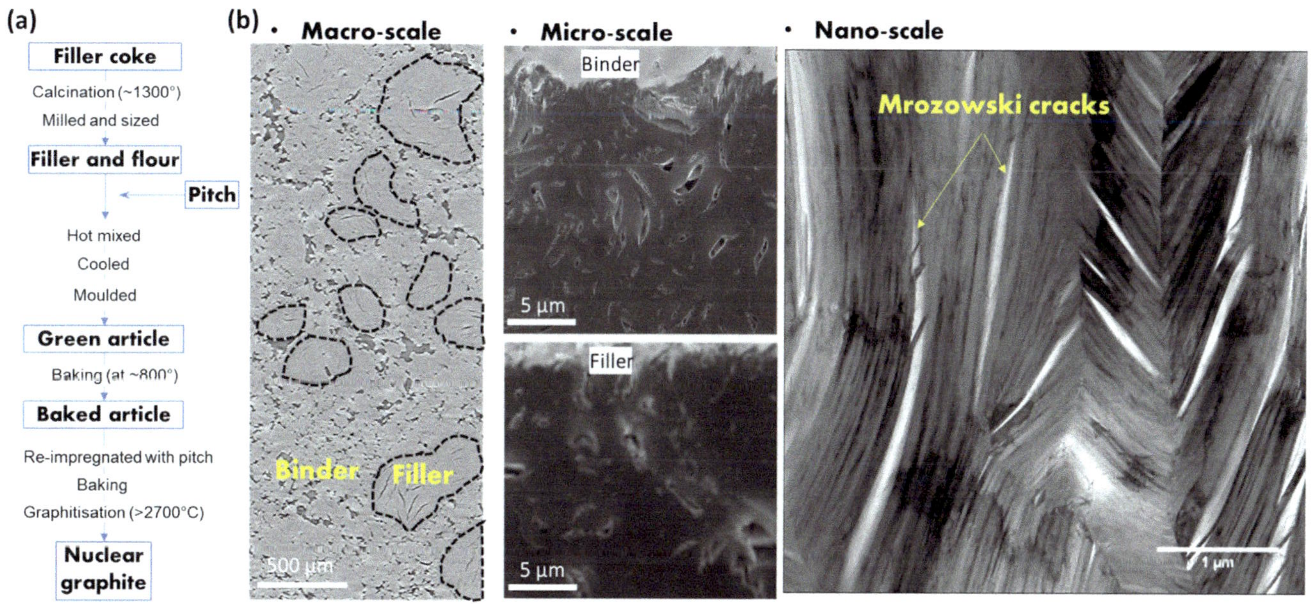

Figure 5.1 (A) Schematic showing the typical procedure used in the manufacturing of synthetic nuclear graphite. The actual number of steps and temperature adopted for a particular graphite may vary depending on the manufacturers and raw materials; (B) using Gilsocarbon graphite as a typical example showing the multiple length-scale structure, including filler particles, binder phase, macro-/micro-/nano-size porosity. A larger area (macro-scale) was imaged by X-ray tomography, showing the random distribution of fillers (some of them are circled by dashed lines) in the binder phase; it is this random mix of filler and binder that results in relatively isotropic bulk properties in Gilsocarbon graphite. The microscale images were taken by a scanning electron microscope from a focused ion beam milled cross-section in the filler and binder, respectively, showing the microsize pores. The last image at the nano-scale is captured by transmission electron microscopy in bright field mode, demonstrating the lenticular Mrozowski cracks formed between the basal planes in the as-manufactured condition.

manufacture, about 30 parts of binder are usually used with every 100 parts of filler particles (by wt.). They are hot-mixed when the binder is in a liquid form (usually at ~100–150°C, depending on the pitch type in the binder) ideally to ensure every one of the filler particles is coated with a layer of binder and is bonded together to form a macrosize graphite block. A relatively uniform distribution of filler particles is achieved in Gilsocarbon graphite by vibrational molding. As a result, when the strength or CTE is measured in samples machined along versus normal to the molding direction, an anisotropy ratio of 1:1.10 is generally seen, indicating the macroscopic Gilsocarbon material is effectively isotropic with respect to mechanical properties such as elastic modulus (10–11 GPa) and flexural strength (~26 MPa) [11,18,19]. More detailed descriptions of the porous structure of graphite can be found elsewhere [9,20–22].

In many graphite grades, finer flour of the coke is also mixed with the binder, and the final product comprises filler particles, pitch binder and filler flour. This is also the case for Gilsocarbon, though the exact amount of flour added to the binder is not readily known. Depending on the size of particles used in the starting mix, the final graphite product can be classified as 'coarse-grained' (particle size >4 mm), 'medium-grained' (particle size <4 mm) or 'fine-grained' (particle size <100 μm) according to the ASTM D7219 and C709-925 standards. Nuclear graphite generally falls into one of the latter two definitions; for example, the above-mentioned UK AGR Gilsocarbon graphite is medium-grained, and the primary graphite of interest to GenIV reactors is fine-grained (note the 'grain' used here is not describing the crystal structure; it is a term for the particle size in the initial mix used to make synthetic graphite).

As a result of complex raw materials and manufacturing processes adopted, long-range order in the 'graphene'-like structure is present only for tens or hundreds of nanometers which is at least one order of magnitude less than a (Highly oriented pyrolytic graphite (HOPG). As reported by Liu et al. [23] in a transmission electron microscope (TEM) study about the Gilsocarbon graphite, within each of these 'graphite'-like regions, the material consists of single crystal platelets, typically 1–2 nm thick, stacked with large relative rotations around the c-axis; individual platelets have both hexagonal and cubic stacking, as shown in Fig. 5.2. Upon cooling from graphitization temperature, various types of cracks form; one type, termed 'Mrozowski' cracks [24] (see the nano-scale graph in Fig. 5.1B) run along the basal planes as the *c*-direction CTE in graphite crystals is ~20 times more than in the *a*- direction. Liu et al. [23] identified that there are other types of voids present at the nano-scale due to crystallographic requirements; this may have a potential impact on the irradiation behavior of the graphite material [23] (Figs. 5.2A and B).

One length-scale up, many such nano-platelet crystals form a 'grain' in graphite [13,25–27]. This is a very different case from metals. For example, each of the regions segmented by the yellow dotted lines in Fig. 5.2A is considered to be a 'grain'. Note this is different from the 'grain' describing the size of the raw material. Within such grains, between layers of nano-platelet, the aforementioned 'Mrozowski' cracks form and are part of these grains [27–32]. However, the complexity of graphite's structure is not simply this. The grains form clusters, or 'domains', that are separated by larger

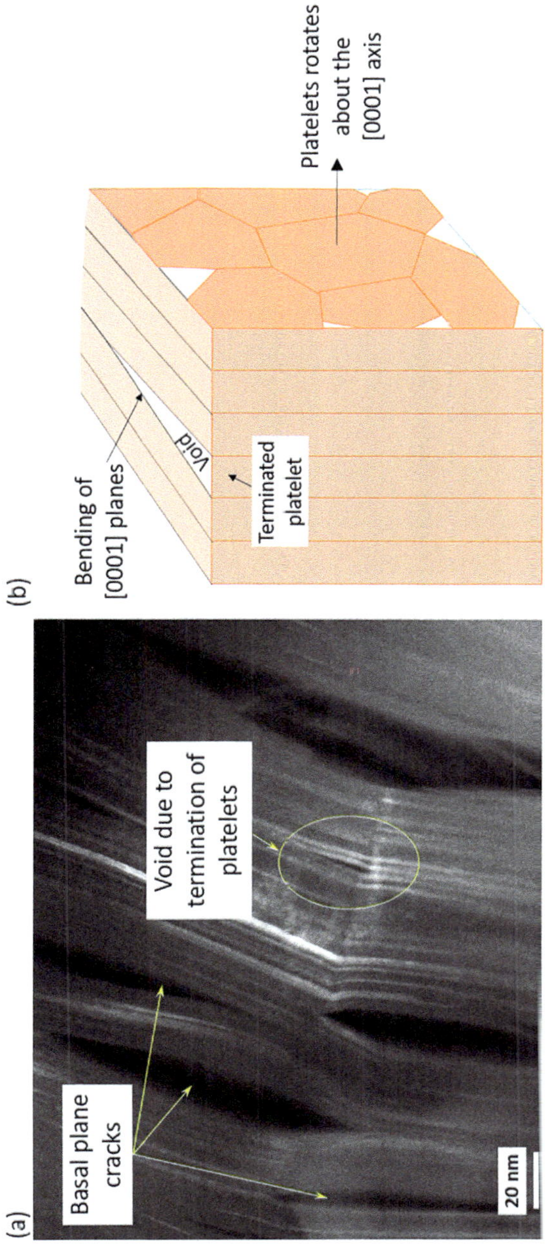

Figure 5.2 (A) Shows the termination of a platelet and the formation of a void as a consequence; (B) a schematic describing this class of nano-voids due to the crystallographic requirement as the [0001] planes bend; the mosaic structure of individual platelets is also illustrated.

nano-pores (marked as 'P' in Fig. 5.3A and these nano-pores have geometries and origins that are different from Mrozowski cracks). These domains form many types of features in nuclear graphite, *e.g.*, well-graphitised structure, nano-sized graphite particles (sometimes referred to as 'rosettes'), quinoline insoluble particles, chaotic structures and nongraphitizing carbon [28]. All these features, together with their complex nanostructures, are all part of the filler particles or the binder phase.

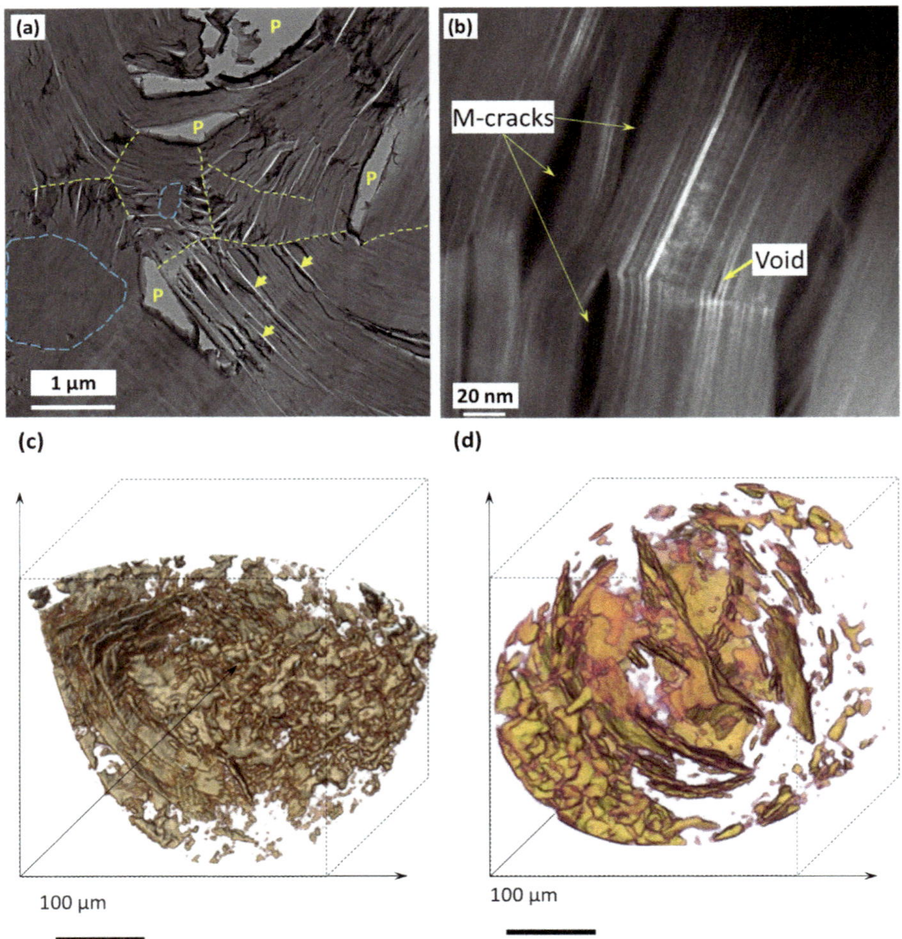

Figure 5.3 (A) A TEM image showing the typical structural features in graphite including pores (P), basal planes edge-on and tilted away (blue circled regions), the formation of nano-cracks along the basal planes, the grain boundaries (doted yellow lines) (B) higher magnification of an area with the basal planes facing edge-on showing the nano-cracks and the stacked platelet crystals in between; (C) and (D) are two 3D tomography images showing the pore structure of Gilsocarbon filler particles where lenticular micro-size Mrozowski cracks distributed circumferentially in a filler particle.

At macro-scale (millimeter or centimetre), macro-pores (either closed or open) are present inside and between filler and binder. In particular, the filler particles (Fig. 5.3C and D) have a crystallite c-axis that is distributed radially outwards from the particle center, and thus the Mrozowski cracks appear radially distributed between the individual folded graphitic sheets (micro-size Mrozowski cracks in the filler particle are shown in the schematic). In terms of the binder, usually spherical pores form, as can be seen from Fig. 5.1B. It is considered that these pores are typically pore-runs resulting from the bubble percolation of volatile gases during the baking stage (~800°C), and micrometer-sized pores formed during the impregnation of liquid pitch.

There is a large amount of research effort in characterizing and modeling porosity in graphite; more in-depth analysis can be found in Refs. [20,33–36]. This is because the behavior of porosity is considered to impact the physical and mechanical properties of graphite as a function of temperature and irradiation dose. For instance, both nano- and micro-scale Mrozowski cracks close up upon heating and with irradiation, and thus can impact the strength and toughness of the graphite material. Most importantly, the closure of these cracks helps to accommodate the irradiation-induced volumetric change along the c-direction.

It is an open question in the area regarding the 3D structure/geometry/network of the nano-crystallites/nano-cracks assembly, as well as the structure at the grain boundaries. There is very little open data on the 3D shape/dimension of the Mrozowski cracks and their population in a meso-scale volume, say a few hundred micrometers. As these cracks are very small, a very high-resolution technique (e.g., transmission electron microscopy) is needed. A recent attempt using TEM tomography was conducted by Arregui-Mena et al. [37], and they imaged a 3D pore, for the first time, in IG-110 graphite (unirradiated and irradiated). Future work is needed to image a larger number of these pores, especially their 3D distribution and surrounding features. As mentioned above, these nano-pores are critical to understand the mechanisms controlling the physical and mechanical properties/behaviors of the graphite with neutron irradiation. The irradiation induced structural and property changes will be elaborated in Section 3.

3. Irradiation-induced structural changes

When graphite components are irradiated in a reactor, significant changes to their dimensions and properties occur. Varying the starting materials, such as the coke and binder, and processing parameters directly impacts the evolution of dimensions and properties with irradiation dose. While the binder is often considered to have a more randomly oriented crystallites/grains, the filler particles and their orientation with respect to the graphite component are thought to be responsible for the anisotropy in graphite dimensions and properties after irradiation [38,39]. The cracks and porosity, depending on their type and relative orientation, will function to 'alleviate' the degree of anisotropy in the bulk material. However, the exact physics behind the irradiation-induced dimensional changes, irradiation anisotropy and creep, and how irradiation damage is transferred from crystallite/grains to fillers/binders to the change in bulk properties is still not well understood.

As a consequence, when new grades of nuclear graphite are put forward, a new irradiation program will be necessary. A modern example of this is the US Next-Generation Nuclear Plant (NGNP) program for graphite selection and qualification. For the NGNP Graphite Technology Development Program (TDP), several candidate graphites were considered, including IG-430 (Toyo Tanso, Pitch Coke), NBG-17 (SGL, Pitch Coke), NBG-18 (SGL, Pitch Coke), PCEA (GrafTech International, Petroleum Coke), PGX (GrafTech International, Petroleum Coke) and 2020 (Garbone of America, Petroleum Coke). A heavy emphasis is placed on the type and source of the coke used in manufacture because the coke's properties largely dictate the properties and behavior of conventionally manufactured graphite. Out of these grades, the Pitch Coke-based NBG-18 from SGL and the Petcoke-based PCEA graphite were considered as two major candidates for qualification in the TDP progam. Burchell, Bratton and Windes [40] produced a comprehensive report detailing the rationales and strategy regarding the graphite selection. It will not be repeated in this chapter. It is emphasized that the research in understanding the fundamental mechanisms controlling the irradiation behavior of graphite is crucial as it can help reduce the need, to some extent, for expensive qualification programmes and to support safety cases during reactor operation, as well as life extensions.

This section of the chapter will provide a review of the state-of-the-art understanding in terms of the irradiation damage in graphite, including historical data and modern results obtained by advanced techniques.

3.1 Point defect irradiation damage model

Carbon and graphite-based materials are probably one of the most studied material groups in nuclear irradiation. Neutron irradiation is expensive and produces materials that are active to handle. The majority of experiments were conducted using ion irradiation/implantation. A benefit of ion irradiation is that it has a high dose rate and low cost compared with neutrons and protons. For instance, neutrons usually have a dose rate of about (~10^{-7} dpa/s) while ions can be in the range of 10^{-10} and 10^{-1} dpa/s[42]. In most facilities, the ion mass can be varied, hence offering the flexibility of changing the recoil spectrum [41]. As such, ion irradiation/implantation is often used for modifying the sample/material surface and also creating disorder to varying depths depends on the incident ion energy [42]. In this section, experimental/modeling results produced by neutrons/protons and ions will all be summarized with the purpose of explaining the irradiation-induced damage mechanisms in graphite/carbon materials, though it is emphasized that there are differences in terms of how different energetic particles interact with carbon atoms/lattice, which will not be elaborated here.

Characterization of bulk property changes in polycrystalline graphite under irradiation dates back to the 1940s, mainly due to its large-scale use in nuclear fission reactors, but pioneering work was carried out on the atomistic irradiation damage models using high voltage electron microscopes around the 1960s–1970s, including the first lattice images of graphitic carbon acquired by Heidenreich et al. in 1968 [43].

The conventional theory about irradiation damage in graphite is the point defect model. In this model, the crystal expands along the *c*-axis and shrinks along the *a*-direction. This is attributed to the formation of interstitial clusters, which eventually rearrange into a new basal plane (contributing to the c-direction expansion) and the collapse of vacant lattice sites (resulting in a-direction shrinkage) [44,45].

Here, some key experimental evidence on this topic is presented based on the irradiation of single-crystal graphite (SCG) materials. This includes the work from Bollman et al. [46], where a single crystal graphite sample was irradiated by several different types of energetic particles. They used neutrons, electrons and carbon ions over a range of temperatures where they observed the formation of interstitials and vacancies after irradiation. Their results showed evidence of the clustering of interstitials and their agglomeration at high temperatures. With increasing irradiation doses (e.g., to 1.5×10^{20} ions/cm^2), using dark field TEM observations and electron diffraction kinematical theory, they suggested the formation of a disc-shaped interstitial monolayer, though they indicated there is no direct evidence found for vacancy clustering [47,48]. Hinman et al. [49] conducted experiments at higher temperatures and observed both vacancy clusters and interstitial clusters with neutron and carbon ion irradiation. Though they proposed the primary fraction of the c-direction crystal expansion may be indeed caused by interstitial clusters. Work from Baker and Kelly [50], where a single crystal graphite was irradiated over a range of temperatures up to 900°C, observed the formation of interstitial dislocation loops. Though these dislocation loops can be eliminated by annealing treatment at higher temperatures than the irradiation temperature, facilitated by the motion of atoms along the c-axis at 2200°C. Reynolds and Thrower [48,51] found that the size of interstitial clusters is dependent on the irradiation temperature and proposed a description of the phenomenon by a simple homogeneous nucleation theory. They supported the argument that vacancy clusters could exist, but only at high temperatures after higher irradiation doses.

As discussed in Section 3.2, nuclear graphite is polygranular and comprise of short range ordered graphite sheets. The size of the crystallite is much smaller than HOPG and single-crystal graphite; in addition, the crystallite boundaries, like many metal grain boundaries, can act as sinks. Therefore, for this graphite in a polycrystalline nature, the size of the defects is much smaller than those formed in an SCG subject to the same irradiation fluence. In particular, the aforementioned Mrozowski cracks, which were formed between the basal planes as a result of the anisotropic thermal shrinkage during cooling from graphitization temperature, are able to accommodate the initial irradiation induced c-axis expansion [52,53]. As such, an initial net bulk volume shrinkage occurs in these polycrystalline graphite materials with irradiation [54,55]. Note that at very low temperatures, for instance at 150°C, many nuclear graphite samples showed an initial volume expansion before the bulk dimensional shrinkage. The cause of this is not clear, but examples will be given in section 4.

3.2 Bending, wrinkling, ruck, and tuck

In recent years, it has been increasingly recognised that the conventional point defect model is insufficient in explaining certain observations. For instance, Tanabe [56] argued that the crystallite c-plane expansions, caused by the discrete interstitial planes distributed between graphite basal sheets, are much less than the bulk volume expansion encountered in neutron irradiated graphite, which can be as high as 200% in some cases [57]. In addition, the point defect model cannot explain the homogeneous dilation of (0002) lattice fringes in highly graphitized carbon fibers [57] as well as the changes in some graphite properties (e.g., the CTE). Tanabe et al. [56] therefore extended the point defect model by including different mechanisms at high irradiation dose (for instance, when the graphite sp^2 structure degrades to sp^3 clusters [58]). When graphite receives a high dose of irradiation, the graphene sheets (i.e., the basal planes) will lose their two-dimensional ordering and will bend, break, and form turbulence, creating new open spaces—these will all contribute to additional dimensional changes.

As described in the point defect model, the defects in sp^2 bonds can be annealed when heated to a temperature higher than the irradiation condition. However, high irradiation fluence will lead to the transition to sp^3 bonds, which is considered highly irreversible [59]. There is an accumulation of evidence showing the mechanisms contributing to the 2D in-plane structure. One of these is marked by Niwase's work [60] where a dislocation dipole model was proposed to explain the irradiation induced amorphization at low temperatures. During the initial stage of amorphization, the ABAB stacking ordering is preserved while in-plane disordering occurs. This is consistent with the decrease in thermal conductivity in graphite crystals due to low dose irradiation induced defects within basal planes [16]. However, the nucleation and growth processes of such dislocations remain unknown. Recent evidence by Karthik et al. [25] showed that the new basal planes, which assisted the c-direction expansion, are likely a consequence of the positive climb of dislocation dipoles; on the other hand, the condensation of the vacancies left behind by the dislocation climb may have contributed to the *a*-direction shrinkage. It is suggested that the breaks, bends, and curls are subsequently introduced to affect the lattice spacing and dimensional change [61–63].

From the modeling aspect, Heggie et al. [64] also noted the inadequacies in the conventional point defect theory and proposed a 'ruck and tuck' defect model including nano-buckling and nano-folding of basal planes. There is some direct experimental evidence for this model in a recent high-temperature irradiation study conducted by Johns et al. [65]. A most recent comprehensive experimental study conducted by Liu et al. proposed a macro-scale ruck and tuck model that induced new porosity, facilitating c-plane expansion and a-direction shrinkage. Vein-like features were observed on the surface of HOPG graphite after ion irradiation at 100°C, even at doses as low as 0.028 dpa, and these veins become more pronounced at 0.339 dpa, Fig. 5.4, and comprise primary veins and secondary ones connecting them. Focussed ion beam milled cross-sectional analysis revealed a multi-layered structure under these veins with pores in-between layers of platelets. These veins in graphite are independent of the graphite crystallite orientations and boundaries, Fig. 5.4B. Note the large crystallite size in the HOPG graphite (Fig. 5.4B) compared with the synthetic nuclear graphite, Figs. 5.1–5.3. Nanosize veins were also found in HOPG graphite thin foils when irradiated in a TEM using in situ ions. These vein-like features

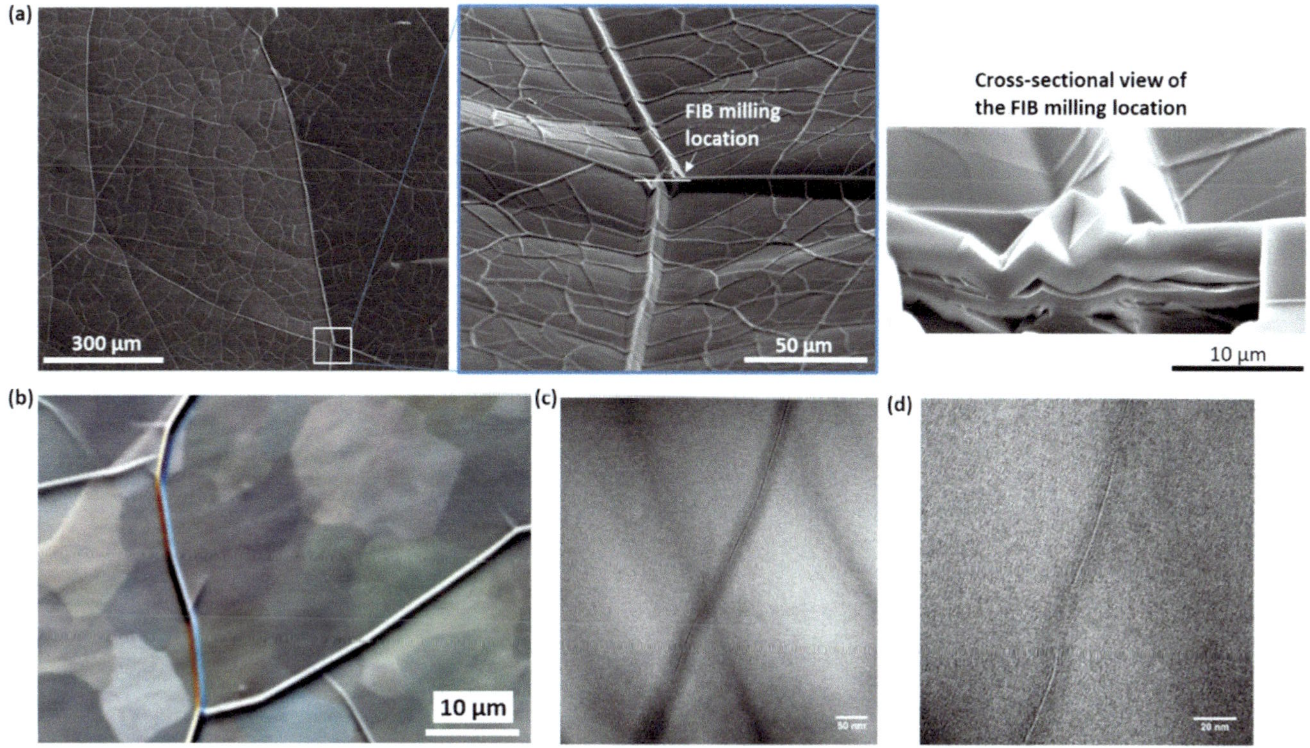

Figure 5.4 (A) Vein like features formed on the surface of HOPG graphite subjected to C^+ irradiation (1.8×10^{17} ions; approximately 0.339 dpa); (B) EBSD map showed the veins are formed independent of the crystal orientation and boundaries; similar veins were observed during in situ Ar^+ irradiation with transmission electron microscopy imaging where (C) bright field and (D) dark field images showing the formation of a thin void tube appears as a discontinuous line.

reported, since publication in 2021, have been repeated and observed in HOPG by other research groups on HOPG.

The importance of these veins lies in the fact that they acted as another mechanism (in addition to what was proposed in the point defect model), causing significant shrinkage along the a-axis due to the contraction of the basal layers, Fig. 5.5A. In the meantime, the c-axis expansion is also observed at the location of the veins (e.g., vein01 and vein02). Vein01 is at a later stage than vein02 whereby a large number of pores or lenticular cracks were formed between the basal layers. Note that this HOPG had no porosity prior to irradiation, and it had well-aligned graphene basal planes showing no disorder D band in a Raman spectrum. The irradiation damage was accommodated in HOPG by the formation of veins. Note the nanostructure at the vein location is extremely similar to that observed in nuclear graphite in as-made condition, albeit a different length-scale, as the nuclear graphite has much smaller crystallites than HOPG.

A detailed investigation of the boundary of the veins showed that such veins have a triangle-shaped zone underneath. The platelets inside the triangular zone are shifted uniformly upwards compared with the zone outside the boundaries; this phenomenon is illustrated by upward arrows in Fig. 5.5B. Inside the triangle, delamination cracks in a lenticular shape formed between the platelets. This is the early stage of vein01 before the formation of a large population of pores. A macro-scale 'ruck&tuck' geometry was thus generated at these triangle structure boundaries. The schematic is shown in Fig. 6A and B. The triangular boundary appears amorphous in TEM, and it was believed to have been caused by out-of-plane folding of the layers. Note that a small increase in c-spacing was observed with irradiation, as expected based on the point defect model. But it is emphasized that the additional deformation caused by irradiation, such as macro-scale ruck and tuck phenomenon described in Figs. 5.4 and 5.5, play a vital role in the dimensional changes in polycrystalline graphite.

Note the dislocation type contributing to the ruck and tuck behavior is considered to be a-type, which means this process, either in the generation of the types of dislocation or its slip process, does not need to involve point defects. This is a different model from the aforementioned point defect theory. Compared with Heggie's modeling where it was proposed that permanent nano-buckling occurs at low temperatures (lower than 250°C) and ruck&tuck behavior takes place at higher temperatures, experimental work showed macro-scale rucking and tucking can occur at low temperatures (100°C), acting as a means to accommodate irradiation induced compressive stress in a layered polycrystalline graphite structure. Further discussion about this ruck and tuck model and how it differs from other existing propositions (e.g., the 'ripplocation' in MAX phases [66]) can be found in Ref. [67].

To conclude this section, the irradiation mechanisms in graphite are still an ongoing topic that requires a significant amount of effort utilizing advanced experimental and modeling techniques. The impact of the irradiation damage on the structure of graphite extends from the nano-crystallite scale to the changes in dimensions and properties at the macro-scale for bulk materials, and they vary with temperature and irradiation fluence.

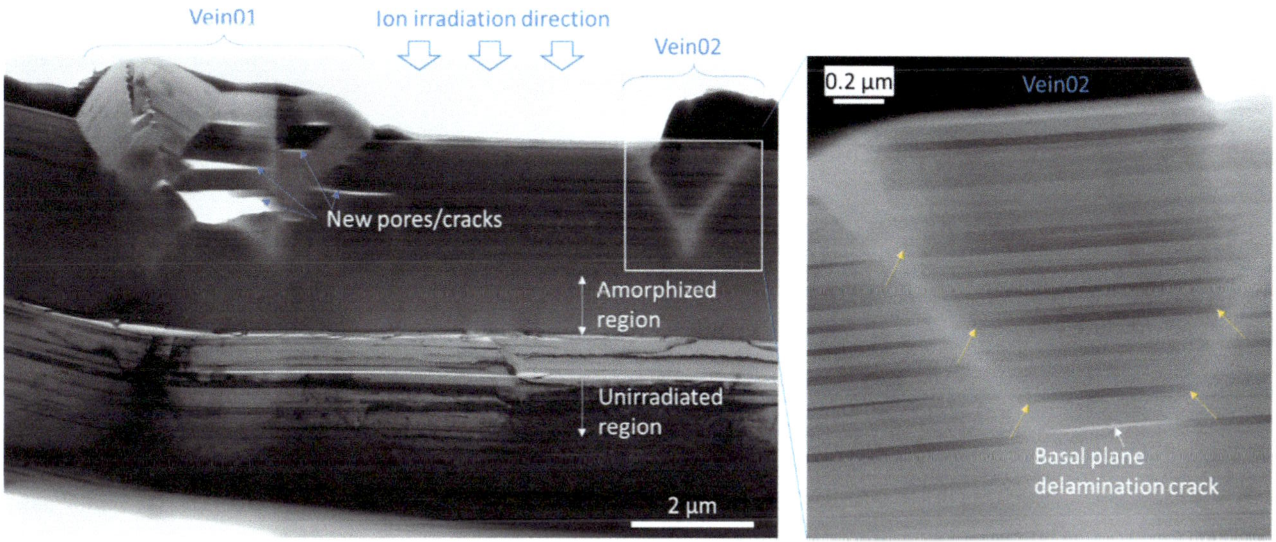

Figure 5.5 A bright field TEM image of a thin foil milled using focussed ion beam across two veins in the HOPG sample, demonstrating the formation of additional pores along the basal planes; a higher magnification image of vein02 revealed the upward shifting of the platelets (marked by orange arrows) between the inclined boundaries of the triangular zone under the vein.

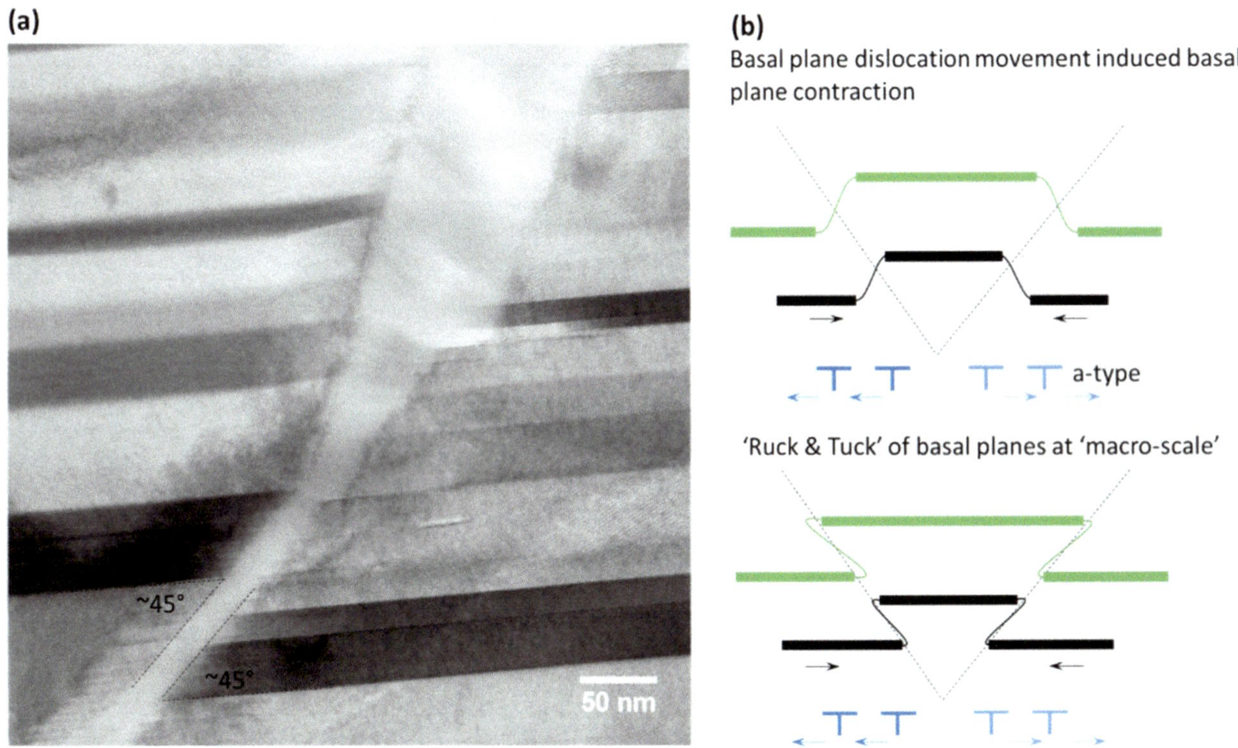

Figure 5.6 (A) A TEM image showing the amorphous appearance of the inclined boundary zone (inclination angle is about 45 degrees). The basal plane platelets are displaced across the boundary; (B) Schematics of the proposed mechanism comprise the dislocation induced contraction of basal planes and the subsequent ruck and tuck configuration at the boundaries of a triangular zone.

4. Irradiation-induced changes in bulk dimensions and properties

This section will first describe typical experimental observations about the graphite crystal dimensional changes (c-spacing and a-spacing), very briefly, as a consequence of irradiation fluence and temperature. This will be used to emphasize the differentiation between crystal volume change and bulk dimensional changes in graphite. The definition of 'With Grain' (WG) and 'Against Grain' (AG) will be elaborated to prepare the reader for the concept of irradiation anisotropy. Finally, typical examples will be given to illustrate the impact of irradiation on a range of properties in graphite and the potential challenges this may cause during service.

4.1 Crystal dimensional change

Irradiation induces changes to the graphite crystallites, as discussed in section 3. An example is given using the HOPG graphite results as shown in Fig. 5.7. The c-plane spacing increases with fluence (irradiation dose), and the rate of the change is a function of temperature. Below and at 200°C, the increase is very steep and the amount of increase can be as high as 20%–25%, Fig. 5.7A. On the other hand, the corresponding a-plane spacing decreases by about 4%–6%, Fig. 5.7B. At temperatures higher than 250°C, both the expansion in c-spacing and a-spacing are reduced significantly, suggesting it is a more stable temperature for defect population. For the operating AGRs and future GenIV reactors, the temperatures are in the range of higher than 250°C.

4.2 Bulk dimensional changes

It is emphasized that the bulk dimensional changes in polycrystalline nuclear graphite are a very different concept from the graphite crystal dimensional changes. This is a point that is often misunderstood by many. To illustrate this, a typical example of the bulk changes in Gilsocarbon graphite is shown in Fig. 5.8. During irradiation, the bulk material will show a volume shrinkage at low fluences first. It will then reach a turnaround point before entering the swelling stage. Usually, when the graphite swells back to its original volume at very high fluences marks the end of life of the reactor in certain designs. Note that the fluence at which the turnaround point occurs generally decreases with the increase in irradiation temperature. The mechanisms for volume shrinkage and swelling are very different and not yet fully understood.

This type of dimensional change introduces stresses in the graphite component, using the UK AGR hollow graphite fuel brick as an example, Fig. 5.9. During early life before the turnaround point, the inner side of the brick at the bore is under tension as the materials there shrink the most, but under constraints applied by surrounding materials resulting in tension. After turnaround, the outer surface of the brick will be under tension resulting in tensile stresses at the keyway root, which become the preferred sites for crack initiation. For different reactor designs, the shape of the graphite core component will differ, but it is critical to have a sound understanding of the irradiation dose, temperature, dimensional change and thermal expansion for the prediction of service stresses.

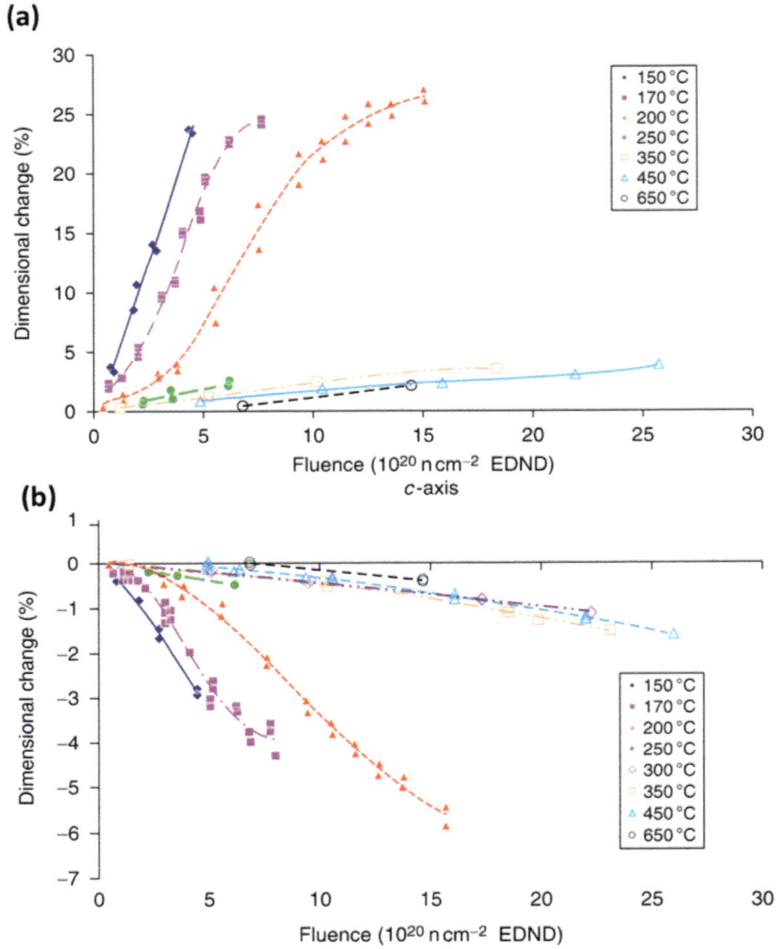

Figure 5.7 Crystal dimensional changes in HOPG illustrated in (A) c-axis and (B) a-axis direction.
Taken from Ref. [.68]

It is not apparent from Fig. 5.8, but there is often an initial swelling at very low fluences, for Gilsocarbon graphite (from 0 fluence to about ~10×10^{20} n·cm^{-2} EDND in Fig. 5.8) which became more pronounced at lower temperatures, Fig. 5.10. This is also the case for many other grades of graphite, including GenIV graphite, demonstrating an initial bulk volume expansion at low irradiation doses before the shrinkage in volume occurs. For instance, Burchell and Snead have observed this in NBG-10 graphite (near isotropic, medium-grained, manufactured by SGL), Fig. 5.11. Kelly studied fine-grain POCO-AXZ-5Q1 graphite, which showed swelling even at high temperatures (Fig. 5.12). The existing literature tends to show finer grain graphite tends to have more pronounced initial swelling both at low and high temperatures. Further complete view on this topic needs more data, and the exact mechanism for this is also unclear. Some attributed it to the relaxation of machining stresses, but there is no experimental evidence to verify.

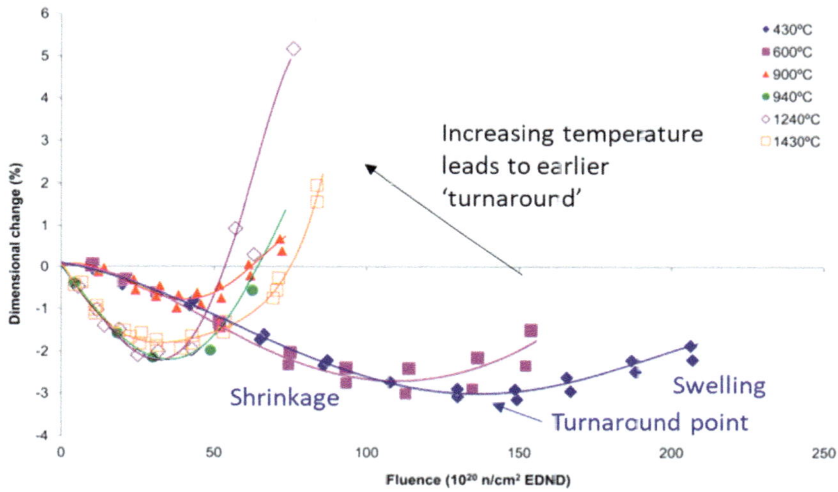

Figure 5.8 Bulk dimensional change in Gilsocarbon graphite showing the shrinkage at low fluence, turnaround point and swelling at high fluences. The point at which the turnaround occurs is dependent on irradiation temperature.
Graph adapted from Ref. [.16]

4.3 Irradiation anisotropy

In general, the nuclear industry prefers graphite with isotropic bulk dimensional changes in all directions; however, many grades of graphite demonstrated anisotropic dimensional/volume change with irradiation. Prior to the discussion of anisotropy in dimensional changes, it is necessary to describe the commonly used terminology: 'with grain (WG)' and 'across/against grain (AG)' - graphite microstructure, bulk anisotropy and thermos-mechanical properties often presented with respect to these orthogonal directions. Specifically, With Grain (WG) is the direction in the graphite that most of the graphite crystals a-axes are aligned, whereas Across/Against Grain (AG) is the direction in which most of the c-axes are aligned. These orientations are generated during the forming methods applied to the green artifact of graphite prior to baking/carbonization. Graphite green artifacts are usually formed by one of the four methods: extrusion, molding/pressing, vibrational molding and isostatic molding/pressing, which will be elaborated below.

(1) Extrusion (Fig. 5.13). This is a commonly used method for historical grades of graphite, where the heated graphite paste (mixture of filler coke particles, filler flour (if any) and liquefied pitch binder) was extruded. In this process, the coke particles are aligned with the

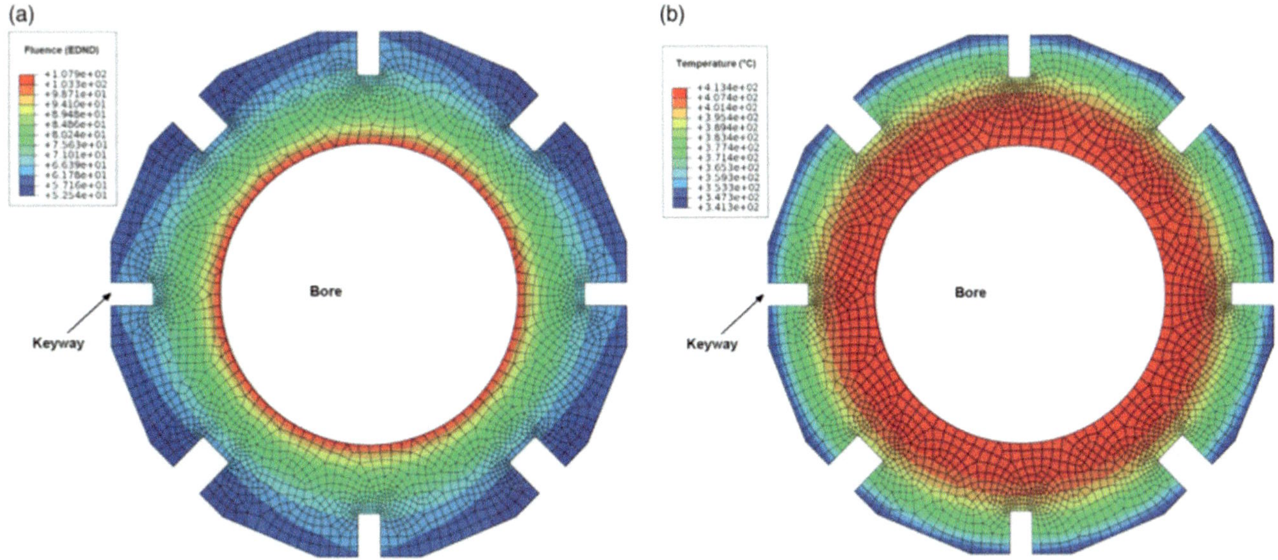

Figure 5.9 The simulated distribution of (A) neutron fluences and (B) temperature gradient across an AGR fuel brick. Taken from Ref. [.16]

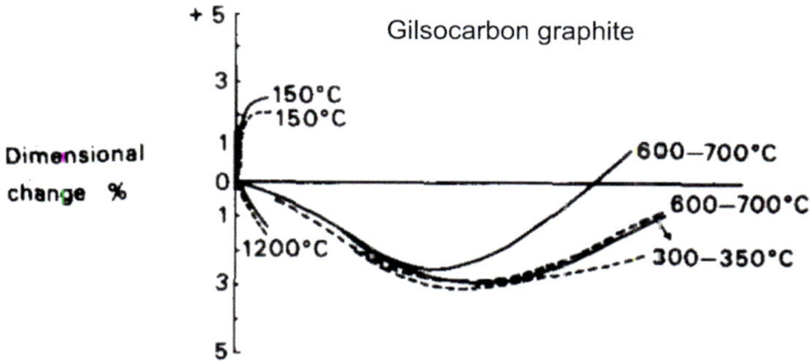

Figure 5.10 Historical data on Gilsocarbon graphite illustrating the swelling at lower temperatures.
image taken from Kelly, 1982 [.1]

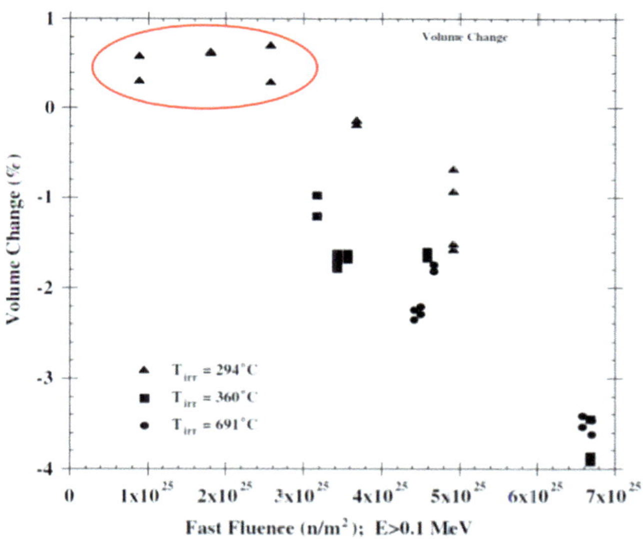

Figure 5.11 Volume swelling at low fluence for NBG-10 graphite in the circled fluence range.
image adapted from Burchell and Snead, 2007 [.44]

extrusion direction. Typical example is medium-grained Pile grade A (PGA) for Magnox reactors and medium-grained PCEA graphite for high temperature gas cooled reactors (HTGRs). For the PGA graphite, a needle-shaped coke was used; therefore, the longer axis of the coke is in line with the extrusion direction, resulting in anisotropies of the material's properties. The needle-like filler particles have inherently lenticular shrinkage cracks formed parallel to the crystal basal plane direction and will tend to align in the direction of extrusion, i.e., the extrusion direction is WG. Therefore, extruded graphite has one WG orientation (parallel to the billet long axis) and two AG orientations. Extrusion is still used on many grades of modern nuclear-grade graphites, for instance, the INNOGRAPH irradiation program (c.a. 2006–07) irradiated a range of extruded graphite (e.g., PCEA, PPEA, NGB-20) to study their irradiation properties (Figs. 5.12 and 5.13).

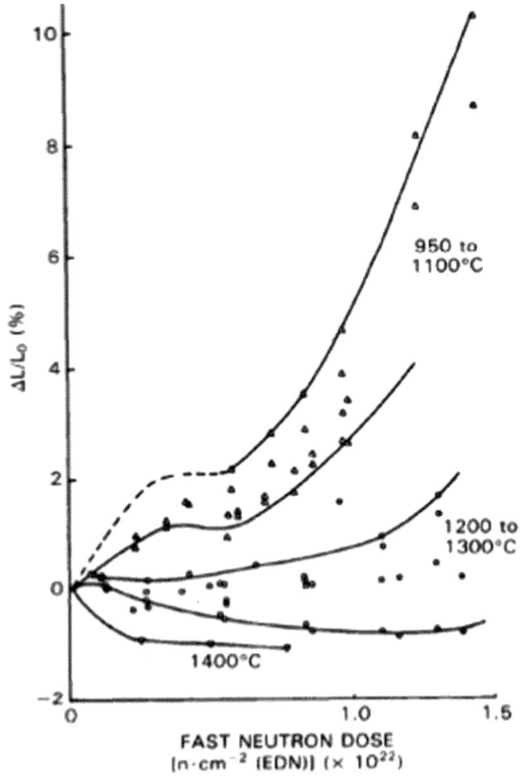

Figure 5.12 Bulk dimensional change of POCO-AXZ-8Q1 graphite as a function of fast neutron irradiation and temperature, showing initial swelling even at very high temperatures [69].

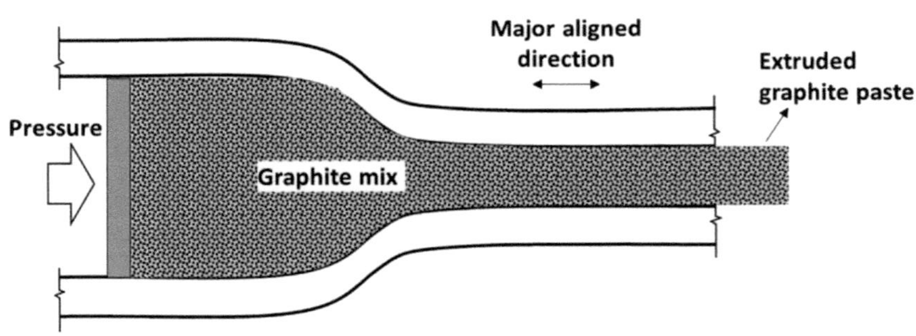

Figure 5.13 Extrusion process of graphite paste mixture of filler particles, binder (and filler flour). The major direction of the grain alignment is parallel to the extrusion direction.

(2) Molding/pressing (Fig. 5.14A) and vibrational molding (Fig. 5.14B). Molding/pressing is a forming method involving the pressing of graphite paste from one or two directions at the same time. A typical example of graphite is the CL 2020 produced by Mersen. For vibrational molding, the mixture is placed in a mold and compacted by vibrating the mold with a load pressing from one side. Typical examples include the Gilsocarbon graphite used for

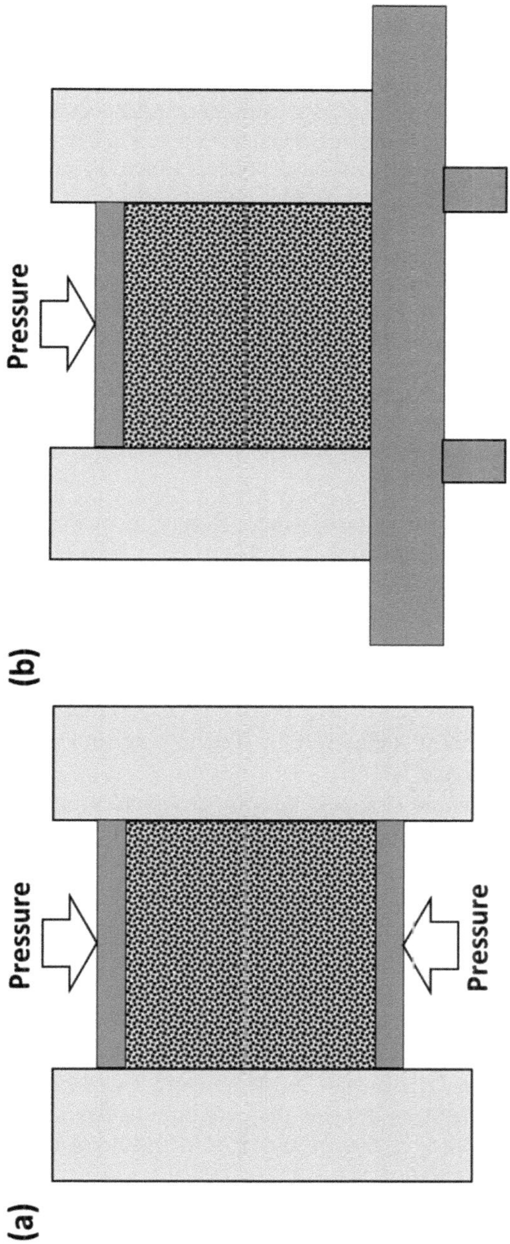

Figure 5.14 Schematics showing (A) molding/pressing and (B) vibrational molding forming methods of graphite paste.

UK AGRs, NBG-17 and NBG-18 (potential candidates for high temperature reactors). Molding and vibration molding tend to align the long axis of the particles in the plane perpendicular to the forming axis. Therefore, these graphites will have two perpendicular WG orientations and one AG orientation. In the case of the producing Gilsocarbon graphite by vibration molding, most of the graphite crystal c-axes tend to align perpendicular (90 degrees) to the pressing direction and therefore, 90 degrees to the brick axis.

(3) Isostatic molding (isostatic pressing) is a more complicated method by which the graphite mixture is molded by applied pressure from all directions in a liquid chamber, resulting in isotropic properties and better uniformity. Typical examples include fine-grain graphite grades such as IG110, IG430, SGL R7650, NBG-25, PCIB-SFG and ETU-10. Note that isostatically molded and isostatically pressed graphite does not have a preferred orientation of filler particles/grains. Therefore, their properties, such as strength, modulus, CTE, and thermal conductivity, are the same in the as-made condition. However, these properties will become very different in AG and WG directions after irradiation. This has been observed in many grades, but the cause of the 'irradiation anisotropy' is unclear. Due to this anisotropy phenomena, even for isotropic graphite, terminologies similar to WG and AG will also be used to fully describe its properties in orthogonal directions. For instance, the axial direction is along the extruding direction if extrusion was used to make the graphite block or is the direction of gravity for iso-molding. A summary of typical terminologies is shown in Fig. 5.15B.

To describe the irradiation anisotropy more clearly, an example is given here based on an irradiation test undertaken at Oak Ridge on a Japanese fine-grain graphite G347A [70,71] (Fig. 5.3A). The 'TR' and 'AX' used in Fig. 5.16 refer to the transverse and axial directions with respect to the processing procedure (see Fig. 5.15B). For the two directions (*e.g.*, at 738°C), the 'cross-over' (where the graphite swells to its original volume marked in Fig. 5.16A and B) happens at neutron fluences in the AX direction (25×10^{25}n/m^2 (Fig. 5.3A)) that are twice of that in the TR direction (12×10^{25}n/m^2 (Fig. 5.3B)). And this anisotropic ratio changes with temperature (Fig. 5.16A and B).

In addition to the volume changes, the properties, such as elastic modulus, also change with neutron fluence and the measurement directions. The modulus changes increase with neutron fluence and reach a plateau before it decreases as a function of neutron fluence (Fig. 5.16C). When it comes to the CTE) (Fig. 5.16D), after an initial increase, it reduced to a constant value regardless of the temperature and direction.

This example given here is to emphasize that although G347A initially has isotropic thermal and mechanical properties, the dimensional change and other properties such as modulus and CTE showed remarkable difference especially after the 'turnaround' point (Figs. 5.16C and D). Moreover, it can be seen from Fig. 5.16 that the irradiation temperature impacts the evolution of these properties differently, and this further complicates the interpretation of the mechanical behavior of graphite during and following irradiation [11,72].

It has been recognised that irradiation-induced volume shrinkage and swelling are not a simple process of density change, as is evident in the slope change in the modulus-fluence relationship before and after the turnaround point (Fig. 5.3C). It was proposed that the change before turnaround was due to *c*-axis expansion

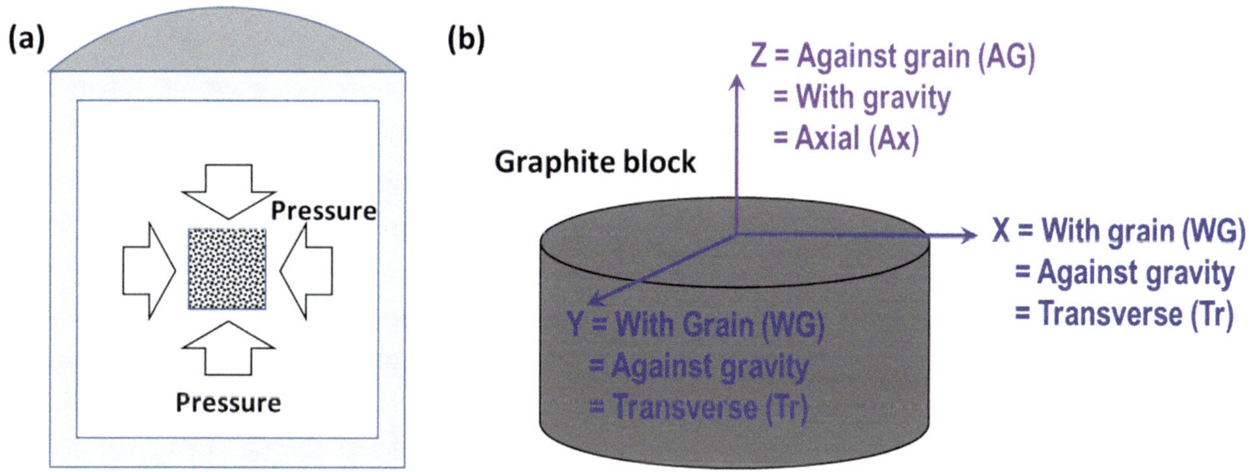

Figure 5.15 Schematics showing (A) the method of isostatically molding/pressing of graphite paste; (B) typical terminologies in describing graphite orientations for those with one AG direction and two WG directions.

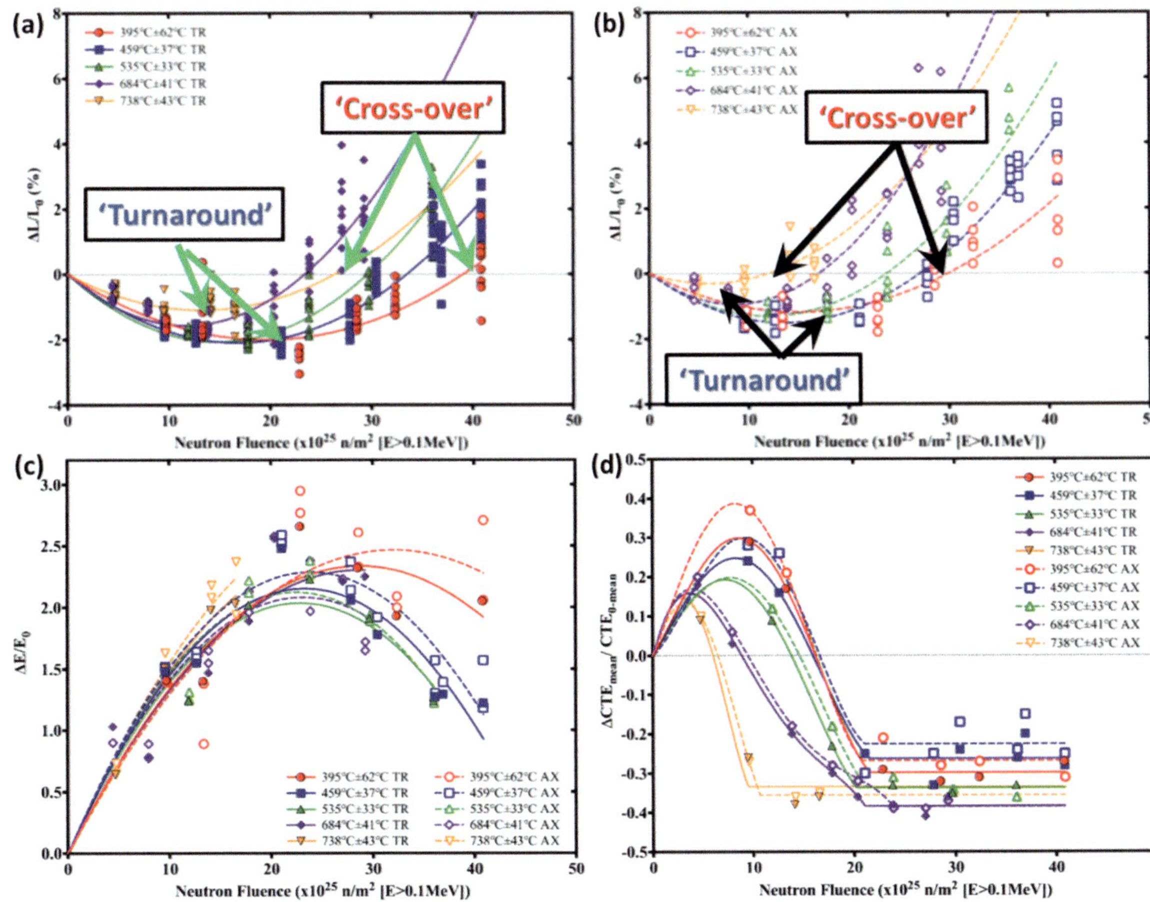

Figure 5.16 Volume changes in (A) transverse and (B) axial direction, (C) modulus and (d) CTE change with neutron fluence in a fine-grain Japanese grade graphite, G347A.
Modified from Refs. [70,71].

accommodated by Mrozowski cracks, and additional accommodation porosity was formed after turnaround to modify the change in modulus. However, there is no consistent experimental evidence to directly support this hypothesis. Other properties, such as CTE (results shown in Fig. 5.16D) and thermal conductivity, do not follow the trend of density either.

When changing the initial raw material mix and manufacturing processes, the trends of change in properties with fluence could differ significantly from the grade shown in Fig. 5.16. Another set of well-known data is from the INNOGRAPH irradiation at Petten (NRG), Figs. 5.17 and 5.18. Four grades of extruded graphite and four grades of isostatic molded graphite were selected for comparison. For all eight grades of graphite, the length shrank in the range of irradiation doses (6–10 dpa) conducted at 750°C, as expected.

The extruded graphite materials showed maximum shrinkage between 2% and 2.7%, and there is no appreciable difference between the measurements conducted at the center and the edge of the sample, Fig. 5.17. The WG direction (symbols in red in Fig. 5.17) has more shrinkage than the AG direction (symbols in blue in Fig. 5.17). This anisotropic behavior in terms of length change is comparable for all four extruded grades.

For the isostatic molded graphite grades, Fig. 5.18, the shrinkage in the WG direction (symbols in red in Fig. 5.18) is significantly higher than the AG direction (symbols in blue in Fig. 5.18), i.e., the anisotropy in length change is much more pronounced than the extruded graphite, which is counterintuitive. Due to this large difference between the two directions, the maximum shrinkage range is also wider, and it is between 1.1% and 2.4%. There is no distinct difference between the results from the samples from the edge and samples from the center of the graphite blocks.

The change in other properties was also measured; the details can be found in Vreeling et al. [73]. In particular, they compared the CTE of an extruded grade and an isostatic molded graphite. The CTE values generally decreased with the increasing neutron irradiation, and the anisotropy for WG and AG directions is still present. They concluded that for graphite with higher CTE in the as-manufactured stage, a larger decrease was observed with irradiation. Fig. 5.19

Irradiation anisotropy is not a preferred property of the nuclear graphite. For example, one direct impact of the anisotropy is that different parts of the core components reach their end of life at different irradiation fluences, and this will cause extreme difficulties in the management and evaluation of the graphite structural integrity. Worse still, such 'irradiation anisotropy' is seen in every grade of graphite, as demonstrated above, especially in modern homogeneous fine-grained graphite. There are no coherent theories, as yet, to consolidate these properties, either within one grade of graphite nor across different grades.

Before ending this chapter, it is worth noting that if graphite is irradiated under load (which is the case for graphite as a reactor core component), irradiation creep can occur. Scientifically, irradiation-induced dimensional change and creep in graphite/carbon materials is an extremely important but very difficult topic. Using Gilsocarbon graphite in the UK Advanced Gas-Cooled Reactor as an example, in a load-free condition with neutron irradiation, it will shrink in volume until reaching a 'turnaround'

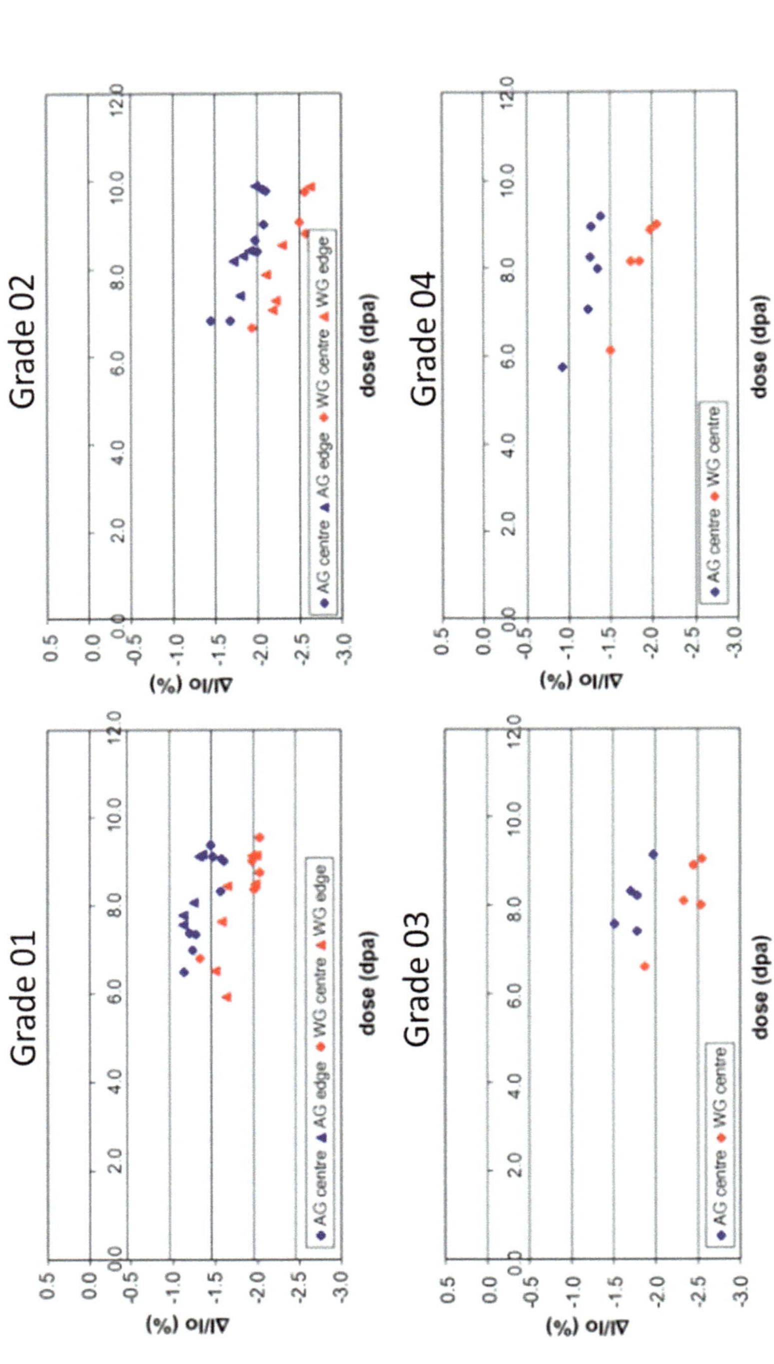

Figure 5.17 The relative length change with irradiation dose from four extruded grades of graphite, irradiated at 750°C, taken from Vreeling et al. [.73]

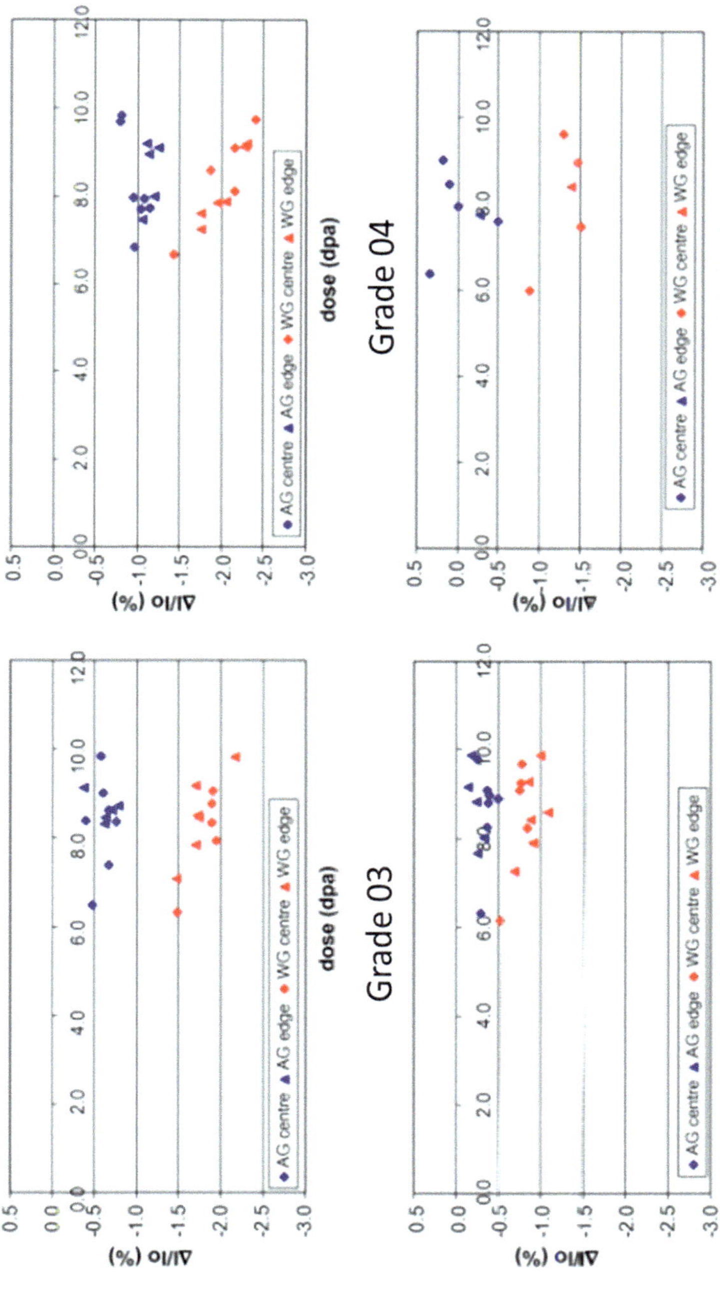

Figure 5.18 The relative length change with irradiation dose from four isostatic molded grades of graphite, irradiated at 750°C. taken from Vreeling et al. [.73]

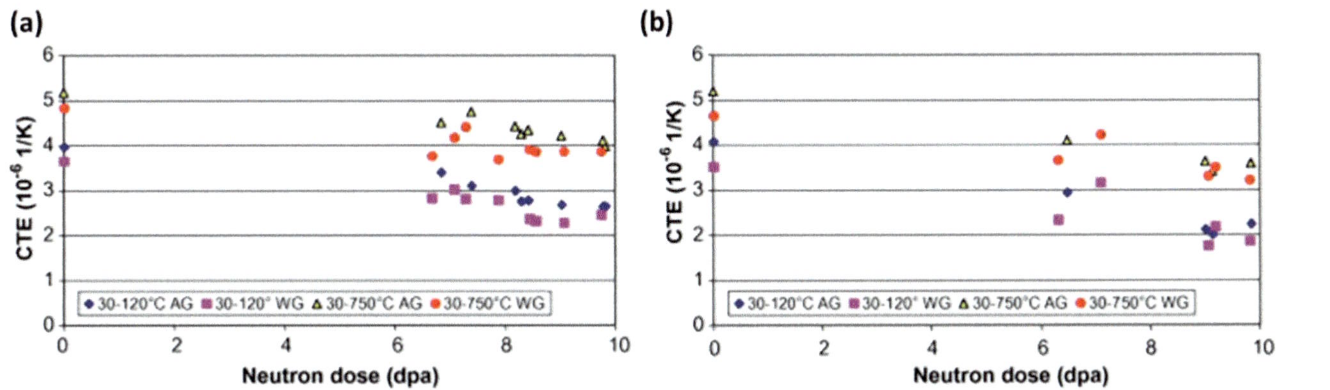

Figure 5.19 Comparison between (A) a typical extruded grade of graphite and (B) a typical iso-molded grade of graphite in terms of the change in coefficient of thermal expansion irradiated at 750°C.
Taken from Vreeling et al. [.73]

point before swelling in volume at higher fast neutron fluences. When irradiated under load, the dimensional change data are modified. The difference between loaded and unloaded dimensional change is referred to as irradiation creep, which is a primary means to delay fracture in these reactor core components by relieving significant thermal, mechanical and dimensional shrinkage stresses. The downside is that irradiation creep introduces additional permanent plastic changes in the dimensions of the component that must be understood and considered in operation. Irradiation creep in nuclear graphite has been mainly attributed to basal plane dislocation sliding due to pinning and unpinning of dislocations, both created and destroyed by irradiation, but as yet there is no convincing microstructural evidence of this mechanism [74]. There are significant more work needed to understand this topic.

5. Concluding comments

This chapter presented a very brief introduction to nuclear graphite, covering several topics including irradiation damage models, irradiation-induced crystal and dimensional changes and touched on the concept of irradiation anisotropy. Due to limited space, this is far from a complete description of these important topics. There is a vast amount of literature in these areas, and some of these were pointed out to the readers in this chapter for further reading. There are many more important topics about nuclear graphite, such as its multiple length-scale mechanical properties [75–77] and its high-temperature deformation and fracture behaviors [78,79]. For instance, due to its complex structure, graphite is stronger in compression than in tension, with its flexural strength usually ~20%-30% higher than in uniaxial tension. As a consequence, the failure of nuclear graphite is not only a function of component size but also of loading conditions [80]. The high temperature strength and fracture toughness of graphite differ from their room temperature values and vary with the raw materials/manufacturing processes. Further, oxidation of graphite is another important topic, not only for the UK AGRs in terms of radiolytic oxidation but also for accident conditions in HTRs. Lately, the waste management of nuclear graphite is another very challenging area, as currently there are more than 200,000 t of graphite waste. A long-term treatment or recycle/reuse strategy is needed [2,4]. In summary, nuclear graphite, as a fascinating material with industrial imperative, has attracted a significant amount of effort and interest from both industry and research institutions. It is hopeful that modern advanced experimental and modeling techniques can shed light on the fundamental mechanisms controlling its in-service behaviors.

References

[1] B.T. Kelly, Graphite—The Most fascinating nuclear material, Carbon N. Y. 20 (1982) 3–11.

[2] Wickham, A. J., Neighbour, G. B. The uncertain future for nuclear graphite disposal: Crisis or opportunity?. Available at: https://inis.iaea.org/collection/NCLCollectionStore/_Public/32/039/32039322.pdf. (Accessed: 6 October 2018).

[3] T.D. Burchell, E.L. Fuller, G.R. Romanoski, J.P. Strizak, Graphite for the nuclear industry, 1991.
[4] A. Wickham, H.-J. Steinmetz, P. O'Sullivan, M.I. Ojovan, Updating irradiated graphite disposal: Project 'GRAPA' and the international decommissioning network, J. Environ. Radioact. 171 (2017) 34–40.
[5] B.T. Kelly, P.A.V. Johnson, P. Schofield, J.E. Brocklehurst, M. Birch, UKAEA Northern division studies of the radiolytic oxidation of graphite in carbon dioxide, Carbon N. Y. 21 (1983) 441–449.
[6] P.J. Hacker, G.B. Neighbour, B. McEnaney, The coefficient of thermal expansion of nuclear graphite with increasing thermal oxidation, J. Phys. D Appl. Phys. 33 (2000) 991–998.
[7] I.M. Pickup, B. McEnaney, R.G. Cooke, Fracture processes in graphite and the effects of oxidation, Carbon N. Y. 24 (1986) 535–543.
[8] E.D. Eason, G.N. Hall, B.J. Marsden, G.B. Heys, A model of Young's modulus for Gilsocarbon graphites irradiated in oxidising environments, J. Nucl. Mater. 436 (2013) 201–207.
[9] G.M. Laudone, C.M. Gribble, G.P. Matthews, Characterisation of the porous structure of Gilsocarbon graphite using pycnometry, cyclic porosimetry and void-network modeling, Carbon N. Y. 73 (2014) 61–70.
[10] J.V. Best, W.J. Stephen, A.J. Wickham, Radiolytic graphite oxidation, Prog. Nucl. Energy 16 (1985) 127–178.
[11] B.J. Marsden, et al., Dimensional change, irradiation creep and thermal/mechanical property changes in nuclear graphite, Int. Mater. Rev. 61 (2016) 155–182.
[12] R. Krishna, A.N. Jones, L. McDermott, B.J. Marsden, Neutron irradiation damage of nuclear graphite studied by high-resolution transmission electron microscopy and Raman spectroscopy, J. Nucl. Mater. 467 (2015) 557–565.
[13] J. Kane, C. Karthik, D.P. Butt, W.E. Windes, R. Ubic, Microstructural characterization and pore structure analysis of nuclear graphite, J. Nucl. Mater. 415 (2011) 189–197.
[14] R. Blackstone, Radiation creep of graphite. An introduction, J. Nucl. Mater. 65 (1977) 72–78.
[15] T.D. Burchell, W.P. Eatherly, The effects of radiation damage on the properties of GraphNOL N3M, J. Nucl. Mater. 179–181 (1991) 205–208.
[16] B.J. Marsden, G.N. Hall, O. Wouters, J.A. Vreeling, J. Van der Laan, Dimensional and material property changes to irradiated Gilsocarbon graphite irradiated between 650 and 750 °C, J. Nucl. Mater. 381 (2008) 62–67.
[17] D.E. Baker, Graphite as a neutron moderator and reflector material, Nucl. Eng. Des. 14 (1971) 413–444.
[18] T.J. Marrow, et al., In situ measurement of the strains within a mechanically loaded polygranular graphite, Carbon N. Y. 96 (2016) 285–302.
[19] J.E. Brocklehurst, M.I. Darby, Concerning the fracture of graphite under different test conditions, Mater. Sci. Eng. A 16 (1974) 91–106.
[20] Z. Mileeva, D.K. Ross, S.M. King, A Study of the porosity of nuclear graphite using small-angle neutron scattering, Carbon N. Y. 64 (2013) 20–26.
[21] Z. Zhou, et al., From nanopores to macropores: Fractal morphology of graphite, Carbon N. Y. 96 (2016) 541–547.
[22] Y. Vertyagina, T.J. Marrow, Multifractal-based assessment of Gilsocarbon graphite microstructures, Carbon N. Y. 109 (2016) 711–718.
[23] D. Liu, D. Cherns, Nano-cracks in a synthetic graphite composite for nuclear applications, Philos. Mag. 98 (2018) 1272–1283.

[24] S. Mrozowski, Anisotropy of thermal expansion and internal stresses in polycrystalline graphite and carbons, Phys. Rev. 86 (1952), 622–622.
[25] C. Karthik, J. Kane, D.P. Butt, W.E. Windes, R. Ubic, In situ transmission electron microscopy of electron-beam induced damage process in nuclear grade graphite, J. Nucl. Mater. 412 (2011) 321–326.
[26] C. Karthik, J. Kane, D.P. Butt, W.E. Windes, R. Ubic, Microstructural characterization of next generation nuclear graphites, Microsc. Microanal. 18 (2012) 272–278.
[27] B.E. Mironov, et al., Electron irradiation of nuclear graphite studied by transmission electron microscopy and electron energy loss spectroscopy, Carbon N. Y. 83 (2015) 106–117.
[28] K.Y. Wen, T.J. Marrow, B.J. Marsden, The microstructure of nuclear graphite binders, Carbon N. Y. 46 (2008) 62–71.
[29] B. März, et al., Mesoscopic structure features in synthetic graphite, Mater. Des. 142 (2018) 268–278.
[30] B. März, et al., Data related to the mesoscopic structure of iso-graphite for nuclear applications, Data Br. 19 (2018) 651–659.
[31] B. März, K. Jolley, R. Smith, H. Wu, Near-surface structure and residual stress in as-machined synthetic graphite, Mater. Des. 159 (2018) 103–116.
[32] R. Krishna, A.N. Jones, R. Edge, B.J. Marsden, Residual stress measurements in polycrystalline graphite with micro-Raman spectroscopy, Radiat. Phys. Chem. 111 (2015) 14–23.
[33] M. Jiang, et al., Multiple length-scale microstructural characterisation of four grades of fine-grained graphite, J. Nucl. Mater. 550 (2021) 152876.
[34] C. Berre, et al., Numerical modelling of the effects of porosity changes on the mechanical properties of nuclear graphite, J. Nucl. Mater. 352 (2006) 1–5.
[35] C. Berre, et al., Failure analysis of the effects of porosity in thermally oxidised nuclear graphite using finite element modelling, J. Nucl. Mater. 381 (2008) 1–8.
[36] A.L. Sutton, V.C. Howard, The role of porosity in the accommodation of thermal expansion in graphite, J. Nucl. Mater. 7 (1962) 58–71.
[37] J.D. Arregui-Mena, et al., Electron tomography of unirradiated and irradiated nuclear graphite, J. Nucl. Mater. 545 (2021) 152649.
[38] G. Haag, Properties of ATR-2E graphite and property changes due to fast neutron irradiation, 2005.
[39] D.K.L. Tsang, B.J. Marsden, Constitutive material model for the prediction of stresses in irradiated anisotropic graphite components, J. Nucl. Mater. 381 (2008) 129–136.
[40] T. Burchell, R. Bratton, W. Windes, NGNP Graphite selection and acquisition strategy, 2007, https://doi.org/10.2172/921767.
[41] C. Abromeit, Aspects of simulation of neutron damage by ion irradiation, J. Nucl. Mater. 216 (1994) 78–96.
[42] L. Venosta, N. Bajales, S. Suárez, P.G. Bercoff, Disorder in H+-irradiated HOPG: Effect of impinging energy and dose on Raman D-band splitting and surface topography, Beilstein J. Nanotechnol. 9 (2018) 2708–2717.
[43] R.D. Heidenreich, W.M. Hess, L.L. Ban, A test object and criteria for high resolution electron microscopy, J. Appl. Crystallogr. 1 (1968) 1–19.
[44] T.D. Burchell, L.L. Snead, The effect of neutron irradiation damage on the properties of grade NBG-10 graphite, J. Nucl. Mater. 371 (2007) 18–27.
[45] B.T. Kelly, W.H. Martin, A.M. Price, J T. Bland, The mechanism of dimensional changes in the crystals of graphites and carbons under fast neutron irradiation, Philos. Mag. 14 (1966) 343–356.

[46] W. Bollmann, G. Hennig, Electron microscope observations of irradiated graphite single crystals, Carbon N. Y. 1 (1964) 525–533.
[47] C. Baker, A. Kelly, Energy to form and to move vacant lattice sites in graphite, Nature 193 (1962) 235–236.
[48] W.N. Reynolds, P.A. Thrower, The nucleation of radiation damage in graphite, Philos. Mag. 12 (1965) 573–593.
[49] G.W. Hinman, A. Haubold, J.O. Gardner, J.K. Layton, Vacancies and interstitial clusters in irradiated graphite, Carbon N. Y. 8 (1970) 341–351.
[50] C. Baker, A. Kelly, An electron microscope study of radiation damage in single crystal graphite, Philos. Mag. 11 (1965) 729–746.
[51] W.N. Reynolds, P.A. Thrower, Irradiation damage in graphite, Radiat. Damage React. Mater. (1962) 553–563.
[52] J. Koike, D.F. Pedraza, Dimensional changes in highly oriented pyrolytic graphite due to electron-irradiation, J. Mater. Res. 9 (1994) 1899–1907.
[53] J.E. Brocklehurst, B.T. Kelly, The dimensional changes of highly-oriented pyrolytic graphite irradiated with fast neutrons at 430°C and 600°C, Carbon N. Y. 31 (1993) 179–183.
[54] J.E. Brocklehurst, B.T. Kelly, Analysis of the dimensional changes and structural changes in polycrystalline graphite under fast neutron irradiation, Carbon N. Y. 31 (1993) 155–178.
[55] D.F. Pedraza, J. Koike, Dimensional changes in grade H-451 nuclear graphite due to electron irradiation, Carbon N. Y. 32 (1994) 727–734.
[56] T. Tanabe, Radiation damage of graphite - degradation of material parameters and defect structures, Phys. Scr. T64 (1996) 7–16.
[57] S. Muto, T. Tanabe, Damage process in electron-irradiated graphite studied by transmission electron microscopy. I. High-resolution observation of highly graphitized carbon fibre, Philos. Mag. A Phys. Condens. Matter, Struct. Defects Mech. Prop. 76 (1997) 679–690.
[58] M. Takeuchi, S. Muto, T. Tanabe, S. Arai, T. Kuroyanagi, Damage process in electron-irradiated graphite studied by transmission electron microscopy. II. Analysis of extended energy-loss fine structure of highly oriented pyrolytic graphite, Philos. Mag. A Phys. Condens. Matter, Struct. Defects Mech. Prop. 76 (1997) 691–700.
[59] B.S. Elman, M.S. Dresselhaus, G. Dresselhaus, E.W. Maby, H. Mazurek, Raman scattering from ion-implanted graphite, Phys. Rev. B 24 (1981) 1027–1034.
[60] K. Niwase, Formation of dislocation dipoles in irradiated graphite, Mater. Sci. Eng. A 400–401 (2005) 101–104.
[61] S. Johns, et al., Fullerene-like defects in high-temperature neutron-irradiated nuclear graphite, Carbon N. Y. 166 (2020) 113–122.
[62] K. Niwase, Irradiation-induced amorphization of graphite: A dislocation accumulation model, Philos. Mag. Lett. 82 (2002) 401–408.
[63] A. Chuvilin, U. Kaiser, E. Bichoutskaia, N.A. Besley, A.N. Khlobystov, Direct transformation of graphene to fullerene, Nat. Chem. 2 (2010) 450–453.
[64] M.I. Heggie, I. Suarez-Martinez, C. Davidson, G. Haffenden, Buckle, ruck and tuck: A proposed new model for the response of graphite to neutron irradiation, J. Nucl. Mater. 413 (2011) 150–155.
[65] S. Johns, et al., Experimental evidence for 'buckle, ruck and tuck' in neutron irradiated graphite, Carbon 159 (2020) 119–121.
[66] M.W. Barsoum, et al., Ripplocations: A universal deformation mechanism in layered solids, Phys. Rev. Mater. 3 (2019) 013602.

[67] D. Liu, D. Cherns, et al., Nano-cracks in a synthetic graphite composite for nuclear applications, Philos. Mag. (2020) 1272–1283.
[68] B.J. Marsden, G.N. Hall, A.N. Jones, Graphite in gas-cooled reactors, Compr. Nucl. Mater. Second Ed. (2020) 357–421, https://doi.org/10.1016/B978-0-12-803581-8.00729-3.
[69] B.T. Kelly, Irradiation behavior of fine-grain graphites, Fusion Technol. 16 (2017) 96–103, https://doi.org/10.13182/FST89-A29099.
[70] A.A. Campbell, Y. Katoh, M.A. Snead, K. Takizawa, Property changes of G347A graphite due to neutron irradiation, Carbon N. Y. 109 (2016) 860–873.
[71] A.A. Campbell, Y. Katoh, A.P. Selby, T. Muto, R. Yamashita, T. Maiazaki, High-fluence neutron irradiation effects in IGS-743NH nuclear graphite. present. 18th Int. Nucl. Graph. Spec. Meet. (INGSM-2017), Sept. 18-21, 2017, 2017. Balt. Maryl.
[72] D.K.L. Tsang, B.J. Marsden, Effects of dimensional change strain in nuclear graphite component stress analysis, Nucl. Eng. Des. 237 (2007) 897–904.
[73] J.A. Vreeling, O. Wouters, J.G. Van der Laan, Graphite irradiation testing for HTR technology at the high flux reactor in petten, J. Nucl. Mater. 381 (2008) 68–75.
[74] B.T. Kelly, A.J.E. Foreman, The theory of irradiation creep in reactor graphite—The dislocation pinning-unpinning model, Carbon N. Y. 12 (1974) 151–158.
[75] D. Liu, P. Flewitt, Deformation and fracture of carbonaceous materials using in situ micro-cantilever bending, Carbon N. Y. (2016).
[76] D. Liu, K. Mingard, O.T. Lord, P. Flewitt, On the damage and fracture of nuclear graphite at multiple length-scales, J. Nucl. Mater. 493 (2017) 246–254.
[77] J.E. Darnbrough, D. Liu, P.E.J. Flewitt, Micro-scale testing of ductile and brittle cantilever beam specimens in situ with a dual beam workstation, Meas. Sci. Technol. 24 (2013) 055010.
[78] C. Malmstrom, R. Keen, L. Green, Some mechanical properties of graphite at elevated temperatures, J. Appl. Phys. 22 (1951) 593–600.
[79] D. Liu, B. Gludovatz, H. Barnard, M. Kuball, R.O. Ritchie, Damage tolerance of nuclear graphite at elevated temperatures, Nat. Commun. 8 (2017).
[80] N.N. Nemeth, R.L. Bratton, Overview of statistical models of fracture for nonirradiated nuclear-graphite components, Nucl. Eng. Des. 240 (2010) 1–29.

Ceramics for nuclear wasteforms

6

Michael J.D. Rushton [1], *S. Middleburgh* [1], *T. Ming* [1,2] *and W.E. Lee* [1]
[1]Nuclear Futures Institute, Bangor University, Bangor, United Kingdom; [2]Department of Materials, Clemson University, Clemson, SC, United States

1. Nuclear waste immobilization mechanisms

Immobilization reduces the potential for migration or dispersion of contaminants, including radionuclides. The product of incorporating waste into a suitable matrix is a wasteform. The IAEA (International Atomic Energy Agency) defines immobilization as the conversion of a waste into a wasteform by solidification, embedding or encapsulation. It facilitates handling, transportation, storage and disposal of radioactive wastes [1]. The processing routes used to make wasteforms are described in Section 2.

The type of wasteform chosen to host the waste depends on the activity of the waste (is it heat generating High Level Waste, HLW, or Intermediate Level Waster, ILW) and how long it needs to be locked up for [2]. Immobilization of HLW, where the radionuclides need to be kept out of the biosphere for thousands of years, is achieved by ***chemical incorporation*** into the atomic structure of a suitable matrix (typically glass or ceramic) so that they are captured and unable to escape (left side Fig. 6.1). Encapsulation is more usually used for low or intermediate level wastes (LILW) where the radionuclides need only be kept isolated for a few hundred years at most. ***Physical encapsulation*** is at the micro- or macro-structural level and is achieved by surrounding the waste in materials (typically bitumen or cement) so it is isolated and radionuclides are retained (right side Fig. 6.1). Encapsulation may include some chemical incorporation as described for cements (via cementation) by Ref. [3] and chemical incorporation may include some physical encapsulation as in glass composite materials and illustrated in Fig. 6.1 by the blue and red lines crossing each other. Physically encapsulating ceramics are described in Section 3, and chemically immobilizing in Section 4.

Vitreous wasteforms represent one end of the spectrum of HLW wasteforms shown in Table 6.1.

The open and random glass structure is able to accommodate most of the Periodic Table, and radioactive species can go into the glass network or other sites. For example, alkalis (^{137}Cs, ^{90}Sr) sit on the large Na site in a sodium borosilicate glass [4]. The key issue during production of vitreous wasteforms is to balance the waste loading (to minimize volume), viscosity (ease of pouring), volatilization (of high vapor pressure species), with wasteform durability and simplicity of the process. For a variety of reasons, some species present in wastes, including ^{99}Tc, ^{36}Cl, ^{129}I, and ^{14}C are inherently difficult to immobilize in glasses [5]. In particular, halogens have low

Ceramics for Nuclear Applications. https://doi.org/10.1016/B978-0-12-823865-3.00006-7

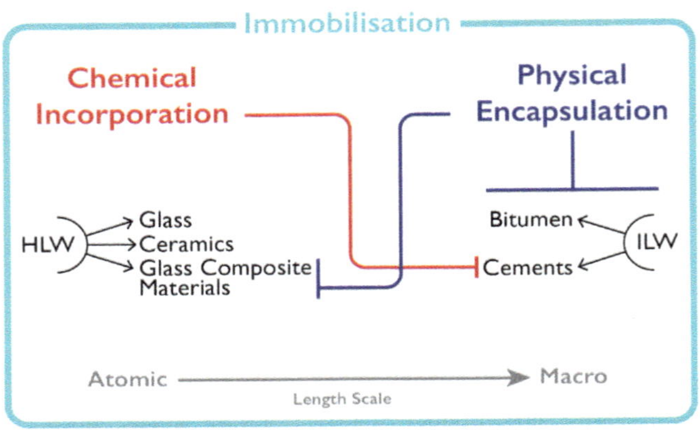

Figure 6.1 Definition of immobilization into chemical incorporation and physical encapsulation mechanisms.

Table 6.1 Classification of types of glass/ceramic wasteforms.

Glasses	Glass composite materials (GCMs)	Cements	Crystalline ceramics
Magnox, Blend, UK DWPF, WVDP, USA	Glass ceramics: zirconolite for MA or Pu	Multiphase composite cements, UK and France	Single phase: zirconolite for MA or Pu, hollandite for Cs.
Aluminophosphate, Russia. R7/T7, France. IR111, India	Crystal waste encapsulated in a glass matrix Crystal-tolerant glasses: U–Mo CCM French wastes, high alkali Hanford wastes	Novel cements, e.g., calcium sulphoaluminates	Multiphase (e.g., synroc) compositions tailored for specific wastestreams Spent fuel waste

DWPF, defense waste processing facility; *WVDP*, west valley demonstration project; *Synroc*, synroc consists of titanates hollandite ($BaAl_2Ti_6O_{16}$), zirconolite ($CaZrTi_2O_7$), perovskite ($CaTiO_3$) and TiO_2. Hollandite mainly fixes the alkali and alkaline earth FP and some process chemicals, whereas actinides and rare earth elements are bound in zirconolite and perovskite.
Adapted from Ref. [2].

solubility in glasses and are volatile at typical glass processing temperatures, while the melting temperatures of phases such as Al_2O_3 are well above glass processing temperatures, leading to bubbles or crystals in the glass.

At the other end of the spectrum shown on the right side of Table 6.1 the use of predominantly crystalline ceramic wasteforms (ceramication) has also been proposed including single-phase ceramics such as zircon or zirconolite [6] to accommodate a limited range of active species such as Pu, and multiphase systems such as Synroc (synthetic rock) to accommodate a broader range of active species [7]. To date, these systems have not been extensively used to immobilize active waste. In addition, spent fuel is directly disposed of in the open fuel cycle and so represents a wasteform, if not deliberately designed to be so. Cements are also used extensively to encapsulate ILW and contain crystalline and amorphous phases.

Like cements, many other wasteforms are intermediate between the "completely" glassy or crystalline materials [8]. Various names have been applied to these glass/crystal mixture wasteforms, including Glass Composite Materials (GCMs) [9], which we shall use here. GCMs can be divided into various types, such as:

- Glass-ceramics where a glassy wasteform is crystallized in a separate heat treatment or which partly crystallize on cooling during, e.g., Cold Crucible Induction Melting (CCIM) [10].
- Materials in which a refractory waste, such as $La_2Zr_2O_7$, is encapsulated in glass to immobilize MA (Minor Actinides, e.g., Np, Am, and Cm) [11] (right side of Fig. 6.1)

These will be discussed in detail in Section 5.

An illustration of nuclear wasteforms used and developed for industrial application is given in Fig. 6.2.

Typically, nondurable crystals are sodium sulfate (Na_2SO_4), sodium molybdate ($Na_2MoO_4.2H_2O$), NaF, and NaCl, while durable crystals include hollandite ($BaAl_2Ti_6O_7$), zirconolite ($CaZrTi_2O_7$), calcium molybdate, powellite ($CaMoO_4$), perovskite

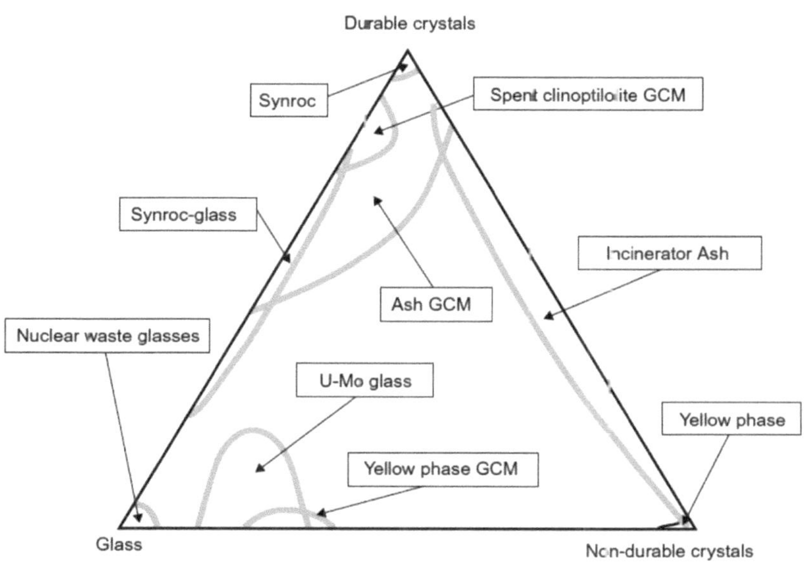

Figure 6.2 Phase compositions of nuclear wasteforms.
From Ref. [9].

($CaTiO_3$), and TiO_2. All efforts are aimed at retaining durable crystals in the wasteform and, if present, isolating nondurable crystals by encapsulation in durable glass.

The general requirements for wasteforms include, but are not limited to:

- High waste loading: the system must be able to accommodate a significant amount of waste (typically 20%–35%) to minimize the wasteform volume, saving space in a potential repository.
- Easy processing: fabrication should be accomplished under reasonable conditions, including low temperatures and ideally an air atmosphere, using well established methods to minimize worker dose and capital cost of the plant.
- High radiation stability: the wasteform should show a high tolerance to radiation effects from decay products to reduce phase changes (which may be detrimental to durability) and volume swelling (resulting in cracking).
- Chemical flexibility: if required, the system should be able to accommodate a mixture of radionuclide and other contaminant species, with minimum formation of secondary phases which may compromise wasteform performance.
- Durability: the wasteform should be resistant to aqueous dissolution under conditions relevant to eventual repository disposal to minimize release of radiotoxic species. Since the half-life of, for example, ^{239}Pu is 24,100 years, a minimum safe containment period of 10 half-lives, i.e., 241,000 years, is required.
- Natural analogs: the availability of natural mineral analogs of the wasteform may give important clues as to the effects on the material of the natural environment, thereby building confidence in the future performance.

2. Wasteform production

This section introduces the various ways of making wasteforms in terms of increasing temperature. The nonthermal (low temperature) techniques are used or being developed predominantly for ILW (Fig. 6.3).

Fig. 6.4 shows the thermal technologies used or under development largely to process HLW, so the higher temperature and hence cost is less problematic than for ILW, although, e.g., radionuclide volatilization and potential to escape to the atmosphere means that expensive filtration systems are often needed.

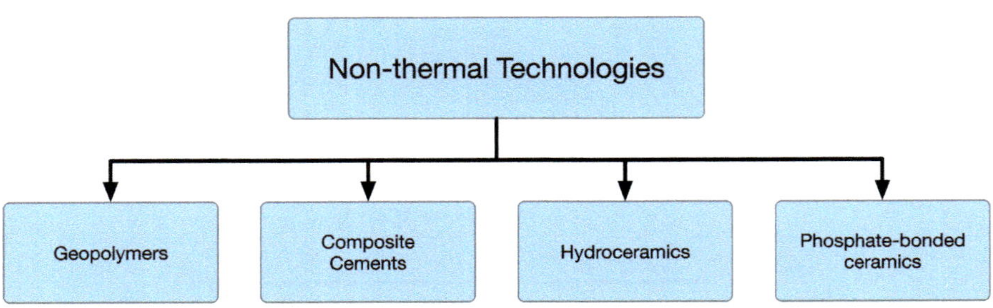

Figure 6.3 Nonthermal technologies used or under development are predominantly for ILW wasteforms. The high volume of ILW material to be immobilized means these techniques must be inexpensive and hence low temperature.

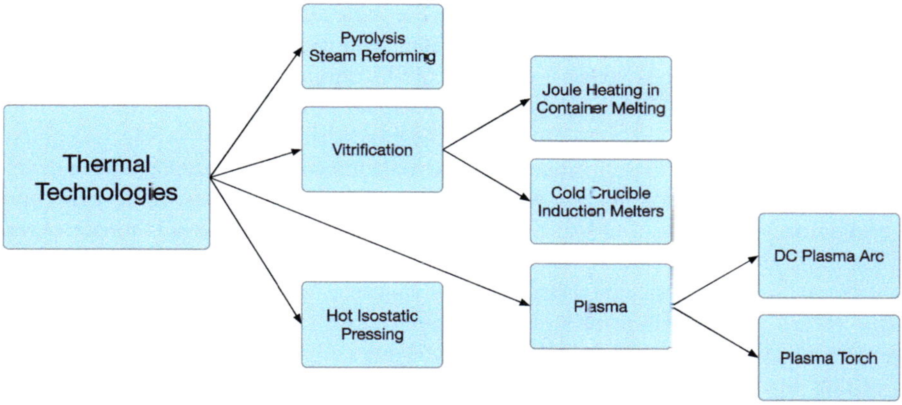

Figure 6.4 Various higher temperature processing technologies are used mostly for HLW immobilization.

The main immobilization technologies that are available commercially and that have been demonstrated to be viable are cementation and vitrification, the latter mainly using borosilicate glasses capable of hosting a range of radionuclides in the open glass structure. Ceramication, in which a range of crystals capable of hosting radionuclides in their structures, such as zirconolite, perovskite, and zircon, is now used commercially. Single-phase systems have been predominantly examined for Pu immobilization, while more complex waste streams need multiphase systems such as the Synroc (synthetic rock) ceramics developed in Australia.

2.1 Room or low temperature processing

Hydraulic cements are inorganic materials that have the ability to react with water under ambient conditions to form a hardened and water-resistant product. Composite systems (Fig. 6.3) based on Ordinary Portland cements (OPCs) are used extensively to encapsulate ILWs and contain crystalline and amorphous phases. OPCs are made up of four main components: tricalcium silicate C_3S (50%–60%), dicalcium silicate β-C_2S (20%–30%), tricalcium aluminate C_3A (8%–12%) and an iron-containing phase C_4AF. These hydrate at room temperature, providing the calcium silicate hydrate (CSH) matrix on setting and are used to encapsulate contaminated solid materials usually contained in a metal drum. Cements offer a number of attributes that make them suitable for encapsulation and immobilization: they are inexpensive and readily available, can be readily prepared and used in remote operations and can be prepared as a fluid grout able to penetrate complex waste shapes. They also assist immobilization of radionuclides by acting as a diffusion barrier, providing sorption and reaction sites and maintaining a high pH that in turn decreases radionuclide solubility. Cements also provide a radiation-stable binder and have controllable permeation and diffusion characteristics over a wide range via selection of their constituents. Drawbacks of composite cements, however, include their alkalinity, which makes them reactive with some wastes (such as Al), their strengths come from hydrated phases, so they cannot usefully be dehydrated, and radiolytic H_2 can be generated along with other gaseous species, causing complications in repository disposal.

Composite cement systems may use additional powders such as Blast Furnace Slag and Pulverized Fuel Ash. These offer cost reduction, energy saving and potentially superior long-term performance. Improved composite cements have been developed using, e.g., radionuclide-sorbing inorganic additives such as bentonite, vermiculite, and zeolites. Organic additives, such as naphthalene sulphonate superplasticizer, are used to reduce water to cement ratio by imparting fluidity and giving a stronger, lower porosity, and permeability concrete. The flexibility of cement systems means that a toolbox of systems made at close to room temperature is being developed, applicable to the complex array of wastes which they must encapsulate [3].

Other low-temperature processed potential hosts for radioactive waste include **Geopolymers,** which are based on aluminosilicate polymers and were first developed by Davidovits and are proving suitable for hosting wastes in chemically demanding situations [12,13]. The ingredients are a soluble alkali silicate and a reactive aluminosilicate precursor such as metakaolinite. The range of materials that can be used indicates flexibility to incorporate a range of waste materials in a medium which ultimately does not rely on a hydrate matrix for strength, as the binder is a 3D amorphous aluminosilicate.

Hydroceramics are another concrete-type material similar to zeolitized rock [5]. They are made by curing a mixture of inorganic waste, calcined clay, vermiculite and Na_2S, NaOH with water under hydrothermal conditions (60–200°C) to form a matrix containing crystalline zeolites embedded in a sodium aluminosilicate matrix.

Phosphate-bonded ceramics [14] are a very different class of inorganic cements and form the basis of glass ionomer cements used in dental and medical applications, as well as being suitable for rapid repairs to damaged concrete, for example, to concrete pavements and airport runways. They rely on acid–base reactions, such as that which occurs between magnesium oxide and an acid phosphate source such as ammonium dihydrogen phosphate, $NH_4H_2PO_4$, or aluminum orthophosphate, $Al(H_2PO_4)_3$. The products are fast-setting with high-early strength development, but the reactions can be controlled to some extent by the use of retarders and various grades of filler such as sand. The product is effectively a composite material with grains of unreacted oxide particles surrounded by a matrix of hydration products. The magnesia can be replaced, partially or fully, by other basic and amphoteric oxides. These materials are sometimes known as chemically bonded ceramics. Because of the ionic nature of the setting mechanism, which gives rise to a rapid reaction with considerable heat evolution, this system is unlikely to be extensively applicable to radioactive waste disposal. It may, however, be useful for the treatment of specific waste streams that require immobilization in a product with a much lower pH than that associated with alkaline cementing systems.

2.2 Melt processing

Vitrification from a molten liquid melt is used in a number of countries (e.g., the United States, United Kingdom, Russia, Japan, and France) to immobilize heat-generating HLW from reprocessing. Vitrification is attractive because of the high capability of glass to reliably immobilize a range of elements; the production technology adapted from glass manufacture is simple and proven over many decades; the small volume of the resulting glassy wasteform; the high chemical durability of glassy wasteforms in contact with natural waters; and the high tolerance of glasses to radiation damage.[35] The open and random glass structure is able to accommodate most of the periodic table, and radioactive species can go into the glass network or other sites. For example, alkalis (^{137}Cs, ^{90}Sr) sit on the large Na site in a sodium borosilicate glass. The borosilicate glass compositions used are very similar to Pyrex (which used to be made by Corning for cookware). The exact compositions of nuclear waste glasses are tailored for easy preparation and melting, avoidance of phase separation and uncontrolled crystallization, and acceptable chemical durability, for example, leaching resistance. Vitrification can be performed at temperatures below 1200°C, thus avoiding excess radionuclide volatilization and maintaining viscosities below 10 Pa s to ensure high throughput and controlled pouring into canisters.

The key issue during the production of vitreous wasteforms is to balance the waste loading (to minimize volume), viscosity (ease of pouring), volatilization (of high-vapor pressure species), with wasteform durability and simplicity of the process. Two vitrification processes are used to make wasteforms in the one-stage process (as used at Savannah River and West Valley in the United States; Fig. 6.5): glass-forming powders are mixed with concentrated liquid wastes in a Joule-heated melter (JHM). The batch is fed to the melter, where water evaporation, calcination, and glass melting occur directly.

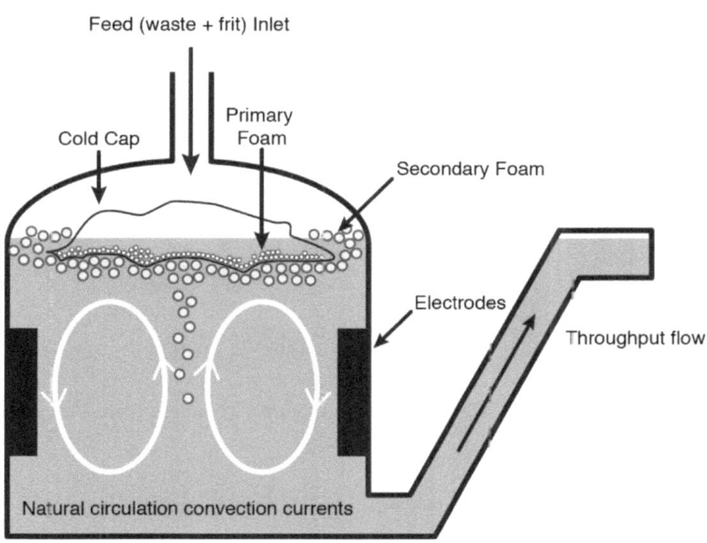

Figure 6.5 Schematic crosssection of a Joule-heated melter.
From Ref. [15].

In the two-stage process (as used at Sellafield, in the United Kingdom, and La Hague, in France; Fig. 6.6), liquid waste concentrate is fed into a separate rotating calcining furnace. After calcination, the calcine product and glass frit are fed into the induction-heated melting furnace.

Figs. 6.5 and 6.6 reveal the development of two quite different technologies that have evolved in different countries, but which both produce an acceptable end product. Radioactive waste management programs and technologies for various countries are examined in detail in Ref. [16].

A more recent development is the application of cold crucible melting to the production of vitrified radioactive wasteforms. The cold crucible is a compact, cooled melter in which the radioactive waste and the glass are melted by direct high-frequency induction heating. This technology can handle highly corrosive solutions and high operating temperatures. Cold Crucible Induction Melter (CCIM) technology was initiated in Russia and France to process radioactive waste since the 1990s [17],

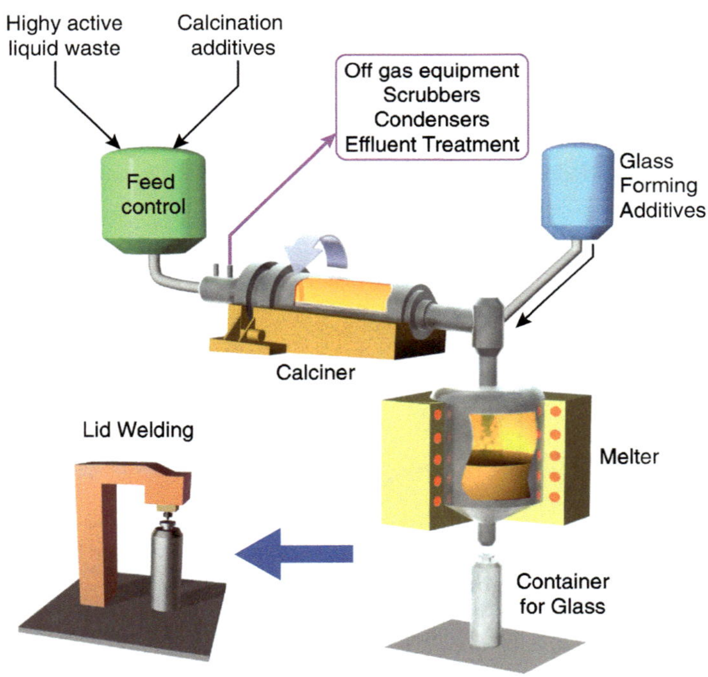

Figure 6.6 Schematic of the two-stage glass-melting process utilizing separate calcining and melting furnaces.
From Ref. [15].

and it has been demonstrated that it can achieve the temperatures required to melt glass wasteforms and ceramic synroc materials [18,19]. The CCIM employs induction heating by coupling a water-cooled high-frequency electrical coil to the materials, causing eddy currents that heat and mix the material. The absence of internal electrodes and refractories allows for increased melter longevity and permits higher-temperature operation. In contrast to established Joule heated melters (JHMs, the reference melter technology used in the U.S. for processing borosilicate HLW glass Fig. 6.5), it better accommodates the formation of ceramic forms given the CCIM's higher temperatures, efficient mixing, tolerance to crystal formation and bottom drain configuration. CCIM also allows for higher throughputs and waste loadings.

An illustration of CCIM operation at US DOE Idaho National Laboratory [20], is shown in Fig. 6.7, (a) is the schematic drawing of the CCIM, (b) is the drain of the CCIM test run. The CCIM design is much less sensitive to wasteform chemistry than JHMs, which allows feeds with more diversified chemistry and lessens the burden put on any pretreatment processes. As such, the CCIM runs at much higher temperatures than is necessary to allow for processing alternative wasteform chemistries such as glass ceramics and ceramics.

In Russia, a pilot plant for vitrifying HLW using a cold crucible melter design is in operation at the Mayak facility [18]. Phosphate glass and borosilicate glass wasteforms have been fabricated using this method (Smelova et al. 2000; [21,22]). In 2010, a CCIM started hot operation for the first time ever in an existing very highly active facility at La Hague, France ([23,24]; Labe et al. 2014). The CCIM has now been operated for essentially processing active effluents from DD (Deactivation and Decommissioning) operations and high-level liquid waste from reprocessed

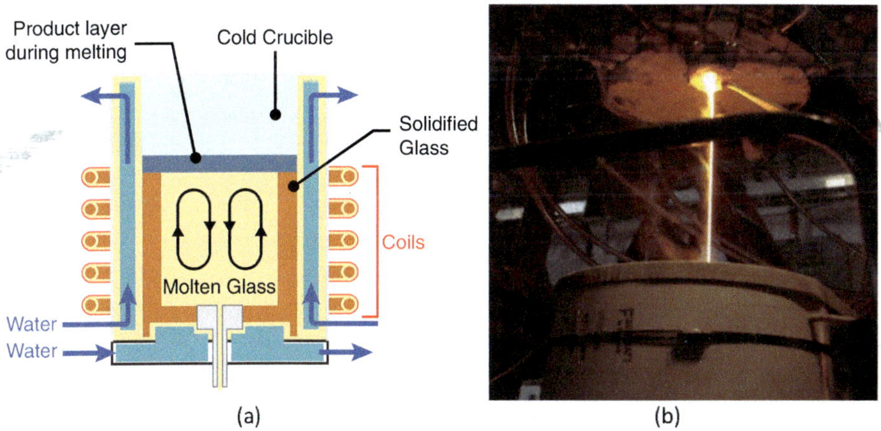

Figure 6.7 (a) Schematic cutaway view of the CCIM. Compared to a JHM, the lack of in-melt electrodes and refractory allows for greater mixing, longer life, greater throughput, less footprint and temperatures exceeding 1700°C. Higher temperature allows for the potential melt production of highly durable HLW forms containing ceramic phases. (b) Drain of the CCIM test run (ambient cooled). Consequently, it was the fastest cooled of all three pours, and the heating mantles for controlled cooling remained off.

U–Mo–Sn–Al spent fuel. These solutions are very rich in molybdenum and phosphorus, whose contents make the molten glass quite corrosive and require a high-temperature glass formulation to obtain sufficiently high waste loading factors (12% in molybdenum oxide). The feasibility of this process has been demonstrated in a full-scale pilot facility with inactive surrogate solutions [25].

2.3 High temperature processing of ceramics

Processing of ceramic wasteforms without complete melting has been extensively studied. In the original Synroc process (Fig. 6.8), the matrix was used to immobilize Fission Products (FPs) and nonextracted actinides from reprocessing of spent fuel by

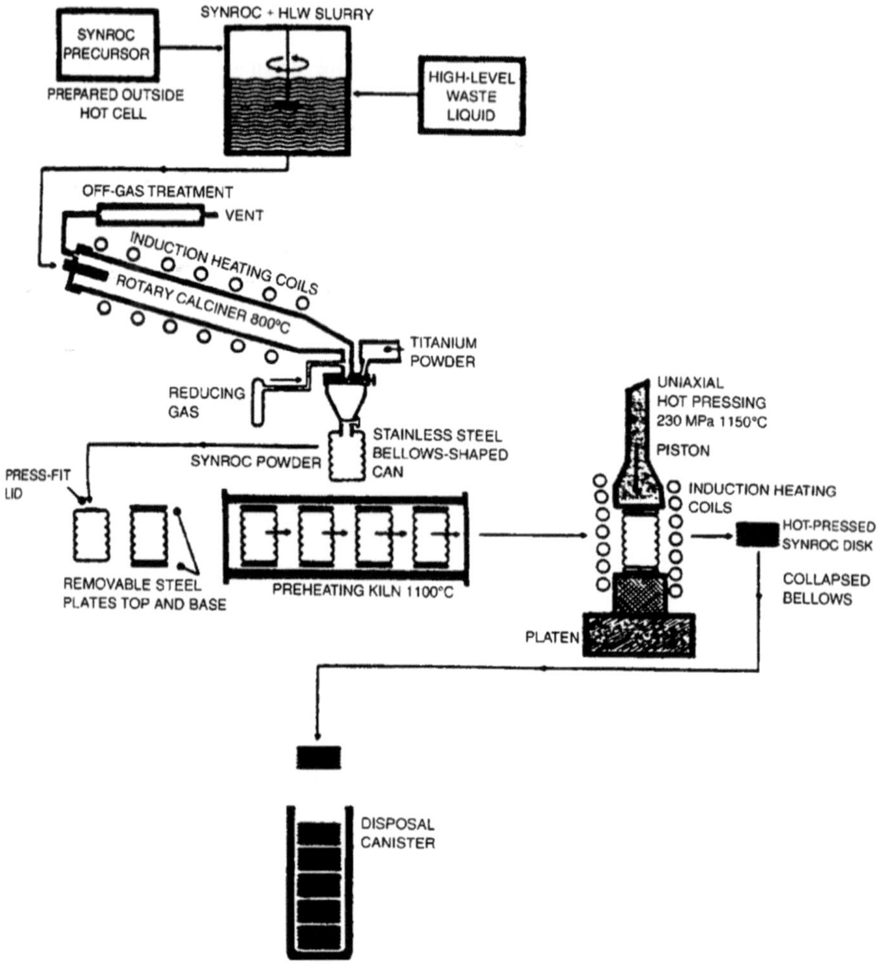

Figure 6.8 Schematic of the Synroc Process developed by the Australian Nuclear Science and Technology Organization (ANSTO).

Hot Isostatic Pressing (HIPing) them into a multiphase ceramic host, leading to direct incorporation into the lattice structure, dividing the actinides into the zirconolite phase and FPs into the hollandite and perovskite phases. The Synroc process uses a relatively simple chemical route involving Ti and Zr alkoxide hydrolysis in the presence of NaOH, followed by drying/calcining under reducing conditions and then hot pressing at 1100–1170°C + 2 wt% Ti to lower the mobility of volatiles and keep Mo metallic, so as to avoid water-soluble molybdates.

Ceramic processing of radioactive wastes is not done at very high temperatures, not only to avoid significant melting but also to avoid uncontrolled release of radioactive volatile species (mostly alkalis). Other key factors when designing ceramic wasteforms include that all host phases are multi-cation and may accommodate several waste cations, and the need to understand charge compensation mechanisms for aliovalent substitutions. Incorporation of separated actinides is an even simpler process, as only a single-phase wasteform is needed as opposed to a multiphase composition such as Synroc. The advantages of this are that it allows a much more precise crystallo-chemical understanding of the incorporation method and, by using a synthetic mineral composition analogous to that found in nature, can provide a means of assessing the long-term behavior of the wasteform.

3. Physically encapsulating ceramics

3.1 Cements

Cements are used to immobilize both liquid (aqueous) and solid radioactive wastes. Solid radioactive waste is embedded into cementitious matrices by pouring grout or mortar into containers with solid waste. Fig. 6.9 shows examples of solid LLW encapsulated via cementation at the waste encapsulation plant in Sellafield, UK.

Figure 6.9 Cement-encapsulated solid, liquid and slurry ILW in 500 L drums. Courtesy of NNL, UK.

Liquid waste-cement mixtures may be prepared either directly in the container (in-drum mixing), which is the final product container or prior to pouring into the container (in-line mixing). After in-drum mixing, the cement–waste mixture is allowed to set, the container is capped with a different composition cement to minimize void space and to avoid surface contamination, and a lid is fitted. In-line mixing processes combine the waste, any additives, water and cement before they are placed into a disposal container.

Composite cement systems of ordinary Portland cement (OPC) with additional waste materials such as pulverized fuel ash and ground granulated blast-furnace slag have been used for the encapsulation of radioactive wastes for many years. These composite systems are cheaper than pure OPC with lower heats of hydration and potentially superior long-term durability. Cements offer a number of attributes that make them suitable for encapsulation and immobilization: they are inexpensive and readily available, can be readily prepared and used in remote operations and can be prepared as a fluid grout able to penetrate complex waste shapes. They also assist immobilization of radionuclides by acting as a diffusion barrier, providing sorption and reaction sites and maintaining a high pH that in turn decreases radionuclide solubility. Cements also provide a radiation-stable binder and have controllable permeation and diffusion characteristics over a wide range via selection of their constituents. Drawbacks of composite cements, however, include that their alkalinity makes them reactive with some wastes (such as Al), their strengths come from hydrated phases, so they cannot be usefully dehydrated, and radiolytic H_2 can be generated along with other gaseous species, causing complications on repository disposal. Cement systems in the United States (where they are termed *saltstone*) are under development for low-activity wastes at Hanford and Savannah River.

Many thousands of the steel drums shown in Fig. 6.9 are expected to be deposited in geological repositories at depths of over 1000 m.

4. Chemically immobilizing glasses and ceramics

4.1 Borosilicate and phosphate glasses

Borosilicate glasses are the first choice of material worldwide for immobilizing both HLW and LILW. This selection is based on the flexibility of borosilicate glass with regard to waste loading and the ability to incorporate many different kinds of waste elements, coupled with good glass-forming ability, chemical durability, mechanical integrity and excellent thermal and radiation stability. The major component of borosilicate glasses is SiO_2 with relatively high B_2O_3, CaO, MgO, Na_2O, and Al_2O_3 contents and minor amounts of many other oxides. SiO_2, B_2O_3, and Al_2O_3 are generally network formers because they form strong covalent bonds involving SiO_4, AlO_4 and BO_4 tetrahedra and BO_3 triangles. Phosphate glasses have been intensively studied in Russia at Eurochemic.

Corporation at Mol, Belgium, at Oak Ridge National Laboratory and at the University of Missouri–Rolla in the US. Russia has been immobilizing HLW from the nuclear fuel reprocessing plant RT-1 in the Ural region in alumina-phosphate glass since 1987. Molten phosphate glasses are highly corrosive to refractory linings, a behaviour that has limited their application. Novel Fe–Pb-phosphate glasses are particularly attractive due to their ability to accommodate enhanced amounts of refractory oxides and their high chemical durability.

The exact compositions of nuclear waste glasses (Table 6.2) are tailored for easy preparation and melting, avoidance of phase separation and uncontrolled crystallization and acceptable chemical durability, for example, leaching resistance. High waste loadings and chemical durability can be achieved in both borosilicate and aluminophosphate glasses. Moreover, such glasses immobilize large quantities of actinides; for example, borosilicate glasses can accommodate up to 7.2 mass% of PuO_2. Phosphate glasses can accommodate large amounts of aluminum oxides; however, in contrast to borosilicate melts, molten phosphate glasses are highly corrosive to refractory linings, behavior that has limited their application. Alumina-phosphate glass has been used in Russia to immobilize HLW from nuclear fuel reprocessing since 1987. Recently, borosilicate glasses have been developed to host Hanford high-Al radioactive waste (SRNL glass, Table 6.2).

4.2 Single- and multi-phase ceramics

Thermodynamically stable crystalline materials sometimes have high natural radionuclide contents that are retained for millions of years. Many durable crystals have a high isomorphic capacity for the incorporation of elements, including radionuclides from nuclear waste. The basic idea is to use as immobilizing wasteforms those durable minerals that have been preserved in natural conditions for geological times, so-called natural analogs. Some potential host minerals are given in Table 6.3 with data on radionuclides that can most efficiently be immobilized, radiation durability and characteristic normalized leaching rates (NR). The radiation durability is given in dpa—the maximum number of displacements per atom caused by the irradiation above which the material loses its crystalline structure, that is, radiation amorphization occurs. The mechanisms by which radionuclides are accommodated vary with species and crystal type but include direct substitution of isovalent species on particular lattice sites, altervalent substitution facilitated by charge compensation on a nearby site, insertion into open channels within the crystal structure and incorporation by generation of planar defects such as twins and crystallographic shear planes.

Single-phase minerals can host a large number of nuclides and can be used as a monophasic wasteform. However, monophase ceramics are difficult to fabricate, and multiphase compositions are more common. The composition of the multiphase ceramic is tailored to that of the waste composition to achieve complete and reliable immobilization of the waste constituents. The most famous multiphase ceramic for nuclear waste immobilization is Synroc. Synroc is short for 'Synthetic Rock', invented in 1978 by T. Ringwood of the Australian National University. Synroc is made of geochemically stable natural titanate minerals, which have naturally immobilized uranium and thorium for billions of years. U/Th-containing natural analogs of the basic constituent of Synroc-zirconolites from Sri Lanka, dating back 550 million

Table 6.2 Typical nuclear waste glass compositions.

Glass/Country oxide (wt%)	SiO_2	P_2O_5	B_2O_3	Al_2O_3	CaO	MgO	Na_2O	Misc	Waste oxides
R7/T7, France	47.2	–	14.9	4.4	4.1	–	10.6	18.8	28
DWPF. USA	49.8	–	8.0	4.0	1.0	1.4	8.7	27.1	33
SRNL, USA	30.5	1.1	15.2	25.0	6.1	0.1	9.6	13.5	45[a]
WVP, UK	47.2	–	16.9	4.8	–	5.3	8.4	17.4	25 (up to 35–38)
Pamela, Germany–Belgium	52.7	–	13.2	2.7	4.6	2.2	5.9	18.7	30
Mayak, Russia	–	52.0	–	19.0	–	–	21.2	7.8	33[b]
Radon K-26, Russia	43	–	6.6	3.0	13.7	–	23.9	9.8	35[c]
P0798, Japan	46.6	–	14.2	5.0	3.0	–	10.0	20.2	
GC-12/9B, China	46.2	–	13.4	4.2	2.5	1.5	9.1	23.1	

DWPF, Defense Waste Processing Facility, Savannah River Site, US; *SRNL*, Savannah River National Laboratory, US; *WVP*, Waste Vitrification Plant, Sellafield, UK.
[a]This glass has been developed to host Hanford high-Al radioactive waste.
[b]$\leq$10 for fission products and minor actinide oxides.
[c]This glass is designed for sodium-containing LLW and ILW.

Table 6.3 Potential single phase ceramic hosts for racioactive waste species.

Mineral	Elements from waste	Radiation durability (dpa)	NR (g/cm^2 day)	Structure
Monazite: (Ce,La,Nd,Th)PO_4	Ln, An	>10	10^{-7}	Monoclinic, *P2$_1$/n*
Zircon: $ZrSiO_4$	Ln, An, Nb, Ta, Hf	0.3–0.4	4×10^{-7}	Tetragonal *I4/amd*
Zirconolite: $CaZrTi_2O_7$	Ln, An, Nb, Sc, Y, Hf	0.2–0.3	4.5×10^{-6}	Monoclinic
Pyrochlore: $A_2B_2O_7$ A: Na, K, Ca, Sr, Ba, Fe^{2+}, RE, U, Th, …; B=Nb, Ta, Ti, Fe^{3+}, …; natural pyrochlore $A_2B_2O_7(OH,F)$	Na, Y, Ln, An, Ti, Nb, Ta, W, Cl, I	0.3–0.4	1.5×10^{-6}	Cubic, *Fd3m*
Zirconia: ZrO_2	Zr, Ln, An	>10		Several polymorphs. Mineral form is Baddeleyite, Monoclinic *P2/c*
Garnet: $A^{VIII}{}_3B^{VI}{}_2[SiO_4]_3$	Cr, Mn, Fe, Co, Ni	0.2		Cubic, *Ia3d*
Hollandite: AB_8O_{16} A = Na, K, Rb, Cs, Sr, Ba, Pb; B=Co, Ni, Fe, Cr, Si, Ti, Mn, $BaA1Ti_6O_{16}$	Na, K, Rb, Cs, Sr, Ba, Ra, Ti, Cr, Mn, Fe, Co, Ni, Mo, Pb, Bi, Ag		10^{-6}	Monoclinic, *I4/m*
Perovskite: ABO_3 $CaTiO_3$	Nb, Fe, Ta, Ln, An, Na, Sr, Y	0.4–1	2.5×10^{-8}	Cubic, *Im3*
Apatite: $Ca_2Nd_8(SiO_4)_6O_2$ Britholite: Ca_2Ln_8 $(SiO_4)_6O_2$	Na, Sr, Ln, An, S, I, Y, Mn	0.24×0.3–0.4	2×10^{-7} 2.5×10^{-6}	Hexagonal
Murataite: $A_4B_2C_7O_{20-22}$ A = Ca, Mn, REE, An; B=Mn, Ti, Zr, U; C=Ti, Al, Fe	Na, Ca, Al, Ti, Mn, Fe, Ni, Ln, Ce, Nd, An	0.2	10^{-9}	Cubic, *Fm3m*
NZP: $NaZr_2(PO_4)_3$	Na, K, Rb, Cs, Sr, Ln, An, Fe			
TiC	^{14}C		10^{-6}	Cubic, *Fm3m*

An, actinides; *Ln*, lanthanides.

years, while amorphized have nonetheless withstood the alteration processes of their natural environment.

Key factors when designing ceramic wasteforms include understanding that host phases may be multi-cation and may accommodate several waste cations, and the need to understand charge compensation mechanisms for aliovalent substitutions. Incorporation of separated actinides such as Pu is an even simpler process, as only a single phase wasteform is needed as opposed to a multi-phase composition such as Synroc. The advantages of this are that it allows a much more precise crystallo-chemical understanding of the incorporation method and, by using a synthetic mineral composition analogous to that found in nature, can provide a means of assessing the long-term behavior of the wasteform.

Synroc can take various forms depending on its specific use and can be tailored to immobilize particular components in the HLW. The original form, Synroc-C, was intended mainly for the immobilization of liquid HLW arising from the reprocessing of light water reactor fuel. The main minerals in Synroc-C are hollandite ($BaAl_2Ti_6O_{16}$), zirconolite ($CaZrTi_2O_7$) and perovskite ($CaTiO_3$). Zirconolite and perovskite are the major hosts for long-lived actinides such as plutonium (Pu), though perovskite is principally for strontium (Sr) and barium (Ba). Hollandite principally immobilizes cesium (Cs), along with potassium (K), rubidium (Rb) and barium. Synroc-C (Fig. 6.10) can hold up to 30 wt% HLW. Synroc-D contains nepheline—(Na, K) $AlSiO_4$ instead of hollandite as host for Cs, Rb and Ba. Synroc-F is enriched in pyrochlores—$(Ca,Gd,U,Pu,Hf)_2Ti_2O_7$—and has been developed for the disposal of unreprocessed spent fuel from light water and CANDU (CANadian Deuterium Uranium) reactors. Zirconolite-rich Synroc has been developed for immobilization of excess plutonium. Zirconolite variants containing hollandite and rutile were produced

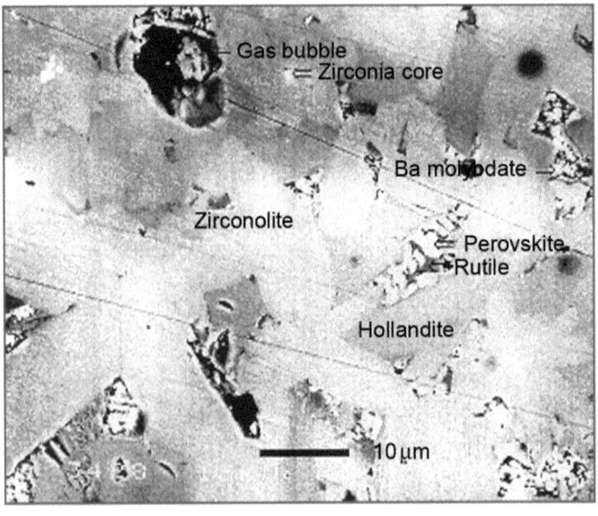

Figure 6.10 Microstructure of multiphase ceramic Synroc containing 30 wt% HLW calcine. Courtesy SV Stefanovsky, SIA Radon, Moscow, Russia.

either by sintering the ingredients in air at 1375°C or by HIPing of calcines at 1280°C and 150 MPa. Pyrochlore-rich Synroc was found to be more efficient for immobilizing the uranium contained in the waste. Pyrochlore is similar to zirconolite and can incorporate up to 50% by mass of PuO_2 and/or UO_2.

5. Glass composite materials

Over the last decade, there has been a trend toward systems intermediate between the 'completely' glassy or crystalline materials. GCMs sit between the fully amorphous glasses and fully crystalline ceramics and contain both glassy and crystalline phases. Depending on the intended application, the major component may be a crystalline phase with a vitreous phase acting as a bonding agent, or, alternatively, the vitreous phase may be the major component, with particles of a crystalline phase dispersed in the glass matrix. GCMs include: (1) glass-ceramics where a glassy wasteform is crystallized in a separate heat treatment or which partly crystallize on cooling often in a cold crucible induction melter (CCIM) and (2) materials in which a refractory waste such as $La_2Zr_2O_7$ is encapsulated in glass to immobilize MA (Minor Actinides e.g., Np, Am, and Cm) [11].

Fig. 6.11 shows the microstructure of CCIM waste glass and glass ceramic samples from Russia and France.

Collaborations between US DOE national laboratories and Russian facilities demonstrated the high efficiency of CCIM with respect to iron phosphate glass [26] and borosilicate glass wasteforms [22]. The US DOE is currently investigating CCIM technology for possible use in its HLW immobilization programs [20]. Scaled proof-of-principle CCIM tests to process multiphase borosilicate glass ceramic and SYNROC-type ceramic wasteforms from a simulated combined (Cs/Sr, lanthanide and transition metal fission products) commercial used nuclear fuel waste stream were recently conducted in the Idaho National Laboratory with Pacific Northwest National Laboratory and Savannah River National Laboratory. Bench scale and melter run results successfully demonstrated the processability of the glass ceramic [27] and ceramic [28] using the CCIM technology. The produced materials were

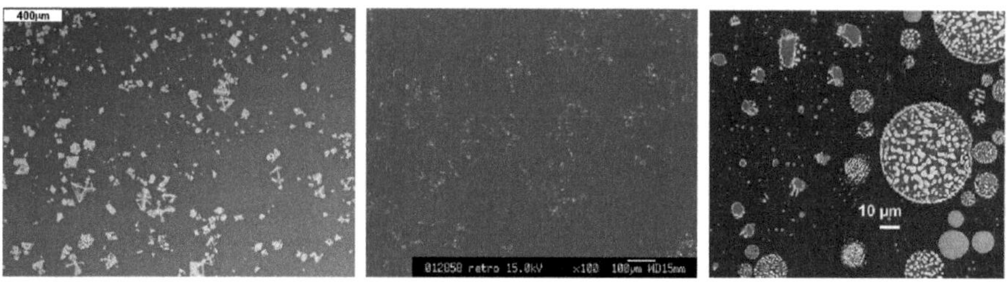

Figure 6.11 (a) Borosilicate glass [22], (b) borosilicate glass (Labe et al. 2014), (c) U–Mo glass ceramic [25].

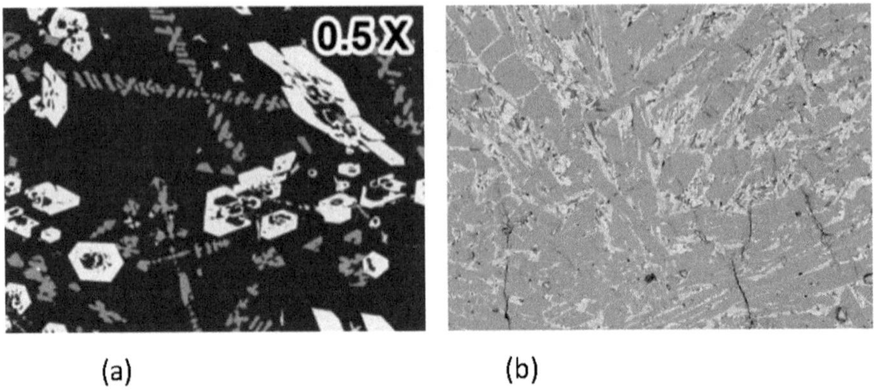

Figure 6.12 (a) CCIM glass ceramic [27], (b) CCIM crystalline ceramic [28].

chemically homogeneous and displayed a uniform phase assemblage with acceptable aqueous chemical durability. Fig. 6.12 shows SEM images of CCIM glass ceramic and ceramic wasteforms from US DOE national laboratories.

Although CCIM products have been successfully fabricated, experience and understanding are limited in comparison to JHM glasses. Furthermore, the design of these melters is challenging in that the material properties, particularly the electrical resistivity of the melt, need to be well known to determine the optimal operation conditions for a given crucible diameter.

6. Modeling of wasteform behavior

High levels of radioactivity combined with difficulties in extrapolating to storage times measured in thousands of years can make experimental studies on nuclear wasteforms challenging. Due to this, computer simulation provides a set of tools which produce data complementing experimental approaches, allowing a more complete description of the long-term behavior of wasteforms.

The challenge of immobilizing energetic radionuclides cannot be understated. In a seemingly alchemical process, the decay of radioactive isotopes can fundamentally change the composition of any waste host. In addition, the energy released during decay can be considerable, leading to significant damage to the material and possible degradation to its physical and mechanical properties. This means there is a risk that, with time, an initially stable wasteform with good performance could become compromised due to these effects. High levels of radioactivity combined with difficulties in extrapolating to storage times measured in thousands of years can make experimental studies on nuclear wasteforms difficult. As a result, it is useful to support experimental studies with computer simulations, which can provide information that improves confidence in predictions of long-term wasteform behavior.

Activation-decay simulation codes are useful for determining waste stream composition and then understanding how a wasteform's isotope content will develop with time. Codes such as Origen (Oak Ridge Isotope Generation) [29] and FISPACT [30] exist for performing inventory calculations. In simple terms, when given an initial nuclide composition and atomic density, these computer programs are able to solve equations governing radioactive decay to provide the material's composition at any time in the future. In fact, they are considerably more sophisticated than this and provide a much wider range of outputs by calling on libraries of nuclear data such as TENDL and ENDF ([31]; Neudecker et al. 2018). These contain information on cross-sections, spectra, fission product yields, photo-atomic and thermal scattering data of particular relevance to neutron-induced reactions. What this means is that Origen and FISPACT can be provided with a neutron flux to predict how material composition is altered by activation and transmutation during irradiation. This can be a real boon to wasteform designers as it allows them to predict the waste-stream composition expected for future reactor designs and fuel cycles, even as they are still being planned. This is especially true given the simulation tools that reactor physicists have at their disposal (such as Serpent [32] and MCNP [33]), which allow them to predict the neutron spectra used as the input to activation-inventory codes. This opens up the real possibility of better integration of the front and back-end of the fuel-cycle, where a reactor's design and operation could be optimized to provide more straightforward waste management.

By tracking the decay processes in a material, activation-decay codes can integrate across all the decay events to output the energy and type of radiation emitted. This provides useful information about the total activity (Bq), α, β and γ activity and the heat produced by the waste. These can be used as source terms in other calculations. For instance, the rate of decay heat production is a useful input to calculations for determining maximum loading and wasteform package design (for instance, to provide the heating rate to a finite element calculation). In this way, the temperature rise in a waste package can be predicted as a function of time after emplacement.

Due to the high cost of nuclear waste repository space, it is important to fill it as efficiently as possible. This cannot be at the expense of safety, which is where the heat calculations provided by activation-decay codes provide a useful tool for repository designers. They allow the thermal load of the repository to be calculated, allowing appropriate countermeasures to be implemented [34]. Parallel to this, neutronics codes can show if there is any criticality risk in placing waste canisters, potentially containing fissile material, in proximity to one another (Fig. 6.13).

6.1 Atomic-scale simulation

Vitrification is a mature industrial process regularly used to immobilize HLW. Despite this and given the utility and widespread use of glass wasteforms, there is still a lot of uncertainty surrounding the atomic structure and behavior of glasses. When applied to crystals, diffraction techniques such as XRD can provide a fairly unambiguous definition of a material's structure. With modern diffractometers and analysis software, determination of a crystal's lattice and motif can be completed in a matter

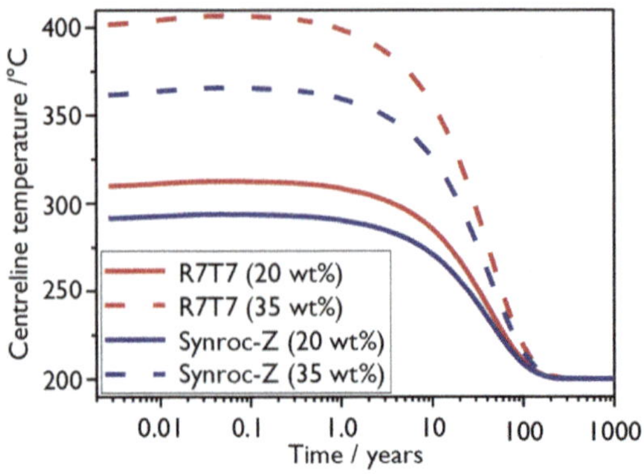

Figure 6.13 Activation decay codes allow the temperature of a waste package to be predicted as a function of time, waste composition, waste loading, waste host type and package geometry. In this example, a cylindrical geometry with a diameter of 36 cm was assumed, and the output of simulations for different waste types was taken from FISPACT and used to predict the centerline temperature of the waste as a function of emplacement time (see Ref. [34] for more details).

of hours. This is not the case for glasses, whose lack of long-range order means that the same diffraction techniques provide only an averaged description of their structure. A typical output of such an experiment would be a radial distribution function, which provides the probability of finding two atoms at a given pair separation. With some more work, individual pair-correlation functions may be obtained. Techniques such as an Extended X-ray Absorption Fine Structure (EXAFS) and Nuclear Magnetic Resonance spectroscopy (NMR) can also be applied to provide more detail regarding the first and second coordination environments around atoms in the structure. This battery of techniques provides a good description of first-nearest neighbor coordination shells existing in glasses; however, beyond second-nearest neighbour distances, experimental assessment of glass structure becomes much less definitive. This is further compounded by the complexity of nuclear glass compositions: even before adding the many species present in typical waste streams, the base compositions for nuclear glasses tend to be complex (e.g., the UK Magnox waste glass contains Si, B, Li, Na, Al, Ca and Zn compounds). Atomic-scale simulation has proved useful for generating model structures for these systems. Methods such as molecular dynamics (MD) and reverse Monte-Carlo (RMC), are able to produce structures containing on the order of 10^2–10^6 atoms. In these, the position of every atom in the simulation box is known, and additionally, for MD, velocities and forces are also known (recently, RMC codes may also output atomic forces, see below). The two methods operate in quite different ways, which will now be described.

In molecular dynamics, Newton's laws of motion are solved to predict how the positions of a set of atoms change with time. Given a set of initial positions, velocities and a description of the forces acting between the atoms, numerical integration of Newton's second law ($F = ma$) allows atomic trajectories to be calculated for a series of steps separated by small time increments. At each step, positions are updated, and forces are recalculated. From these, the acceleration and velocities are obtained, which are then used to update positions for the next time-step. In this way, the MD algorithm proceeds until a sufficient period of time has been simulated to capture the phenomena of interest.

The size of the MD integration time-step is necessarily small: it needs to be small enough to accurately describe atomic trajectories smoothly, but needs to be balanced against the desire to simulate as long a timescale as possible. Evaluating interatomic forces is the most computationally demanding part of the MD process, and this imposes a limit on how many time-steps a computer can calculate in a day, meaning there is a limit on how much time can be simulated during a given project. In practice, time-steps of between 1 and 4 fs are most often used. At higher temperatures, or when high-energy events such as radiation damage occur, much shorter time-steps may be required to accurately represent the motion of the fast-moving particles in these simulations. In such cases, a time step of 0.1 fs would not be unusual. As a result, due to the sheer number of steps required, even long MD runs simulate timescales that are on the order of nanoseconds, which means MD is ill-suited to studying slow processes and infrequent events (although methods such as temperature accelerated dynamics, parallel replica and par-splice have found some success in extending timescales, see Refs. [35,36] for more).

Glass structures can be generated from MD by using a melt-quench approach. This aims to reproduce the experimental procedure where a glass-forming liquid is rapidly cooled to produce an amorphous solid. To simulate this, the initial simulation box is populated with atoms to give the desired composition and its volume is set to give a density appropriate for the molten state.

During the "melt" phase, molecular dynamics is performed at a temperature above the melting point of the glass to induce a liquid like state. To accelerate this process, a temperature well in excess of the melting point is often chosen, typically allowing this step to be completed over a run lasting tens of picoseconds. The desired starting temperature is obtained by initializing atomic velocity vectors with magnitudes drawn at random from the appropriate Boltzmann distribution. During the "quench" phase of the "melt-quench", the system's temperature is reduced incrementally over a time period lasting between a few hundred picoseconds and several nanoseconds by scaling atomic velocities using an algorithmic device known as a thermostat.

RMC methods can also be used to produce representative atomic-scale structures of glass. In its simplest form, these are generated by filling a simulation box with atoms at initially random positions and using the trial-and-error approach typical of other Monte-Carlo optimization methods. These positions are randomly displaced with the intention of nudging the atoms into positions consistent with a set of experimentally determined pair correlation functions. This is achieved over a series of iterations with the following algorithm applied at each step:

1. An atom is chosen at random and displaced along a random vector.
2. Pair-correlation functions are calculated for the model system.
3. The model pair-correlation functions, $g_{\text{model}}(r_{ij})$, are compared with experimental pair-correlation functions provided as inputs to the RMC code, $g_{\text{expt}}(r_{ij})$. In this way, a residual

value (χ^2) is obtained between the desired structural data and the current state of the model. This is weighted by σ, allowing the significance of a dataset to be increased or decreased in the RMC fitting process. A χ^2 of zero would be obtained if the match between the experiment and the model data was perfect:

$$\chi^2 = \frac{\sum \left[g_{expt}(r_{ij}) - g_{model}(r_{ij}) \right]^2}{\sigma^2}$$

4. The χ^2 values are summed to obtain a measure of the quality of fit between the model and experimental data. The difference between the current and last iteration is calculated to give (χ^2). If this is negative, then displacing the atom has improved the fit to the experimental data and the move is accepted. If it has led to an increase in (χ^2) then the move is accepted only if a random number in the range (0,1) is less than $\exp(-\Delta(\chi^2)/2)$. If not, the displacement is reversed, and another displacement is attempted. By sometimes allowing displacements that decrease the quality of the match with experiment, this allows the RMC process to escape shallow local minima, thereby allowing more of the search space to be considered in the hope that an even better solution can be found.

It should be noted that there is considerable complexity in calculating values from the model (step 2) that are suitable for comparison with experimental data. For instance, the experimental correlation functions may have been obtained using X-ray diffraction or neutron diffraction. This means the simulated correlation functions must be appropriately weighted by the appropriate neutron scattering length (in the case of neutrons) or atomic form factors, before comparison. Similarly, other effects such as instrument resolution and choice of windowing function must be included in the simulated shape-factors from which the pair-correlation functions are obtained, to take account of the artifacts these introduce into the experimental data.

The RMC method has been improving continually since it was first developed. The general thrust of these efforts has been to increase the speed at which the algorithm converges to a representative structure and also to improve the quality of the structures produced. This has been achieved primarily by allowing a wider range of experimental data to be included in the fit. Information such as Raman, EXAFS and NMR spectra, bond angle distributions can all be included. Certain codes also allow classical potential models to be specified, so that the system energy can be included in the RMC process with the intention of producing lower energy structures than would be obtained if this were not considered.

The atomic configurations produced by RMC and MD melt-quenches can be analyzed to provide detailed information about the atomic environments of glass constituents and their structural role in the glass. Understanding the distribution of network modifiers. The original suggestion by Zachariasen's continuous random network model from the 1930s [37] was that network modifiers would be homogeneously distributed through the glass network. In the early 1980s, Neville Greaves used EXAFS and NMR to argue for his modified random network theory [38,39] in which the modifiers are arranged into modifier-rich channels. This was soon supported by molecular dynamics simulations [40]. The same calculations also demonstrated that, for mixed alkali compositions, species of different types were

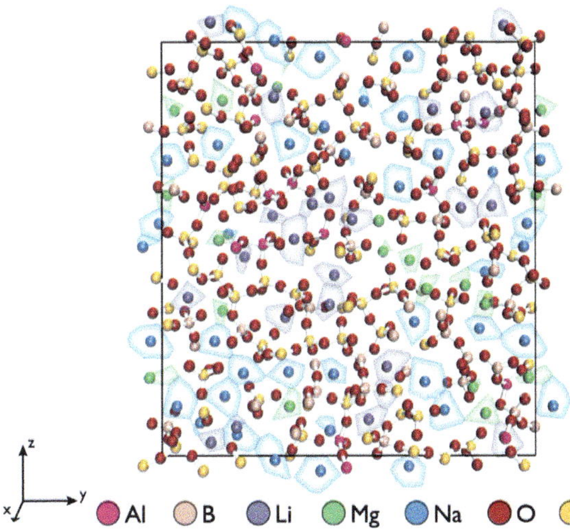

Figure 6.14 A slice through a glass structure generated from a molecular dynamics melt-quench simulation. This is for a glass used for immobilizing Magnox waste and is a magnesium aluminoborosilicate composition. Voronoi domains are shown around network modifiers to show where they have modifiers as nearest neighbors.
Image from Ref. [42].

intimately mixed in the modifier-rich regions. This result was used to explain aspects of the mixed alkali (the nonlinear relationship between property values when a single alkali is gradually substituted for another to create a mixed system [41]). An example of an MD-generated nuclear waste glass showing these effects is shown in Fig. 6.14.

The diffusion of alkali species is important in the behavior of glass wasteforms, and their migration is important in the dissolution of glasses by water. This is especially true in the rapid, early stages of dissolution, where there is ion exchange between protons in the water and the relatively loosely bound alkali atoms [43]. The channel-like structures predicted by the modified random network would be expected to provide routes into the glass from which the second phase of dissolution, network hydrolysis, can proceed. As the alkali content of a glass is increased and the modifier-rich regions grow, a point is reached, known as the percolation threshold, where these regions become fully interconnected. This well-connected network provides ready access to the interior of the glass for water during the early stages of dissolution. Higher-level models for glass dissolution increasingly include structural information as inputs (see Jantzen's discussion of medium-range order [4]), which can be extracted from simulated MD and RMC structures [44].

These structures are also useful as the starting point for other simulations. Following the melt-quench procedure, molecular dynamics can be used to interrogate glass systems. For instance, more mobile species diffuse within glasses on timescales accessible via MD simulations. A number of studies have measured mean squared displacements from MD simulations of nuclear waste compositions and used these to calculate diffusivities and activation energies, which are themselves useful inputs for kinetic waste dissolution models. For instance, Ren, Deng and Du used this method for International Standard Glass (ISG) and Simplified ISG compositions (sodium borosilicate and boroaluminosilicate compositions, respectively) to provide diffusion coefficients for Na, O, B, Al, and Si [45].

6.2 Radiation damage and stopping range

Computer modeling is a useful tool in understanding how radiation interacts with wasteform materials. When an energetic particle emitted by radioactive decay travels through it may cause damage as it gives up its energy to the material. Generally speaking, energetic particles initially slow down through electronic stopping. In this first phase, the particle feels an electronic drag force, as some of its energy is transferred to the electrons associated with the atoms it travels past. Wasteforms are typically glasses or ceramics and, as a result, are almost all electronic insulators; consequently, this electron heating also leads to localized temperature increase in the region surrounding the particle's track due to electron-phonon coupling effects. The particle may also lose energy through the creation of electronic defects via processes such as ionization and excitation. Interactions due to photon irradiation (e.g., gamma rays) are also predominantly with the system's electrons through the photoelectric effect, Compton scattering and pair production.

As their velocity decreases, through electronic stopping, the probability of a direct collision between the incident particle and a wasteform atom greatly increases. In a process akin to atomic billiards, higher mass particles (such as alpha, heavy ions and neutrons) can transfer sufficient momentum to atoms in the material that they are displaced from their sites. These primary knock-on atoms may themselves have sufficient energy to cause a cascade of subsequent collisions, each creating atomic defects in the material. Finally, when there is no longer sufficient kinetic energy to cause displacements, the energy is dispersed as heat through the material.

Following radiation damage, some recovery can take place. With time, Frenkel pairs can recombine, and other defects may also have time to anneal out. This is important when considering different approaches for modeling radiation damage. Some methods are only able to predict defect creation and do not take into account recovery. Meaning they are able to predict peak damage but not the defect population, sometimes after a damage event.

The radiation damage processes outlined in the previous few paragraphs describe three phases: an initial electronic stopping phase, a displacive damage phase and subsequent recovery. Some methods are better suited than others for modeling each stage of this process.

The Stopping Range of Ions in Matter (SRIM, formerly TRIM) is in widespread use by both modeling specialists and experimentalists alike. The latter often uses it before ion implantation experiments to predict the distribution of implanted ions in their samples. SRIM is a Monte-Carlo code that calculates the partition of an incident particle between electronic and nuclear stopping. The code contains a comprehensive database of electronic stopping powers, parameterized from an extensive range of stopping experiments. Whilst nuclear stopping is described using the binary collision approximation with the interactions between nuclei described using a shielded Coulomb pair potential known as the Ziegler-Biersack-Littmark (ZBL) potential, after the authors of SRIM/TRIM. The ZBL potential aims to provide a universal potential for any ion pair [46].

SRIM has certain limitations. Key amongst these is its simplistic representation of materials: incident particles interact with the target based on the density of each species they contain. In other words, SRIM sees materials as probability fields and includes no real concept of atomic structure; the atomic densities are used to calculate electronic drag and the probability of nuclear-nuclear collisions. This means that crystallographic effects (e.g., channeling) or interaction with microstructural features such as grain boundaries, bubbles/voids or stacking faults, cannot be considered explicitly with SRIM. The code also has a tendency to overestimate electronic stopping for heavy ions incident on targets containing light elements [46].

Molecular Dynamics can be used to simulate the displacive and recovery phases of radiation damage. MD allows the time evolution of atom positions, velocities and forces to be predicted. This means that a very detailed picture of how a system responds to radiation damage can be tracked through the damage phase, through to recovery.

MD collision cascades proceed as follows. A simulation box containing the atomic structure under consideration is defined and equilibrated at the desired irradiation temperature. An atom in the structure is then chosen to be the primary knock-on atom (PKA). This defines the initial point of the collision cascade and represents the first atom to be hit by the incident particle. The PKA is then given the necessary kick to initiate the cascade by assigning it a high kinetic energy (typically in the keV range) in a chosen direction. The MD algorithm is then run, and the progress of the collision cascade can be tracked. In this way, the effects of choosing PKAs with different species, initiating PKAs in different crystallographic directions and over a range of energies can be investigated.

The recovery phase of a cascade is simulated by conducting a long MD run. Due to the limited timescale accessible through conventional molecular dynamics and the large number of atoms typical of cascade simulations, the maximum equilibration time possible is on the order of nanoseconds. Although this is sufficient for many defects to recombine, using molecular dynamics alone, it remains difficult to establish whether any retained damage would also recover over the seconds, minutes and years available to a wasteform. For this reason, methods such as temperature-accelerated dynamics, parallel replica dynamics and kinetic Monte-Carlo approaches have been used to extend timescales in order to study longer-term recovery.

6.3 Damage cascade simulations

Jan et al. performed a series of collision cascade simulations on sodium borosilicate and sodium aluminoborosilicate glasses. These had compositions related to those used by the French high-level waste immobilization program and also international simples' glass [47]. They initiated collision cascades in systems produced using an MD melt-quench procedure by assigning 4 keV of kinetic energy to an atom (given the mass of uranium) in the corner of the system and firing it toward the center of the system. MD was then used to monitor the progress of this cascade. After the first cascade's energy had dissipated, another was set in train, with this procedure being repeated to give 120 runs in total. This allowed the system's evolution to be tracked as a significant damage level was accumulated. This work provides a worthwhile case study in how MD damage modeling can be used to better understand the effects of radiation on wasteform evolution and behavior.

Jan's damage simulations were able to reproduce the reduction in density also observed during experimental irradiation of glass. Despite increasing damage, they observed that the basic tetrahedral Si and Al building blocks of the glasses were maintained; an important change in boron coordination was observed: the number of three-coordinate boron atoms was seen to increase. Jan suggested this was caused by high temperatures immediately adjacent to the damage track, which promoted the creation of high-temperature features, such as three coordinate boron. The super-heated region within the cascade core represents a very small volume, meaning that it experiences a rapid quench as its heat dissipates into the far larger and cooler bulk material. It is thought that this explains the retention of these high-temperature features in the glass following a damage event.

A decrease in network polymerization was also reported, accompanied by increased mixing of chemical species. This, together with the decrease in boron coordination are thought to have an impact on increasing glass dissolution. Qualitatively, the aluminium-bearing glasses in this study showed similar trends in the effects exhibited following damage. However, the aluminum was shown to reduce the magnitude of the damage, possibly indicating why it is beneficial to include it in waste glass compositions. The results of the MD damage simulations are consistent with experiments showing increased leaching rates and alteration depth for irradiated glass.

7. Summary

This chapter introduced the various types of glass, ceramic and glass composite material wasteforms being used to immobilize radioactive wastes, their processing and microstructures and recent developments in modeling to examine their behavior. Here, high-level simulation codes with only very simplistic models of materials have been described, which allow radiation damage and the effects of decay to be predicted over geological timescales. Atomic scale simulation methods have also been shown to allow incredibly detailed models of wasteform materials to be considered, but at the expense of only being able to be applied over short time-scales. By combining various modeling techniques across length and timescales, it is possible to obtain a more complete image of how the various parameters available to scientists and engineers can be adjusted to give wasteforms with improved performance. Before concluding, it should be emphasized that modeling results cannot fully replace experiments, but they can facilitate a deeper understanding of nuclear wasteforms and identify areas where experimental effort should be focused.

References

[1] M.I. Ojovan, W.E. Lee, S.N. Kalmykov, An Introduction to Nuclear Waste Immobilisation, third ed., Elsevier, 2019.
[2] W.E. Lee, M.I. Ojovan, M.C. Stennett, N.C. Hyatt, Immobilisation of radioactive wastes in glasses, glass composite materials and ceramics, Adv. Appl. Ceram. 105 (1) (2006) 3–12.
[3] N.B. Milestone, "Reactions in cement encapsulated nuclear wastes: need for toolbox of different cement types, Adv. Appl. Ceram. 105 (1) (2006) 13–20.
[4] C. Jantzen, K.G. Brown, J.B. Pickett, Durable glass for thousands of years, Int. J. App. Glass Sci. 1 (2010) 38–62.
[5] W.E. Lee, M. Gilbert, S. Murphy, R.W. Grimes, Opportunities for advanced ceramics and composites in the nuclear sector, J. Am. Ceram. Soc. 96 (7) (2013a) 2005–2030.
[6] M. Gilbert, C. Davoisne, M. Stennett, N. Hyatt, N. Peng, C. Jeynes, W.E. Lee, Krypton and helium irradiation damage in Neodymium-Zirconolite, J. Nucl. Mater. 416 (2011) 221–224.
[7] I.W. Donald, B.L. Metcalfe, R.N.J. Taylor, The immobilisation of high level radioactive wastes using ceramics and glasses, J. Mater. Sci. 32 (1997) 5851–5887.
[8] M.I. Ojovan, W.E. Lee, New Developments in Glassy Nuclear Wasteforms, Nova Science Publishers, New York, 2007.
[9] M.I. Ojovan, J.M. Juoi, A.R. Boccaccini, W.E. Lee, Glass composite materials for nuclear and hazardous waste immobilisation, Mater. Res. Soc. Symp. Proc. 1107 (2008) 245–252.
[10] D. Caurant, P. Loiseau, O. Majerus, V. Aubin Chevaldonnet, I. Bardez, A. Quintas, Glasses, Glass-Ceramics and Ceramics for Immobilization of Highly Radioactive Nuclear Wastes, Nova, New York, 2009.
[11] A.R. Boccaccini, S. Atiq, R.W. Grimes, Hot-pressed glass matrix composites containing pyrochlore phase particles for nuclear waste encapsulation, Adv. Eng. Mater. 5 (2003) 501–508.
[12] J. Davidovits, Geopolymer Chemistry and Applications, fifth ed., Institut Geopolymere, Saint-Quentin, France, 2020.
[13] J.L. Provis, J.S.J. van Deventer, Geopolymers: Structure, Processing, Properties and Industrial Applications, Woodhead, 2009.
[14] A.R. Wagh, Chemically-bonded Phosphate Ceramics, second ed., Elsevier, 2016.
[15] D.W. Richerson, W.E. Lee, Modern Ceramic Engineering, fourth ed., CRC Press, 2018.
[16] W.E. Lee, M.I. Ojovan, C.M. Jantzen (Eds.), Radioactive Waste Management and Contaminated Site Clean-Up: Processes, Technologies and International Experience, Woodhead, 2013b.
[17] M.L. Elliott, Letter Report: Cold Crucible Melter Assessment, PNNL-11018, Pacific Northwest National Laboratory, Richland, Washington, USA, 1996.
[18] A.I. Demine, N.V. Krylova, P.P. Poluektov, I.N. Shestoperov, T.N. Smelova, V.F. Gorn, G.M. Medvedev, Mater. Res. Soc. Symp. Proc. 663 (2001) 27–34.
[19] S.V. Stefanovsky, A.G. Ptashkin, O.A. Knyazev, S.A. Dmitriev, S.V. Yudintsev, B. S. Nikonov, J. Alloys Compd. 444 (2007) 438–442.
[20] V. Maio, Initiating the validation of CCIM processability for multi-phase ceramic (SYNROC) HLW forms: plan for test BFY14CCIM-C, US Department of Energy Report INL/EXT-14-32724 (FCRD -SWF-2014-000444), Idaho National Laboratory, Idaho Falls, Idaho, USA, 2014.

[21] F.A. Lifanov, M.I. Ojovan, S.V. Stefanovsky, R. Burcl, Cold crucible vitrification of NPP operational waste, Mat. Res. Soc. Symp. Proc. 757 (2003) 201–206.

[22] A.P. Kobelev, S.V. Stefanovsky, O.A. Knyazev, T.N. Lashchenova, A.G. Ptashkin, M. A. Polkanov, E.W. Holtzscheiter, J.C. Marra, Results of a 50% Waste Loading Vitrification Test Using the Cold Crucible Melter for Savannah River Site Waste, WM2006 Conference Tucson, AZ, 2006.

[23] S. Naline, F. Gouyaud, V. Robineau, C. Girold, B. Carpentier, Vitrification 2010 – A Challenging French Vitrification Project to Retrofit a Cold Crucible Inductive Melter at the La Hague Plant, WM2010 Conference Phoenix, AZ, 2010.

[24] R. Didierlaurent, J.F. Hollebecque, O. Pinet, E. Leveel, C. Veyer, I. Trevisan, C. Roussel, S. Philippe, UMo Solutions Processing in La Hague Cold Crucible Induction Melter: The Feedback from the First Years of Operation, WM2016 Conference Phoenix, AZ, 2016.

[25] O. Pinet, J.F. Hollebecque, I. Hugon, V. Debono, L. Campayo, C. Vallat, V. Lemaitre, J. Nucl. Mater. 519 (2019) 121–127.

[26] C.W. Kim, C.S. Ray, D. Zhu, D.E. Day, D. Gombert, A. Aloy, A. Mogus-Milankovic, M. Karabulut, J. Nucl. Mater. 322 (2003) 152–164.

[27] J.V. Crum, V.C. Maio, J. McCloy, C. Scott, B. Riley, B. Benefiel, J.D. Vienna, K. Archibald, C.P. Rodriguez, V. Rutledge, Z. Zhu, J.V. Ryan, M. Olszta, J. Nucl. Mater. 444 (2014) 481–492.

[28] J.W. Amoroso, J. Marra, C.S. Dandeneau, K. Brinkman, Y. Xu, M. Tang, V. Maio, S. M. Webb, W.K.S. Chiu, J. Nucl. Mater. 486 (2017) 283–297.

[29] W.A. Wieselquist, R.A. Lefebvre, M.A. Jessee (Eds.), SCALE Code System, ORNL/TM-2005/39, Version 6.2.4, Oak Ridge National Laboratory, Oak Ridge, TN, 2020.

[30] J.-C. Sublet, J.W. Eastwood, J.G. Morgan, M.R. Gilbert, M. Fleming, W. Arter, FISPACT-II: an advanced simulation system for activation, transmutation and material modelling, Nucl. Data Sheets 139 (2017) 77.

[31] A.J. Koning, D. Rochman, J.-C. Sublet, N. Dzysiuk, M. Fleming, S. van der Marck, TENDL: complete nuclear data library for innovative nuclear science and technology, Nucl. Data Sheets 155 (1) (2019) 1.

[32] J. Leppänen, M. Pusa, T. Viitanen, V. Valtavirta, T. Kaltiasenaho, The Serpent Monte Carlo code: status, development and applications in 2013, Ann. Nucl. Energy 82 (2015) 142–150.

[33] C.J. Werner, J.S. Bull, C.J. Solomon, F.B. Brown, G.W. McKinney, M.E. Rising, D. A. Dixon, R.L. Martz, H.G. Hughes, L.J. Cox, A.J. Zukaitis, J.C. Armstrong, R.A. Forster, L. Casswell, "MCNP6.2 Release Notes", Los Alamos National Laboratory, report LA-UR-18-20808, 2018.

[34] Y.-H. Hsieh, M.J.D. Rushton, P.C. Fossati, W.E. Lee, Thermal footprint of a geological disposal facility containing EURO-GANEX wasteforms, Prog. Nucl. Energy 118 (2020) 103065.

[35] D. Perez, B.P. Uberuaga, Y. Shim, J.G. Amar, A.F. Voter, Chapter 4: Accelerated molecular dynamics methods: introduction and recent developments, in: Annual Reports in Computational Chemistry, vol 5, 2009, pp. 79–98.

[36] D. Perez, E.D. Cubuk, A. Waterland, E. Kaxiras, A.F. Voter, Long-time dynamics through parallel trajectory splicing, J. Chem. Theor. Comput. 12 (2015) 18.

[37] W.H. Zachariasen, The atomic arrangement in glass, J. Am. Chem. Soc. 54 (1932) 3841–3851.

[38] G.N. Greaves, A. Fontaine, P. Lagarde, D. Raoux, S.J. Gurman, Local structure of silicate glasses, Nature 293 (1981) 611.

[39] G.N. Greaves, EXAFS and the structure of glass, J. Non-Cryst. Solids 71 (1985) 203–217.

[40] B. Vessal, G.N. Greaves, P.T. Marten, A.V. Chadwick, R. Mole, S. Houde-Walter, Cation microsegregation and ionic mobility in mixed alkali glasses, Nature 356 (1992) 504.
[41] J.O. Isard, The mixed alkali effect in glass, J. Non-Cryst. Solids 1 (1969) 235–261.
[42] M.J.D. Rushton, Simulations of Glass and Ceramic Systems for Nuclear Waste Applications, Imperial College London, 2006. PhD Thesis.
[43] S. Gin, A. Abdelouas, L.J. Criscenti, W.L. Ebert, K. Ferrand, T. Geisler, M.T. Harrison, Y. Inagaki, S. Mitsui, K.T. Mueller, J.C. Marra, C.G. Pantano, E.M. Pierce, J.V. Ryan, J. M. Schofield, C.I. Steefel, J.D. Vienna, An international initiative on long-term behavior of high-level nuclear waste glass, Mater. Today 16 (2013) 243–248.
[44] C. Le Losq, D.R. Neuville, W. Chen, P. Florian, D. Massiot, Z. Zhou, G.N. Greaves, Percolation channels: a universal idea to describe the atomic structure and dynamics of glasses and melts, Sci. Rep. 7 (2017) 1–12.
[45] M. Ren, L. Deng, J. Du, Bulk, surface structures and properties of sodium borosilicate and boroaluminosilicate nuclear waste glasses from molecular dynamics simulations, J. Non-Cryst. Solids 476 (2017) 87–94.
[46] W.J. Weber, Y. Zhang, Predicting damage production in monoatomic and multi-elemental targets using stopping and range of ions in matter code: Challenges and recommendations, Curr. Opin. Solid State Mater. Sci. 23 (2019) 100757.
[47] A. Jan, J.M. Delaye, S. Gin, S. Kerisit, Molecular dynamics simulation of ballistic effects in simplified nuclear waste glasses, J. Non-Cryst. Solids 505 (2019) 188–201.

Further reading

[1] D.A. Brown, M.B. Chadwick, R. Capote, A.C. Kahler, A. Trkov, M.W. Herman, A. A. Sonzogni, Y. Danon, A.D. Carlson, M. Dunn, D.L. Smith, G.M. Hale, G. Arbanas, R. Arcilla, C.R. Bates, B. Beck, B. Becker, F. Brown, R.J. Casperson, J. Conlin, D. E. Cullen, M.-A. Descalle, R. Firestone, T. Gaines, K.H. Guber, A.I. Hawari, J. Holmes, T. D. Johnson, T. Kawano, B.C. Kiedrowski, A.J. Koning, S. Kopecky, L. Leal, J.P. Lestone, C. Lubitz, J.I. Márquez Damián, C.M. Mattoon, E.A. McCutchan, S. Mughabghab, P. Navratil, D. Neudecker, G.P.A. Nobre, G. Noguere, M. Paris, M.T. Pigni, A.J. Plompen, B. Pritychenko, V.G. Pronyaev, D. Roubtsov, D. Rochman, P. Romano, P. Schillebeeckx, S. Simakov, M. Sin, I. Sirakov, B. Sleaford, V. Sobes, E.S. Soukhovitskii, I. Stetcu, P. Talou, I. Thompson, S. Van Der Marck, L. Welser-Sherrill, D. Wiarda, M. White, J. L. Wormald, R.Q. Wright, M. Zerkle, G. Žerovnik, Y. Zhu, ENDF/B-VIII.0: the 8th major release of the nuclear reaction data library with CIELO-project cross sections, new standards and thermal scattering data, Nucl. Data Sheets 148 (2018) 1.

Ceramics for nuclear medicine

7

Lee J. Evitts[1], *Michael J.D. Rushton*[2] *and Fiona Pearce*[3]
[1]United Kingdom Atomic Energy Authority (UKAEA), Abingdon, United Kingdom; [2]Nuclear Futures Institute, Bangor University, Bangor, United Kingdom; [3]Nottingham University Hospitals NHS Trust, Nottingham, United Kingdom

1. Diagnostic imaging

1.1 PET, SPECT, and LSO

Utilizing radioactive materials for internal diagnostic imaging has been a staple of nuclear medicine for decades. In the most common procedure, the short-lived metastable ^{99m}Tc is attached to a pharmaceutical to create a radiopharmaceutical that can be introduced intravenously into the patient. The radiopharmaceutical then attaches to a particular biological target. In one example, ^{99m}Tc can be coupled with medronic acid and injected into the patient, carrying the radioisotope to regions of growth in bones, as the acid replaces bone phosphate [1]. The radioisotope decays with a half-life of 6 hours, emitting a single 140 keV γ-ray in the process. The energy of the γ-ray is large enough to escape the body without depositing a significant dose, yet small enough to be detected by an external gamma camera (scintigraphy). As such, a picture can be constructed of the source of radiation and can inform clinicians of any abnormalities, e.g., in the previous example, the clinician can diagnose areas of fractures or unusual bone growth. The 2-D images can be taken from multiple angles in order to produce a 3-D scan of the patient by means of Single-Photon Emission Computed Tomography (SPECT), schematically shown in Fig. 7.1(A).

An alternative to SPECT is offered in Positron Emission Tomography (PET), where the radioisotope attached to the pharmaceutical is a positron emitter. One of the most common procedures involves attaching radioactive ^{18}F to a glucose analogue, which, when injected into a patient, is picked up in areas where glucose uptake is greatest, e.g., in the brain, kidneys and cancerous cells [2]. The ^{18}F decays with a half-life of just under 2 hours and emits a positron in the process. The positron immediately annihilates with an available electron and produces two 511 keV γ-rays emitted in opposing directions. The two γ-rays are absorbed by the ring of scintillator detectors that surround the patient, as shown in Fig. 7.1(B), allowing a 3-D picture to be constructed.

The imaging methods described above both utilize scintillating materials to detect the γ-ray radiation. The incident radiation excites the electrons of the scintillator to higher states, and as the electrons quickly relax, they emit a number of visible or UV photons proportional to the energy of the original radiation. The scintillator

Ceramics for Nuclear Applications. https://doi.org/10.1016/B978-0-12-823865-3.00002-X

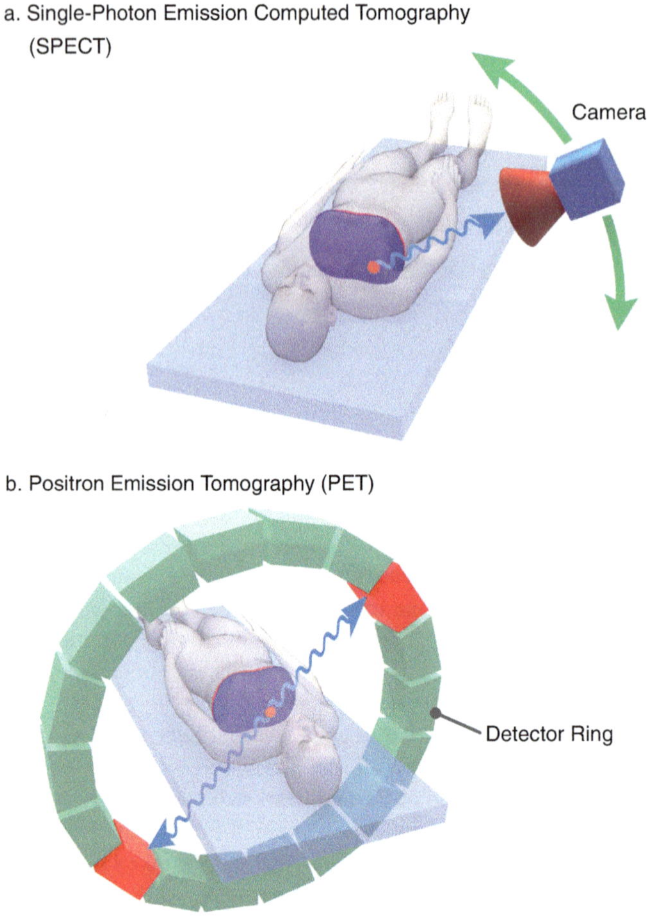

Figure 7.1 Schematic diagrams of (A) single-photon emission computed tomography (SPECT) and (B) positron emission tomography (PET).

material is coupled to a photomultiplier tube (PMT) or a complementary metal oxide semiconductor (CMOS), such as a silicon photomultiplier (SiPM), which collects the emitted light and ultimately converts the radiation interaction to an electrical signal.

Ideally, the scintillating material in a PET scanner would be cheap to produce, be highly sensitive (i.e., have a high efficiency for the detection of γ-rays), and have a short relaxation time to allow for good coincidence timings between the detector elements. The scintillating material should also be transparent to the light it emits and possess an emission wavelength that matches the detection window of the attached PMT. The most common scintillator for PET is cerium-doped lutetium oxyorthosilicate (LSO:Ce); a single crystal inorganic scintillator with a high density (i.e., γ-ray detection efficiency), a short relaxation time and high emission intensity [3].

The LSO:Ce single crystals are grown via the Czochralski process, but as they have a high melting temperature of around 2100°C, specialized high-temperature crucibles are required. The Ce^{3+} also has a low segregation coefficient in the LSO lattice, causing the composition of the liquid to change as the crystal grows from the melt, leading to the production of small crystals of high quality. These factors limit the sizes and shapes of the single crystals that can be produced, in turn limiting their applications in medical imaging [4]. The cost and length of fabrication can be reduced by instead forming LSO:Ce ceramics. To accomplish this, nanopowders of LSO:Ce are synthesized by a sol-gel method and heat-treated. After sieving and dry pressing the powders to pellets, they undergo cold isostatic pressing at a pressure of 200 MPa. The residual porosity of the ceramic is further reduced through hot isostatic pressing with argon at 150 MPa, and then the ceramic is annealed in air to reintroduce lost oxygen. This fabrication method can produce an LSO:Ce ceramic that has a shorter decay time and produces ~95% the light yield of its single crystal counterpart [5].

1.2 X-ray imaging

X-ray radiation is used in the most common form of internal diagnostic imaging. The radiation produced by a generator passes through a patient and on to a screen behind, producing an internal image of the patient due to the different attenuation factors between tissues and bones. The generator includes a tungsten cathode filament, which emits electrons when heated, which are accelerated from the cathode to a copper anode to produce the X-rays. Recently, it has been found that a film of carbon nanotubes can also emit electrons when exposed to an electric field at room temperature. This type of field emission source also offers additional benefits, including near-instantaneous emission, a greater uniformity of electron energy and a more compact size, enabling a greater resolution in X-ray imaging [6,7].

Ceramic materials possess a number of properties that are beneficial for a scintillator material in X-ray imaging, e.g., ceramics are easy to produce in a variety of shapes and sizes. However, ceramic materials typically have poor transparency and so must be made thin enough to allow the scintillating light to escape. The slim profile provides a significantly poor detection efficiency for γ-ray energies, but can be used for low-energy X-rays.

As ceramic scintillators can be made cheaply into varying sizes, ceramic phosphors tend to be used in intensifying screens in X-ray imaging [8]. A schematic of an X-ray screen is shown in Fig. 7.2, where a layer of light-sensitive film or a digital imaging plate is affixed to one or two thin layers of intensifying screens made of a scintillating material. The film or plate is exposed to the light emitted by the surrounding scintillators, which absorb a fraction of the incident X-ray radiation. The central film or plate is more sensitive to light than X-ray radiation, so the scintillator screens intensify the X-ray image (at the cost of some blur). As a result of the intensification process, the exposure received by the patient is approximately 100 times less than using the film or digital plates alone.

Terbium-activated gadolinium oxysulfide (Gd_2O_2S:Tb, or simply GOS:Tb) has become a widely used ceramic phosphor for intensifying screens owing to its high

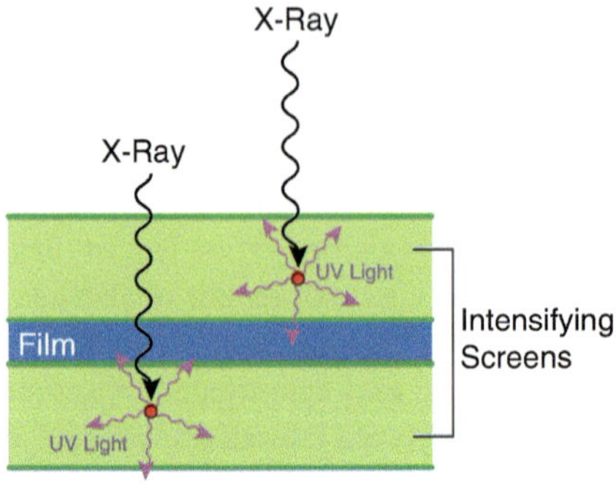

Figure 7.2 A schematic diagram of an intensifying screen used in X-ray imaging.

density, wide band gap and significant radiation resistance [9]. A variety of synthesis methods for the synthesis of GOS:Tb powders have been considered, including combustion synthesis [10] and a water-bath method [11], but traditionally the powder is synthesized by first mixing oxide precursors (i.e., Gd_2O_3 and Tb_4O_7) with a flux, mineralizing additives and sulfur suppliers before firing the mixture at 1200 °C in air [12]. The ceramic is then fabricated by means of one-step hot-pressing or hot isostatic pressing.

Other potential alternatives for intensifying screens include glass-ceramic composites, where scintillating crystallites sit within a host glass matrix, e.g., scintillating rare-earth halides like BaF_2 within an alumino-borosilicate host glass [13]. The size of the crystallites is made nanoscale so that the light emitted by the scintillators is not absorbed or scattered, and thus the host remains transparent at the desired wavelengths.

2. Radiotherapy

2.1 Brachytherapy

In brachytherapy, small sealed sources of radiation can be surgically inserted into the body at the site of a tumor, either inside or adjacent to it. The 'seeds' can be a few millimeters in length and contain a short-range emitting radioisotope, e.g., ^{125}I, which can be used for the treatment of prostate cancer, emits many low-energy electrons and X-rays which have a small range in tissue (typically less than a cm). One of the benefits of brachytherapy is therefore the limited exposure to surrounding healthy tissue.

The seeds come in a variety of shapes and sizes, but are usually made of a metal containing the desired radioisotope. There are alternative materials being explored that can positively utilize the properties of ceramics and glasses. For example, while the electrons emitted from ^{125}I have a small range in tissue, the X-rays have a mean free path of up to 2.6 cm and can therefore potentially damage healthy tissue. Alternatives exist in the rare earth oxides containing β^- emitting radioisotopes (e.g., ^{90}Y,

^{166}Ho, ^{142}Pr) which can be contained in large quantities (up to 70 wt%) in rare earth aluminosilicate (REAS) glasses. The REAS glasses are also biologically inert with no leaching of the radioisotopes during treatment. During the fabrication stage, the glasses can also be loaded with a stable precursor isotope and activated by an external beam, thereby reducing exposure to personnel.

The REAS glasses can be made by melting oxide powders in a crucible (with an example given in the following section) or by sol-gel methods. When creating microspheres, the sol-gel route offers higher purity REAS glasses with a uniform size at a lower processing temperature. The precursor sol typically contains tetraethoxysilane (TEOS), an acidic catalyst, water, and nitrate salts of the aluminum and rare earth isotope. The sol is partially hydrolyzed and gelled into microspheres by means of either an emulsion technique or through the liquid drop process, wherein sol droplets are passed through a heated vertical tube. The gelled microspheres are then heat-treated between 800°C and 1000°C to form REAS glasses [14].

The applications of REAS glasses in brachytherapy have been considered, including ^{142}Pr microspheres for the treatment of arteriovenous malformations (i.e., an abnormal formation of blood vessels) [15] and seeds for prostate cancer [16]. The most common application of radioisotope-containing glasses is in radioembolization.

2.2 Radioembolization

The liver is particularly sensitive to radiation, and severe liver toxicity may occur for typical doses received by external beam therapy [17], which causes problems for some specific liver cancers that cannot be removed through surgical procedures. The dose received by surrounding healthy liver tissue can be minimized through the application of radioembolization, or selective internal radiation therapy (SIRT). In SIRT, microspheres containing yttrium-90 are introduced intravascularly and transported to the site of the tumor. The ^{90}Y particles decay into ^{90}Zr by emitting a β^- particle with a mean energy of ~930 keV (equivalent to a penetration depth of ~2.5 mm in tissue). The microspheres lodge themselves in the smaller blood vessels in the proximity of the tumors, simultaneously irradiating them and blocking the blood flow to the tumor to prevent further growth.

The first applications of ^{90}Y were in the form of a yttrium oxide powder suspended in a viscous medium, but the high density of the powder created difficulties during injection, and the powder tended to settle in the bloodstream [18]. A solution to the problem was found by attaching the radionuclide to the surface of ceramic or plastic spheres, which varied from 1 to 200 microns in diameter [19,20]. However, the application of radionuclides to the surface of a microsphere represents a risk wherein the two may be separated within the body, depositing radioactive material in healthy tissue. The direct application of ^{90}Y to the surface of microspheres also presents an exposure risk to any technicians involved in the application process.

Currently, there are three commercially available microspheres for the purpose of radioembolization. Sirtex Medical (Sydney, Australia) offers biocompatible SIR-spheres, which are coated with a polystyrene resin containing ^{90}Y. The TheraSphere glass microspheres offered by MDS Nordion (Toronto, Canada) incorporate stable ^{89}Y in the production process; the glass microspheres are then activated by an external source of neutrons to produce ^{90}Y. The TheraSphere microspheres were originally produced by the Mo-Sci Corporation (MO, United States) and continue to be produced bioactive glasses for medical purposes that, for example, bind tissue and promote regeneration of bone tissue [21,22]. In recent years, Quirem Medical B.V. (Netherlands) have developed QuiremSpheres, which are made of a poly-L-lactic acid containing ^{166}Ho, which emits both a β^- particle and a low-energy γ-ray, allowing for simultaneous imaging and therapeutic treatment [23,24].

The glass TheraSphere is made by first mixing pure yttria (ideally isotopically pure ^{89}Y), alumina and silica powders into a homogeneous mixture. The mixture is melted in a furnace between 1500°C and 1600°C, depending on the exact composition of the initial powders. The melted mixture is quenched, crushed into a fine powder and passed through a gas/oxygen flame to remelt and form spherical liquid droplets. The droplets are rapidly cooled to prevent agglomeration and preserve the spherical shape [18,25]. The glass microspheres are passed through sieves to filter out beads with a diameter between 20 and 30 microns; the size window allows for convenient delivery through the hepatic artery to the liver and enables the permanent lodging mechanism that defines radioembolization.

3. Devices

3.1 Dosimetry

During every part of a medical procedure involving the application of radiation, it is important to ensure that the dose received by every person involved is as low as reasonably practicable. Careful consideration is made at every stage of the treatment plan to ensure that the patient only receives the necessary dose, ideally at the site of the disease, while leaving the surrounding healthy tissue unaffected. The clinicians performing the procedures must also consider the mechanism of treatment to ensure they receive minimal dose, particularly as they could potentially perform many procedures over the course of a day, e.g., when taking an X-ray, the clinicians will shield themselves behind a lead-lined window or wall.

The amount of radiation that a clinician or patient is exposed to is monitored, usually with a thermoluminescence dosimeter (TLD), to ensure procedures are being followed carefully. A TLD is a passive dosimeter that absorbs incident radiation, exciting electrons to higher states. In the scintillator materials discussed previously, the excited electrons rapidly decay back down to the ground state. However, in a TLD the electrons remain trapped in higher states until they're heated up to 300°C at some later stage; the electrons are then able to relax and emit light in the process. The intensity of the emitted light is proportional to the total amount of radiation absorbed by the TLD, giving a reading of the absorbed dose or exposure to the wearer. So the TLD that is worn during a treatment is analyzed weeks or months later by heating the material and measuring the emitted light.

The most common materials used for TLD's are calcium fluoride and lithium fluoride, depending on the environment in which they're used, i.e., the LiF can indirectly measure neutron irradiation through the $^{6}Li(n,\alpha)$ reaction. The fluorides are doped with impurities like Mn, Tm or Dy to trap the electrons in higher states. However, there is a difficulty in growing the doped single crystals due to the high vapor pressure of the fluoride dopant, leading to variations in dopant concentration throughout the crystal. The difficulties can be avoided by using optically transparent ceramics with the same compound. The fabrication routes and characterization of CaF_2 transparent ceramics with various dopants have been investigated, including Yb [26,27], Eu [28] and Mn [29]. In order to produce CaF_2:Mn transparent ceramics, nano-powders can first be synthesized by coprecipitation, then hot pressed under 20 MPa of pressure at 1000 °C in a vacuum chamber [29]. Through this method, a CaF_2 transparent ceramic, with a doping concentration of 1.5 at.%, can be produced, which possesses a linear thermoluminescence response to the intensity of incident radiation and can measure a minimum dose of 3 mGy [29].

4. Nuclear batteries for implantable cardiac pacemakers

4.1 A brief history of cardiac pacemakers

Atomic cardiac pacemakers are an interesting medical application of nuclear ceramics. The human heart contains its own pacemaker, called the sinus node, which generates electrical impulses that are transmitted through the heart, causing its muscle to contract and pump blood. Working in concert with the autonomic nervous system and the atrioventricular node, a regular heartbeat is maintained with the different chambers of the heart contracting and relaxing in sequence. Sometimes, however, this system doesn't work properly, leading to abnormal heart rhythms. In such cases, an artificial device may be implanted into patients to take over the job of the sinus node and provide the regular impulses required by the heart. For a relatively brief period, nuclear batteries were used to power these artificial pacemakers. This section will describe the history of these devices and the nuclear materials they contained.

Devices for heart pacing were developed as early as the 1930s (the Hyman-Otor device [30]); however, the long-lived implantable pacers that we are now familiar with really started to take shape during the 1950s. In collaboration with Dr. C. Walton Lillehei, engineer Earl E. Bakken developed a portable battery-powered pacemaker in 1957. Although the fundamental design is still in use today, this had the disadvantage of being an external device requiring leads to be passed from the outside into the body (these were made of multi-strand braided stainless steel surrounded by a Teflon sleeve [31]). These could become infected, making such external designs unsuitable for long-term use.

The first fully-implantable pacemaker was received by Arne-Larsson on October 8, 1958 [30]. His pacemaker was designed by Rune Elmqvist and implanted by surgeon

Åke Senning in Stockholm (Sweden) [31]. The device used a nickel-cadmium battery that was designed to be recharged through the skin at regular intervals. This first pacemaker failed after only 3 hours, requiring another to be implanted the following day, which itself failed soon after. In all, the patient required another 21 pacemakers throughout his life. The small battery capacity and poor durability of the Elmqvist-Senning design prevented it from being widely adopted. However, it had shown that fully implantable devices were worth pursuing.

A more successful cardiac pacemaker was invented by Wilson Greatbatch following a lucky mistake in 1956. He was an electrical engineer working as an Associate Professor at the University of Buffalo, New York. It was whilst working on an oscillator designed to record fast heart sounds that he installed a resistor with the incorrect value in his circuit. The oscillator produced a repetitive pattern of 1.8 ms electrical pulses separated by one-second pauses. Realising this was similar to the human heartbeat, and recalling an earlier conversation with two visiting surgeons who described the difficulty in treating blockage of the atrioventricular node, he set to work on adapting his oscillator circuit into an implantable device [30]. Over the following 2 years, and using $2000 of his own funds, Greatbatch created his pacemaker and hand-built the first 50 examples in his home workshop [30]. His device was first implanted during an operation on June 6, 1960, at the Millard Filmore Hospital in Buffalo and extended the life of the 77-year-old patient by 18 months.

The Greatbatch pacemaker represented a considerable step forward in pacemaker technology as it had a practical lifespan of 1–5 years. For a large number of patients, this was sufficient for it to be considered a permanent implant, for others, such as younger patients, long-term use would require multiple devices and associated medical procedures to replace them. In terms of patient wellbeing there was clearly a benefit to be had in extending the useful life of pacemakers.

The Achilles' heel of these pioneering devices was the chemical batteries used as their power sources. The Greatbatch pacemaker used a battery of mercury-zinc cells, which had the tendency to fail prematurely. Although designed for a life of 5 years, in the late 1960s, early battery exhaustion meant that devices typically failed between the 13th and 18th months after implantation [32,33]. This was due to acid leaks from the mercury-zinc cells and ingress of saline body fluids into the pacemaker capsule [32]. These limitations led to the idea of using nuclear radioisotope generators to power cardiac pacemakers. Starting in the 1960s, several countries invested in the development of nuclear pacemakers. Although not all led to commercial devices, these included Alcatel in conjunction with the CEA in France, the Atomic Energy Research Establishment (AERE) at Harwell in England, the Australian Atomic Energy Commission, Atomic Energy of Canada Ltd., Siemens AG in Germany and several organizations in the United States (the AEC, Numec, Douglas Labs, Hitman Associates and Gulf General Atomic) [33,34].

Some of these looked to existing research that had been working toward creating long-lived power sources for other applications. When in 1965 the US Atomic Energy Commission (AEC, later to become the DoE) were asked to consider developing a radioisotope power source capable of sustaining a pacemaker for the entire life of its recipient [32], they were able to use knowledge gained through their SNAP program (Systems for Nuclear Auxiliary Power, also known as Space Nuclear Auxiliary Power) which ran from 1955 to 1973 [35]. Through which they had been developing compact, lightweight and reliable nuclear electric power sources for naval, space and land uses. This has seen considerable research and development activity into various nuclear sources and energy conversion methods [35,36]. These ranged from compact fission reactors to radioisotope thermal generators, with the latter being most suitable for implantable medical devices.

Similarly, the UK's RIPPLE program provided the basis for the AERE's early nuclear pacemaker batteries. RIPPLE stands for Radio Isotope Powered Prolonged Life Equipment, and without the demand from a significant British or European space program it focused on unmanned marine applications such as marine navigational aids, remote communication equipment (radio and submarine cable repeaters), sonar beacons and weather stations [37]. The applied physics group at Harwell had been working on direct methods for converting nuclear heat to electricity since 1960, and this work led to the first RIPPLE devices being commissioned in March 1965 [37].

This kind of research led to viable pacemaking devices that started to enter clinical trials in the late 1960s. The US AEC pacemaker, designed in 1966, became the first pacemaker implanted in animals in 1968 [38]. The first device implanted in humans used the French G.I.P.S.I.E. One nuclear battery with Medtronic-9000 pacemaker electronics; this was implanted at Broussais Hospital in Paris in April 1970 [38].

4.2 Principle of operation

Radioisotope power sources aim to capture energy emitted during radioactive decay and convert it into electricity. Thermionic emission, thermoelectric and beta-voltaic conversion systems have all been considered for application in pacemakers [33,34].

4.2.1 Thermionic emission

In thermionic emission, decay heat is used to heat an electrode to temperatures in the 1500–2000K range where it can emit electrons. These electrons can travel from the an electrical current. To lower the work function for thermionic emission, nuclear batteries based on this principle often use a reservoir of Cs ions between the positive and negative terminals. McDonnell Douglas developed their Isomite, miniature nuclear thermionic devices, that were potentially small enough for use in pacemakers, but ultimately this did not lead to a viable device [34].

4.2.2 Thermoelectric effect

Devices based on the thermoelectric effect also make use of the heat from nuclear decay to produce an electrical current. This was the most widely used energy conversion method for nuclear pacemakers. Thermoelectric materials rely on the Seebeck effect for their operation. This was first observed in 1821 when the German scientist Thomas Johann Seebeck noticed that a compass needle was deflected when the junction between dissimilar metals, which formed a closed loop, was heated [39]. At the

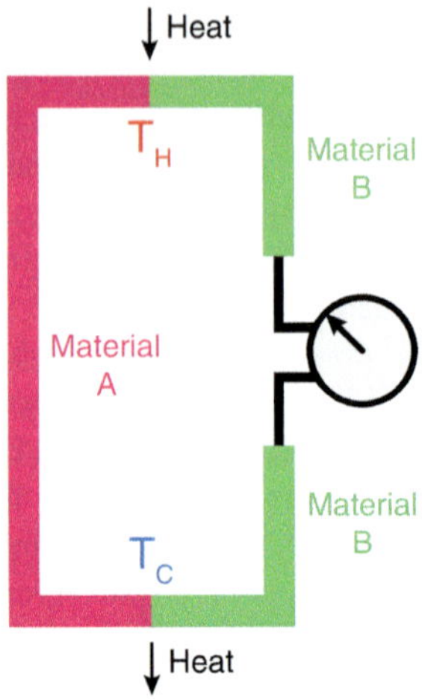

Figure 7.3 A loop of dissimilar materials containing two junctions, a configuration that gives rise to the seebeck effect.

time, he suggested this response was a thermo-magnetic effect, but it was later shown that the magnetic field he observed was due to a thermally induced electric current flowing in the circuit.

This configuration is shown in Fig. 7.3, which contains a hot junction at a temperature T_H between materials A and B and another, cold junction, at T_C. This temperature drives charge carriers (electrons/holes) to diffuse from the hot side to the cold side, resulting in a flow of current.

A pair of materials, such as that shown in Fig. 7.3 is called a thermocouple and is often used to measure temperatures as a thermometer. This is because the open circuit voltage (V_o) is proportional to the temperature difference between the two junctions:

$$V_O = (\alpha_A + \alpha_B)\,(T_H - T_C)$$

Where α_A and α_B are the Seebeck coefficients for the two materials in the pair. For an implanted device, the low temperature side (T_C) is at body temperature, which is normally 37.5°C. The hot-side temperature is determined by the geometry, material and radioisotope density of the fuel pellet. For reasons of durability and reliability, within the compact dimensions of the pacemaker capsule and the desire not to cook patients from the inside, there is a practical upper limit for T_H. Within thermoelectric pacemakers, this was typically set to fall in the range 110–120°C [40], although the ARCO pacemaker battery ran at 200°C to compensate for the low efficiency of its cupron-tophel metallic thermocouples [40].

The Seebeck voltage for a metallic thermocouple was 10–60 μVK^{-1} and 200–500 μVK^{-1} for the semiconductors available in the 1970s [40]. This is too low to use on its own, so thermocouples are attached in series, so their voltages add up. Connecting

Figure 7.4 The Meander like structure of connected thermocouples gives a thermopile.

thermocouples in this way produces what is known as a thermopile, which has the meander structure shown in Fig. 7.4.

The pacing circuits of the time required continuous power at 100–200 μW at a voltage of 5 V. This led to two design philosophies for the energy conversion systems in thermoelectric pacemakers - shaped by the conversion efficiency of their thermocouple materials.

Those based on metals used many couples to reach the desired output voltage (achieving heat to electricity conversion efficiencies of 0.1%–0.3%). The higher efficiency of semiconducting materials means that their thermopiles contain a lower number of couples (between 30 and 100). Although providing good power, the voltage from this system only produces a voltage of 0.3–1V, meaning semiconductor thermopiles require additional circuitry in the form of a DC–DC converter to increase the voltage to the necessary level. This meant that, although the heat to electricity conversion efficiency of the thermopile was higher than metal thermocouples with a value between 0.5% and 1%, the electronics of the time meant that the DC–DC converter robbed some of this to give an overall system efficiency of 0.25%–0.5%. An exception to this was the Syncal corporation battery, which used 300 silicon-germanium couples. This large number avoided the need for a DC–DC converter. However, it is not apparent that this ever saw clinical use [40].

Once the temperature difference (T_H-T_C) is set, the output of the thermoelectric device is determined by the thermoelectric materials it uses. These can be assigned a figure of merit, Z, which is defined as $Z = \alpha^2\sigma/\lambda$. Where α is the Seebeck coefficient, σ is the electrical conductivity, and λ is the thermal conductivity. The higher the value of Z, the better. It can be seen from this that the development of thermoelectric materials is a question of optimizing these three parameters (α, σ, andλ). To obtain a high Z, the Seebeck coefficient (α) and electrical conductivity (σ) must be large, but thermal conductivity (λ) must be minimized.

This requirement is contradictory due to the Wiedemann-Franz law, which requires that the electronic part of thermal conductivity (σ) is proportional to electrical conductivity (λ) whilst the Pisarenko relation limits the simultaneous increase of α and σ [39]. Ultimately, this limits the capacity of thermoelectric materials to convert heat into electricity. During the period relevant to nuclear pacemakers, the best materials had Z values of 1.0, which represents a best-case conversion efficiency of 4%–5% (50 years of development has since yielded Z values just over 2.5) [39].

Beyond the germanium conversion system described above and a handful of devices based around metallic thermocouples, most thermoelectric pacemaker batteries used bismuth telluride for heat-to-electricity conversion.

Bismuth telluride has the chemical formula Bi_2Te_3 and possesses the tetradymite crystal structure in the $R\bar{3}m$ space group [41]. The material has a layered structure as shown in Fig. 7.5; this can be considered as repeated two-dimensional quintuple layers of $Te^{(2)}$-Bi-$Te^{(1)}$-Bi-$Te^{(2)}$, stacked along the *c*-axis of the unit-cell. The numbers in parentheses identify two nonequivalent chalcogen sites.

Bonding between atoms in a quintuple layer group is of the covalent-ionic type. The bonding between quintuple groups is through van der Waal's interactions, meaning that the $Te^{(2)}$ atoms terminating each quintuple form bonds in this way across the vertical gaps between the blue atoms in Fig. 7.5). Within the layers themselves, the $Te^{(2)}$ atoms are covalently bonded to three other tellurium atoms.

Bismuth telluride is widely used as a thermoelectric material at room temperatures due to its low lattice thermal conductivity and high electronic weighted mobility [42]. Its low band-gap of 0.14 eV ultimately limits its performance with the production of thermally generated minority carriers, reducing thermoelectric efficiency [42,43]. Attempts have been made to mitigate this by doping more heavily than would be considered ideal for the majority carriers [43].

As mentioned above, good thermoelectric properties are favored by high electronic conductivity but low thermal conductivity. These are normally correlated, which makes it difficult to decrease thermal conductivity without damaging the electronic

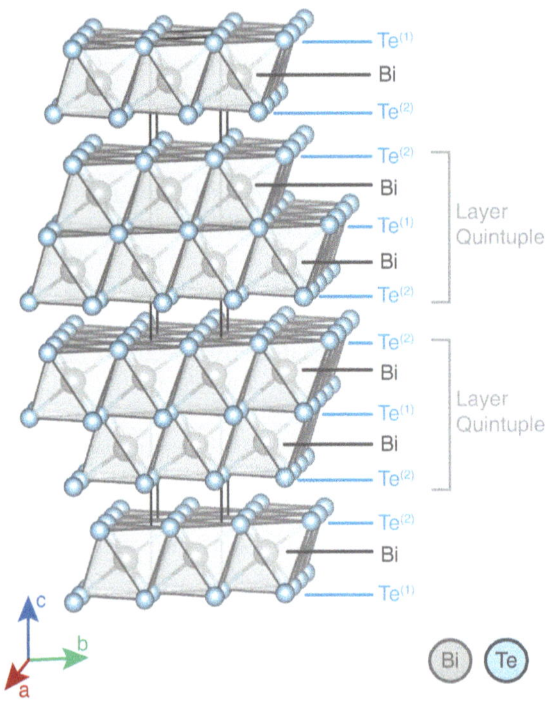

Figure 7.5 Bismuth telluride crystal structure.

properties. Early improvements in the performance of Bi_2Te_3 were achieved by disrupting the lattice component of thermal conductivity through the formation of solid solutions with large mass variation [43,44]. Introducing dopant atoms with a large mass difference to those in the bismuth telluride host lattice allows reduction of thermal conductivity by scattering high-frequency phonons [43]. The use of solid solutions of the compounds, antimony telluride and bismuth selenide, for this purpose is common [44]. These also have the advantage that they do not usually result in a reduction in the mobility of the charge carriers [44]. Adding antimony to bismuth telluride dopes it *n*-type, whilst selenide doping produces a *p*-type material.

To give an idea of how the bismuth telluride material was used in pacemakers, it is informative to consider Renner's description of the Alcatel radiothermal battery. He described the thermoelectric module as consisting of thin rods of bismuth-telluride glued together. With the rods being manufactured from "adequately *p*- and *n*-doped bulk materials by applying conventional techniques such as sawing and grinding" [40]. The fuel capsule was then glued on top of this bismuth telluride module to form the battery. The heat from the fuel capsule enters the bismuth telluride, which then produces an electrical current to power the associated pacemaker circuitry.

4.2.2.1 Fuel pellets

Plutonium-238 is the favored isotope for use in thermoelectric pacemakers. Its long half-life of 89 years, power density of 3.5 W/cm^3 and low radiation output of easily shielded alpha particles, coupled with low cost (at least in the 1960 and 70s) made it the logical choice for devices based on the thermoelectric effect [33].

Although the choice of isotope may have been straightforward, there was a difference of opinion regarding the form in which it should be used in radiothermal batteries, with both metal and ceramic fuels being used. In terms of ceramics, PuO_2 was the material of choice in almost all the devices that ended up in patients' chests; however, PuN was also seriously considered. In addition, beyond the use of the ^{238}Pu isotope, there were key differences between medical-grade PuO_2 and the material sometimes used in fission reactors, with pacemaker fuel requiring isotopic enrichment of the oxygen it contained. The plutonium-bearing ceramic fuels, used in thermoelectric pacemakers, will now be described.

As described above ^{238}Pu is well suited for use in pacemaker batteries, it is not without issues. The isotopic composition of the Pu produced for radiothermal generators at Oak Ridge National Laboratory in the USA is shown as 'high-grade' in Table 7.1. As can be seen, there are a number of contaminants alongside the ^{238}Pu which can increase the dose received by patients from their pacemakers. The ^{239}Pu is of low activity, meaning it does not contribute much to the energy released by the plutonium. The ^{236}Pu, by comparison, is more undesirable as it results in the emission of high-energy gamma rays from its thallium-208 decay product [38]. Gamma rays are a particular problem as their penetrating nature is at odds with the requirement for a compact and lightweight device, which means there is limited scope for any significant shielding. By comparison, the α emissions of the primary ^{238}Pu are more easily blocked by a pacemaker's metal housing. Despite this, these alpha emissions did have

Table 7.1 Plutonium may be produced in an isotope reactor by irradiating ^{237}Np targets. The isotopic composition of the Pu typically obtained via this route (conventional) is given. With further care and attention, the high-grade composition is produced, and this is better for use in pacemakers.

	Abundance wt.%	
Isotope	**Conventional [45]**	**High-grade [38]**
^{236}Pu	1×10^{-4}	$<3 \times 10^{-5}$
^{238}Pu	80.0	90.14
^{239}Pu	16.3	9.21
^{240}Pu	3.0	0.607
^{241}Pu	0.6	0.029
^{242}Pu	0.1	<0.01

implications for the ceramic PuO_2 fuel pellets, which became the main fuel for nuclear pacemakers, as will now be described.

Some early pacemakers used plutonium metal fuel; however, ceramic fuel pellets had obvious advantages, making their adoption desirable. PuO_2 is a stable ceramic with a much higher melting point than Pu metal (2744°C vs. 640°C [46]), which reduces the chances of accidental radioactive release in cases where pacemakers were not removed before cremation. It has low solubility in water and is biologically inert, which is beneficial in the case of pacemaker capsule leaks or accidental penetration, allowing the ingress of body fluids [33].

One disadvantage of the oxide form in comparison to metal fuel is the interaction of the α particles emitted by ^{238}Pu with the oxygen in its structure, which can give rise to neutron radiation. As is the case for gamma rays, neutron emission is difficult to shield in the small confines of an implantable pacemaker and leads to a patient receiving an increased radiation dose (there is also the small chance of these neutrons causing fission, which would multiply the neutron radiation further). To reduce the effect of these (α,n) reactions, the PuO_2 used in pacemakers is of a medical-grade enriched in ^{16}O, which distinguishes it from the PuO_2 fuel which may be used in fission reactors. There are three stable isotopes of oxygen with the following natural abundance: 99.76% ^{16}O, 0.04% ^{17}O and 0.20% ^{18}O. Although only present in small quantities, ^{17}O and ^{18}O have (α,n) reaction cross-sections of ~0.3 b and 0.5 b for the 5.6 MeV α particles emitted by ^{238}Pu [47]. By comparison, the ^{16}O has a negligible (α, n) cross-section, meaning that ^{16}O enrichment can reduce these reactions in PuO_2 by 4.4 times [38], bringing them down almost to the same level as ^{238}Pu metal fuel (see the PuO_2^{15} entry later in Table 7.2). For this reason, the NRC required ^{16}O enrichment for the PuO_2 used in pacemakers.

The other ceramic fuel material used in thermoelectric pacemakers was plutonium nitride. PuN has a higher plutonium density than PuO_2, allowing its power density to be much closer to Pu metal (see Table 7.2). This is desirable, particularly for implantable devices, as more heat can be produced from the same volume of material, which

Table 7.2 Property and dose-rate comparison for ^{238}Pu fuel materials in compounds with natural isotope abundance (nat) and isotopically enriched forms. Dose rates are for 3/8×3/8 inch cylinders measured at a distance of 10 cm through the Ta window. Pu is assumed to contain 80% ^{238}Pu and was obtained by electro-refining, and power-density assumes 90% pellet density.

		Equivalent dose rate/μSv•h$^{/1}$ g$^{/1238}$Pu	
Material	**Power density/W•cm^3**	**Gamma**	**Neutron**
Pu metal	7.2	2.4	3.1
PuN^{nat}	5.6	5.2	3.1
PuN^{15}		2.7	3.1
PuO_2^{nat}	4.2	3.5	15.0
PuO_2^{16}		2.8	3.4

Data from references [52] and [53].

could allow a smaller battery. PuN can also tolerate high temperatures, decomposing at 2587°C, making it safer than Pu metal in case of accidental cremation (although this is lower than the melting point for PuO_2, 2744°C) [48]. Plutonium nitride also has a higher thermal conductivity than the oxide, this is less important in radiothermal generators than when used as fuel is high temperature fission reactors.

The AERE at Harwell in England and Gulf General Atomic in the United States both developed working thermoelectric batteries based on PuN suitable for use in pacemakers [34]. Although the Gulf battery did not make it into any patients, the AERE's did. These were implanted into two patients in the summer of 1970 and weren't a great success: one device was removed after 3 months due to a malfunctioning power-supply and the second was also removed not long after implantation [49]. These failures are thought to have been due to poor shock resistance of the battery [40].

The nuclear battery was based around 180 mg of PuN purchased from the United States Atomic Energy Commission, which was coupled to a bismuth telluride energy conversion system [49]. Its dimensions were 17 mm in diameter × 49 mm long and came to an overall weight of 48g [50]. This could deliver 400 μW of electrical power at 0.4V to an asynchronous pacing circuit manufactured by Devices Implants Ltd [49].

Although PuN avoids the higher neutron emission from the ^{17}O in PuO_2, when used with its natural isotope mix, ^{14}N undergoes a reaction which increases the gamma dose considerably. The decay of $^{17}O^*$ produced from the $^{14}N(\alpha, P)$ reaction results in an 871 keV peak in the gamma emission spectrum, 1.8 times higher than the 761 keV peak resulting from the decay of ^{238}Pu [51]. This leads to the nitride fuel producing a gamma dose over two times higher than metal Pu fuel and 1.9 times higher than ^{16}O-enriched PuO_2 [52,53].

The gamma emission from ^{14}N could also be addressed by isotopic enrichment, producing PuN rich in ^{15}N. This type of PuN^{15} material was certainly synthesized and underwent a program of experiments in the late 1960s at Los Alamos National

Laboratory in the USA, although this was not applied in the AERE device [52–54]. Table 7.2 gives the gamma and neutron dose rates from Pu metal fuel, PuO_2 and PuN (data for the ceramics are provided for both natural and isotopically enriched forms). This shows that with ^{14}N, removed, the gamma dose produced by PuN^{15} is almost the same as that of the metallic fuel. A possible disadvantage of this approach is that nitrogen occurs as 97% ^{14}N, with only 0.4% being the more desirable ^{15}N. By comparison, enrichment for medical-grade PuO_2 only needs to remove the much smaller proportion of ^{17}O that occurs naturally (0.04%). Given that the dose emitted by PuO_2^{16} is very similar to PuN^{15}, the additional effort in enriching nitrogen does not provide a distinct advantage in reducing radiation exposure to patients and is therefore not worthwhile. This may go some way to explaining why PuN was abandoned as a power source for pacemakers, as subsequent AERE batteries employed PuO_2.

4.2.3 Betavoltaics

Almost 300 pacemakers, using the Betacel nuclear battery, were implanted in patients [55]. Unlike the devices based around ^{238}Pu these were not radiothermal generators, instead relying on the principle of betavoltaic conversion to produce the electricity required by the cardiac pacing circuitry.

In terms of betavoltaic nuclear pacemakers, the McDonnell Douglas Betacel was the most important. Betacel was the name given to a series of betavoltaic batteries, some of which were used in pacemakers that were implanted into patients. A clinical evaluation program had started in November 1972 to implant Biotronik pacemakers powered by model-400 versions of the Betacel into patients from Germany and the Netherlands [55]. At that time, over 60 pacemakers of various types had also been implanted in the USA [55].

A diagram of a Biotronik IKP pacemaker, of the type used with the Betacel is shown in Fig. 7.6 - this illustrates the battery encased inside its capsule, with shielding above and below (brown in the figure) at the center of the pacing device. The Model-400 Betacel used here measured 2.3 cm in diameter and was 2.44 cm high with a weight of 98 g, from which a maximum power of 400 μW at 4V was obtained [56]. The efficiency of this battery was 1.7% [56]. Development of the Betacel started in 1967 at the Donald W. Douglas Laboratories of the McDonnell Douglas Corporation based in Richland, Washington, USA [57], with the program being led by Larry Olsen [58]. The concept of the Betacel is shown in Fig. 7.6C. This shows the basic building block of the battery, which is a *p-n* junction, irradiated by a β emitter. These are layered on top of each other to produce the cell. In the case of the Betacel the radioisotope was ^{147}Pm contained in the ceramic Pm_2O_3. Silicon was used to form the *p-n* junctions. The fuel layers were formed by depositing promethia powders onto a metal substrate [57]. Oxide powders were provided to the lab by the US Atomic Energy Commission, having previously been purified by Batelle-Northwest [57]. The thickness of Pm_2O_3 in the Model-400 gave a mass of 4.05 mg/cm^2 with an activity of 770 C/g.

In Betavoltaic batteries, semiconducting materials capture the energy of β particles emitted when certain isotopes decay to produce electricity. In much the same way that the photovoltaic cells are used to produce electricity in solar cells, β-particles produce

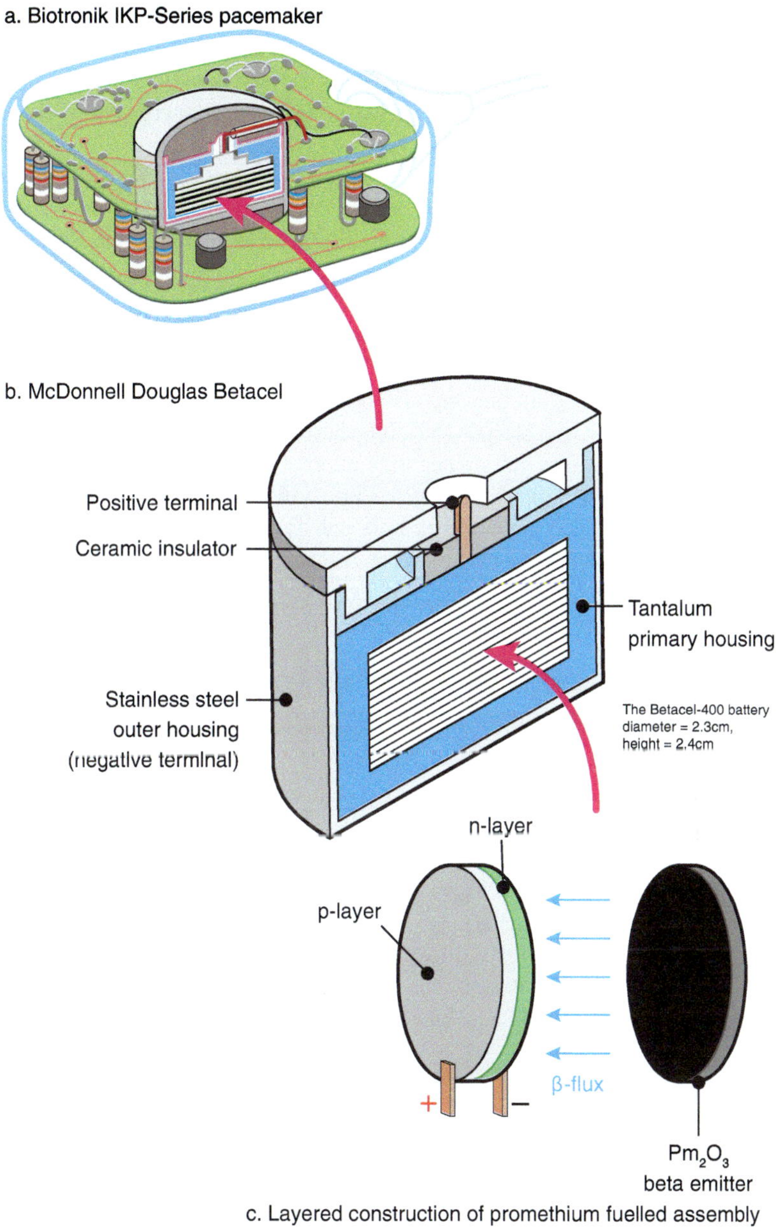

Figure 7.6 The McDonnell douglas betacel and how it was housed in a shielded capsule inside the biotronik IKP-54 pacemaker.

hole-electron pairs in the material, as electrons are promoted from the semiconductor's valence band into the conduction band. Irradiating a *p-n* junction (or other rectifying junction) in this way can then drive an electrical load such as a pacemaker circuit.

In a photovoltaic cell, hole-electron pairs are created by incident photons and, for the solar spectrum, most of these have energies less than 3.5 eV [58]. Creating the hole-electron pair uses an amount of energy equal to the semiconductor's bandgap. Even if sufficient energy is left over, it is very rare for multiple pairs to be created by a single photon and the excess leads to the creation of acoustic and optical phonons - heating the material. Although this reduces the efficiency of light-to-electricity conversion, useful amounts of electricity can be produced as the band-gaps of common semiconductors are of the same order as the energy of the incoming photons (as an example, the band gap of Si is 1.1 eV) [58].

By comparison, β particles produced during nuclear decay are released with energies in the keV range. Not only are these often sufficiently large to cause radiation damage (discussed below), but they are also orders of magnitude higher than the band gap of even wide band-gap materials [59]. It is therefore useful that β particles are able to create many hole-electron pairs when they interact with a semiconducting material [58]. Klein developed an empirical relationship showing that the average amount of radiation energy consumed per electron hole-pair is 14/5 E_g + 0.5 eV (where E_g is the bandgap). Of this, 1.8 E_g is dissipated as acoustic phonons and 0.5 eV as optical phonons [60]. This can be used to define a theoretical upper limit for the conversion efficiency of the semiconductor, which is plotted in Fig. 7.7 as a function of bandgap

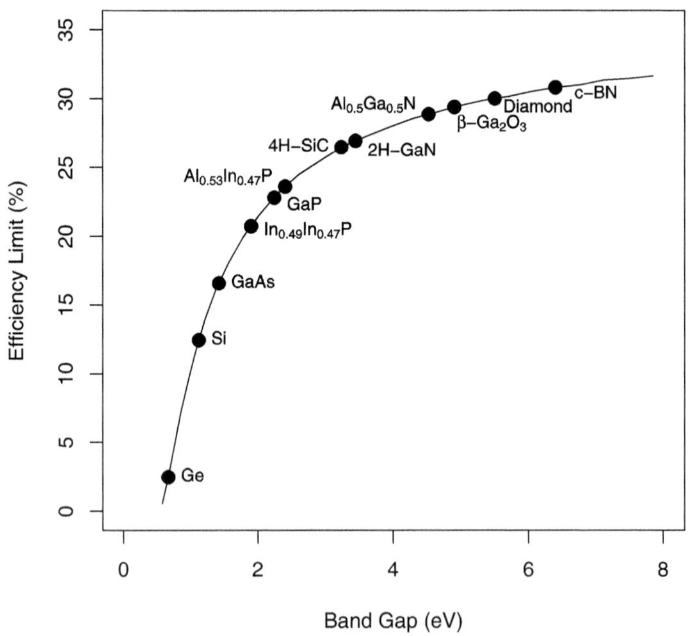

Figure 7.7 Theoretical betavoltaic semiconductor conversion efficiency for various materials. The efficiency curve is taken from Ref. [63], and band-gap data is given in Table 7.3.

and shows that efficiency increases with bandgap size, converging to a value around 33%. This suggests that, based solely on this, wide band-gap materials are preferable in betavoltaic applications.

Olsen proposed that the efficiency of a betavoltaic system should be characterized by the source efficiency (η_B), coupling efficiency of the semiconductor (η_C) and semiconductor conversion efficiency (η_S) [56]. The discussion in the previous paragraph concerned η_S, the overall efficiency of the battery is dependent on several other additional factors, each requiring consideration to maximize its performance. The source efficiency is a measure of how many of the β particles produced by the radioisotope reach the surface of the semiconductor. Some of these are lost through self-absorption within beta-emitting material. This is largely governed by radioisotope choice, its density in the host fuel material and the composition of other species in the fuel. To minimize leakage of β particles out of the battery, fuel laminates alternate with semiconductor layers. The thickness of these fuel layers is a useful design tool for the battery designer. There is a trade-off between self-absorption and how much isotope in each layer. Thick layers contain more isotope, but emission is less efficient as more β particles are absorbed before reaching the semiconductor. Thin layers decrease the amount of material the particles need to travel through before reaching the energy conversion material, but may not produce a high enough flux to yield an efficient device. To give an impression of optimal layer thickness, a study by Maximenko et al. showed that a ^3H titanium tritide film reached a maximum semiconductor incident power of 4.25 μW/cm^2 with a fuel layer thickness of ~2 μm whilst the generating rate of 17.6% purity ^{63}Ni film started to saturate at ~5 μm at 0.937 μW/cm^2 [59].

Coupling efficiency defines how much of the β energy entering the semiconducting material is collected for electron-hole pair creation. Scattering of electrons is the primary mechanism through which energy is lost. However, high-energy electrons can also produce X-rays, alongside secondary and Auger electrons (although the probability of this alternative mechanism is low, meaning they can be ignored to a first approximation) [61]. The scattering coefficient of the semiconductor, and consequently the coupling coefficient η_C, is directly related to the atomic number of its constituent elements [62]. This means that semiconductors with a high effective atomic number have lower coupling efficiency than those with lower atomic numbers. For this reason, the Ga containing compounds from Table 7.3 have η_C values that are notably lower than the lighter materials. For a tritium source, the η_C values were predicted by Maximenko et al. as GaAs = 63%, 2H-GaN = 66%, Ga_2O_3 = 66% and $In_{0.49}In_{0.47}P$ = 58% and were much higher for c-BN and diamond at 86% and 87% respectively [59].

Another parameter relevant to the design of betavoltaic batteries is the thickness of the semiconductor layers. The job of these is to absorb the majority of β-particles. Too thin, and they can pass straight through with the attendant loss of energy. The energy spectrum of the β-particles incident on these absorber layers is determined by the choice of radioisotope, with high-energy particles requiring thicker layers to stop. This difference can be significant: take the example of an Si absorber, for a ^{63}Ni source with its 67 keV β-particles the layer thickness required to deliver 87% of maximum output power has been calculated as 11.5 μm but for a titanium-tritium

Table 7.3 Properties of some semiconducting materials considered for application in betavoltaic devices.

Semiconductor	Bandgap/ eV	Direct/Indirect bandgap	Density/ g/cm^3	Minority carrier diffusion length/μm
Ge	0.67	I	5.5	–
Si	1.12	I	2.33	>40
GaAs	1.42	D	5.32	~6
$In_{0.49}In_{0.47}P$	1.9	D	4.4	–
GaP	2.24	I	4.14	–
$Al_{0.53}In_{0.47}P$	2.4	I	2.4	–
4H–SiC	3.23	I	3.21	~12
2H-GaN	3.44	D	5.88	~1
$Al_{0.5}Ga_{0.5}N$	4.52	D	4.68	<1
β-Ga_2O_3	4.9	D	6.1	0.4
$Al_{0.75}Ga_{0.25}N$	5.25	D	3.9	<1
Diamond	5.5	I	3.52	~40
2H–AlN	6.05	D	3.26	–
c-BN	6.4	I	3.48	<0.2

Data is from reference [59].

alloy emitting β particles with a maximum energy of 18 keV and an average of 7.5 keV the required thickness was only 1.1 μm [59]. It should also be considered that longer minority carrier diffusion lengths also tend to increase collection efficiency and that indirect band gaps tend to provide these.

Before continuing to describe the Betacel, which was actually used in pacemakers, it should be noted that when the Betacel was developed, the range of available semiconducting materials and analysis tools was considerably more limited than today. This is particularly true of the wide band-gap materials shown in Fig. 7.7 and listed in Table 7.3. A number of these materials, whilst promising, have not reached a level of maturity that would allow their use in safety-critical medical devices. It is not enough to be able to make a small amount of semiconducting material in the lab; to make useful devices, it is necessary to be able to successfully dope the materials to produce the rectifying interface (*p-n* junction or Schottky) on which betavoltaic conversion relies. In addition, betavoltaic batteries tend to be based around a layered construction. Thin films of semiconducting material are deposited onto suitable substrates that are then brought into close contact with the beta-emitting material. This requires a good understanding of substrate interactions and having the ability to produce films of sufficient thickness whilst minimizing the population of undesirable defects, such as dislocations. That is to say, the work required to take a semiconductor out of the research lab and into a working device is significant.

The choice of isotopes for betavoltaic batteries will now be discussed. Paul Rappaport produced the first studies where a beta source was coupled to semiconductors to produce a recognizable betavoltaic device whilst at RCA [64]. In his 1954 letter, he reported using a 50 mCi (1.85 GBq) ^{90}Sr–^{90}Y to irradiate silicon and germanium *p-n* junctions. In the case of the Si junctions, this resulted in 0.8 μW of electrical power being produced at a beta to electrical conversion efficiency of 0.4%. During Rappaport's program of work, it became apparent that the use of these specific isotopes presented a challenge when he observed the power of his Si cell halving every 14 h [65]. This precipitous degradation in performance was due to the creation of radiation damage-induced defects in the Si. Although the average energy of β particles was 196 keV, ^{90}Sr emissions have a maximum energy of 546 keV, whilst ^{90}Y is much higher at 2284 keV. Given that the threshold energy for atomic displacements is in the region of 300 keV for β-particles in Si, these energies are well above this. The defects produced result in the semiconductor having a higher back/dark current, as the mid-gap states they induce encourage electron-hole pair recombination [58]. All of which leads to higher energy losses. As alluded to earlier, high-energy particles also tend to cause other problems; high penetration depths into the semiconductor require thick junctions with very long minority carrier diffusion lengths [66]. In addition, the bremsstrahlung radiation emitted increases dose rates [66]. This gives us our first criterion for selecting a radioisotope for use in a betavoltaic device: isotopes should have maximum beta energies less than 300 keV.

The radioisotope must also have a reasonable half-life in order that the battery has good endurance, imposing our threshold energy criterion, to exclude high energy emitters, also eliminates many of the available long half-life isotopes. Table 7.4 shows the candidate isotopes with maximum energies less than 1 MeV. It should be remembered that the Betacel was developed in the late 1960s when silicon was the favored energy conversion semiconductor, consequently ^{204}Tl and ^{85}Kr must be dismissed as they produce betas above this material's threshold energy (Olsen reported that experiments with capsules of ^{85}Kr gas coupled to Si, resulted in a device that could only operate at full power for less than a month [66]). This left tritium, ^{63}Ni and ^{147}Pm as the possible isotopes. Also, at the time of the Betacel's development, it was concluded that ^{3}H and ^{63}Ni would not provide the necessary power for most applications. This left ^{147}Pm as the chosen isotope for use in the Betacel. The promethium isotope produces beta particles with energies that are below the damage threshold but are still able to provide a good specific power output [66]. The main disadvantage of ^{147}Pm is its relatively short half-life. However, by overpowering

Table 7.4 Candidate isotopes for use in betavoltaic batteries.

Isotope	Maximum β energy/keV	Half-life/years
^{3}H	18	12.3
^{63}Ni	67	92
^{147}Pm	230	2.62
^{204}Tl	765	3.75
^{85}Kr	670	10.9

Data from [66].

devices at the start of their life, it was possible to produce batteries that could last between 5 and 10 years [66].

Promethium sesquioxide is a ceramic material. It exhibits a body centered cubic structure at room temperature. This undergoes a transformation to a monoclinic form (B) somewhere between 700 and 760°C. On further heating, this will undergo a transformation to the hexagonal close-packed, A, form at 1740°C. Above 2000°C, Pm_2O_3 exhibits H and X forms before reaching its melting point at 2310°C [67]. These are believed to be modifications of the hexagonal and cubic phases, respectively [68]. Properties of Pm_2O_3 are given in Table 7.5.

In Fig. 7.6, the primary container for the stack of Pm_2O_3 and semiconductors is tantalum. However, for the compatibility with Pm_2O_3 and its containment, it is necessary that it is very pure. Material interactions have been extensively studied up to 1100°C, and these have shown that impurities such as water, carbon or organics cause most metals to be attacked by the oxide [45]. In the presence of these impurities, only rhenium was found to be compatible. When purified, a wider range of materials, including Hastelloy X, tungsten alloys and tantalum, showed no significant degradation after 18 months [45].

Promethium is a fission product of ^{235}U. The thermal neutron fission yield of ^{147}Pm is 2.25% [70]. Some also being formed from the β decay of ^{147}Nd to ^{147}Pm with a half-life of 11 days. Small amounts of ^{146}Pm, ^{148}Pm and ^{148m}Pm are also formed during the fission process. The ^{148}Pm and ^{148m}Pm are of concern due to the high-energy gammas they emit. These would mean that heavy shielding would be needed if fresh Pm was used, which would be particularly problematic for an implantable device. Luckily, the half-lives of ^{148}Pm and ^{148m}Pm are 42 and 5 days, respectively. These are much shorter than that of the desired ^{147}Pm (2.62 years), meaning that if the fuel is aged for 2–3 year before use, this problem largely disappears, without too much loss of the desired product (^{147}Pm).

The ^{146}Pm gamma energy is considerably lower, meaning it can be effectively shielded. This isotope was present at 0.25 ppm in fresh fuel and could be further decreased by reirradiation of the fuel if necessary to give 0.15 ppm or less at additional cost (although it is not clear if this was ever performed for pacemaker fuel) [67].

Table 7.5 Properties of the various forms of promethium oxide.

Crystal structure	Cell parameters (Å) [69]	Density (g/cm³) [67]
C-form (bcc)	a = 10.99	6.85
B-form (monoclinic)	a = 14.22 b = 3.65 c = 8.91 β = 100.1 degree	7.48
A-form (hcp)	a = 3.81 c = 6.03	7.49
H-form	–	–
X-form	–	–

Earlier, the method of ^{238}Pu production was described. This involved irradiating reactor fuel from which ^{237}Np was extracted via a solvent extraction route, followed by further irradiation and subsequent extraction to give the ^{238}Pu isotope. Promethium production follows part of the same route, meaning it is possible to extract the ^{147}Pm from the same irradiated material providing the ^{237}Np targets used in ^{238}Pu production. During the solvent extraction step, the Pm enters what would normally be considered the waste stream (with the actinide species being obtained from the other side of the process). Lanthanide fission products are then extracted from this highly active waste via precipitation and solvent extraction. The ^{147}Pm is then separated from the other lanthanides by an ion exchange step [45].

Current review articles on betavoltaic devices tend to dismiss ^{147}Pm as a power source on the grounds of availability and expense, as it is no longer routinely produced and separated, with tritium and ^{63}Ni being promoted instead [58,61]. During the period relevant to the development of the Betacel however, this was not the case, and this isotope was more readily available. The Hanford Isotope Production Plant aimed to produce 30 MCi of purified Pm, starting in 1968 [71]. Before this facility came online, the Strontium Semiworks (also at Hanford) was able to produce 10 MCi of Pm alongside other isotopes (and did so between 1961 and 1967) [71,72].

4.2.4 The decline of nuclear pacemakers

Nuclear pacemakers of the type described over the preceding pages are no longer implanted into patients, even though the technology was generally successful. Over 3000 devices were implanted worldwide [73], with some patients receiving nuclear pacemakers as late as 1989 [32] (although many hospitals abandoned the technology well before this). Ultimately, improvements in lithium-ion battery technology, electronics and the ability to monitor the state of a pacemaker nonintrusively meant that the use of chemical cells as a power source became more than adequate. With this competition, the exceptional longevity of nuclear batteries was no longer enough to outweigh the inconvenience of their manufacture, handling, disposal and the slight additional health risk to patients from the radiation dose received from their use. Despite this, nuclear thermoelectric and betavoltaic pacemakers performed impressively during their period in the sun. Chauvel et al. reported proper pacemaker function of 97% after 18 years [73]. Whilst in a study of 139 patients, Parsonnet found 82% were functioning correctly after 15 years and reported that early fears of increased cancer rates were unfounded [32,74]. He also reported that a Numec NU-5 pacemaker he had implanted into a 20-year-old patient in 1973 was still working well in 2007 after 34 years of operation [74]. It was estimated that, although initially more expensive than a conventional pacemaker, when including follow-up costs (hospitalization, medical fees) the nuclear device cost $36,000 over this period compared to $55,000 if a standard device had been used. A saving of $1050 per annum [74]. Finally, doctors in Lille (France) reported the case of a patient who received their nuclear pacemaker as a 19-year-old in 1978, with the device only failing after 36 years of reliable service [75].

References

[1] O. Buckley, S. O'Keeffe, T. Geoghegan, I.D. Lyburn, P.L. Munk, D. Worsley, W. C. Torreggiani, 99mTc bone scintigraphy superscans: A review, Nucl. Med. Commun. 28 (2007) 521.

[2] G.J. Kelloff, J.M. Hoffman, B. Johnson, H.I. Scher, B.A. Siegel, E.Y. Cheng, B. D. Cheson, J. O'Shaughnessy, K.Z. Guyton, D.A. Mankoff, L. Shankar, S.M. Larson, C. C. Sigman, R.L. Schilsky, D.C. Sullivan, Progress and promise of FDG-PET imaging for cancer patient management and oncologic drug development, Clin. Cancer Res. 11 (2005) 2785.

[3] C.L. Melcher, J.S. Schweitzer, Cerium-doped lutetium oxyorthosilicate: A fast, efficient new scintillator, IEEE Trans. Nucl. Sci. 39 (1992) 502.

[4] O.A. Graeve, A. Yazdani, J.P. Kelly, R. Kanakala, J. Tinsley, Effect of SiO2 on the sintering of cerium-doped lutetium oxyorthosilicate, Opt. Mater. 100 (2020) 109650.

[5] L. Fan, M. Jiang, D. Lin, D. Zhou, Y. Shi, Y. Wu, H. Yao, F. Xu, J. Xie, F. Lei, L. Zhang, J. Zhang, Densification of cerium-doped lutetium oxyorthosilicate scintillation ceramics by hot isostatic pressing, J. Alloys Compd. 720 (2017) 161.

[6] H. Sugie, M. Tanemura, V. Filip, K. Iwata, K. Takahashi, F. Okuyama, Carbon nanotubes as electron source in an X-Ray tube, Appl. Phys. Lett. 78 (2001) 2578.

[7] R.J. Parmee, C.M. Collins, W.I. Milne, M.T. Cole, X-Ray generation using carbon nanotubes, Nano Converg. 2 (2015) 1.

[8] P. Lecoq, Development of new scintillators for medical applications, Nucl. Instrum. Methods Phys. Res. Sect. Accel. Spectrometers Detect. Assoc. Equip 809 (2016) 130.

[9] I. Kandarakis, D. Cavouras, Experimental and theoretical assessment of the performance of Gd2O2S: Tb and La2O2S: Tb phosphors and Gd2O2S: Tb–La2O2S: Tb mixtures for X-Ray imaging, Eur. Radiol. 11 (2001) 1083.

[10] T. Xia, W.-H. Cao, X.-X. Luo, Y. Tian, Combustion synthesis and spectra characteristic of Gd2O2S: Tb3+ and La2O2S: Eu3+ X-Ray phosphors, J. Mater. Res. 20 (2005) 2274.

[11] X. Huang, J. Ding, X. Liu, X. Li, H. Chen, D. Hu, D. Zhu, T. Xie, J. Zhou, X. Jiang, Z. Sun, J. Li, Fabrication of Gd2O2S: Tb scintillation ceramics from the uniformly doped nanopowder, Opt. Mater. 117 (2021) 111192.

[12] E.-J. Popovici, L. Muresan, A. Hristea-Simoc, E. Indrea, M. Vasilescu, M. Nazarov, D. Y. Jeon, Synthesis and characterisation of rare Earth oxysulphide phosphors. I. Studies on the preparation of Gd2O2S: Tb phosphor by the flux method, Opt. Mater. 27 (2004) 559.

[13] M.B. Beckert, S. Gallego, Y. Ding, E. Elder, J.H. Nadler, Medical imaging scintillators from glass-ceramics using mixed rare-earth halides, Opt. Mater. 60 (2016) 513.

[14] J.E. White, D.E. Day, Rare Earth aluminosilicate glasses for in vivo radiation delivery, Key Eng. Mater. 94–95 (1994) 181.

[15] S.-W. Lee, W.D. Reece, Dose calculation of 142Pr microspheres as a potential treatment for arteriovenous malformations, Phys. Med. Biol. 50 (2004) 151.

[16] J.W. Jung, W.D. Reece, Dosimetric characterization of 142Pr glass seeds for brachytherapy, Appl. Radiat. Isot. 66 (2008) 441.

[17] T.S. Lawrence, J.M. Robertson, M.S. Anscher, R.L. Jirtle, W.D. Ensminger, L.F. Fajardo, Hepatic toxicity resulting from cancer treatment, Int. J. Radiat. Oncol 31 (1995) 1237.

[18] D.E. Day, G.J. Ehrhardt, United States patent: 4789501-Glass microspheres, 1988, 4789501.

[19] I.M. Ariel, Treatment of inoperable primary pancreatic and liver cancer by the intra-arterial administration of radioactive isotopes (Y90 radiating microspheres, Ann. Surg. 162 (1965) 267.
[20] I.M. Ariel, Treatment of inoperable primary and metastatic cancer of the lung by internal irradiation, Surg. Clin. North Am. 47 (1967) 1483.
[21] L. Hench, J. Jones, Biomaterials, artificial organs and tissue engineering, Elsevier, 2005.
[22] L.L. Hench, M.B. Fenn, J.R. Jones, New materials and technologies for healthcare, World Scientific, 2012.
[23] C.G. Radosa, J.C. Radosa, S. Grosche-Schlee, K. Zöphel, V. Plodeck, J.P. Kühn, J. Kotzerke, R.-T. Hoffmann, Holmium-166 radioembolization in hepatocellular carcinoma: Feasibility and safety of a new treatment option in clinical practice, Cardiovasc. Interv. Radiol. 42 (2019) 405.
[24] F.X. Sundram, J.R. Buscombe, Selective internal radiation therapy for liver tumours, Clin. Med. 17 (2017) 449.
[25] M.A. Westcott, D.M. Coldwell, D.M. Liu, J.F. Zikria, The development, commercialization, and clinical context of Yttrium-90 radiolabeled resin and glass microspheres, Adv. Radiat. Oncol. 1 (2016) 351.
[26] P. Aubry, A. Bensalah, P. Gredin, G. Patriarche, D. Vivien, M. Mortier, Synthesis and optical characterizations of Yb-Doped CaF2 ceramics, Opt. Mater. 31 (2009) 750.
[27] A. Lyberis, A.J. Stevenson, A. Suganuma, S. Ricaud, F. Druon, F. Herbst, D. Vivien, P. Gredin, M. Mortier, Effect of Yb3+ concentration on optical properties of Yb:CaF2 transparent ceramics, Opt. Mater. 34 (2012) 965.
[28] P. Samuel, H. Ishizawa, Y. Ezura, K.I. Ueda, S.M. Babu, Spectroscopic analysis of Eu doped transparent CaF2 ceramics at different concentration, Opt. Mater. 33 (2011) 735.
[29] S.G. Singh, S. Sen, G.D. Patra, S. Bhattacharya, A.K. Singh, S. Shinde, S.C. Gadkari, Fabrication, properties and thermo-luminescent dosimetric application of CaF2:Mn transparent ceramic, Nucl. Instrum. Methods phys. Res. Sect. B Beam Interact. Mater. At. 287 (2012) 51.
[30] H. Beck, W.E. Boden, S. Patibandla, D. Kireyev, V. Gupta, F. Campagna, M.E. Cain, J. E. Marine, 50th anniversary of the first successful permanent pacemaker implantation in the United States: Historical review and future directions, Am. J. Cardiol. 106 (2010) 810.
[31] O. Aquilina, A brief history of cardiac pacing, J. Appl. Cardiol. 8(2) (2006) 17–81.
[32] V. Parsonnet, J. Driller, D. Cook, S.A. Rizvi, Thirty-one years of clinical experience with nuclear-powered pacemakers, Pacing Clin. Electrophysiol. 29 (2006) 195.
[33] F.N. Huffman, J.J. Migliore, W.J. Robinson, J.C. Norman, Radioisotope powered cardiac pacemakers, IEEE Trans. Nucl. Sci. 21 (1974) 707.
[34] V. Parsonnet, Power sources for implantable cardiac pacemakers, Chest 61 (1972) 165.
[35] W.R. Corliss, R.L. Mead, Power from radioisotopes (U.S atomic energy commission division of technical information, 1964.
[36] S.S. Voss, Snap Reactor Overview, Airforce weapons laboratory, 1984.
[37] F.W. Yeats, The United Kingdom atomic energy authority RIPPLE program, IEEE trans, Aerosp. Electron. Syst. AES-3 862 (1967).
[38] D.L. Purdy, Nuclear batteries for implantable applications, in: B.B. Owens (Ed.), Batteries for implantable biomedical devices, Plenum Press, New York, 1986, p. 285.
[39] X. Zhang, L.D. Zhao, Thermoelectric materials: Energy conversion between heat and electricity, J. Materiomics 1 (2015) 92.
[40] T. Renner, Thermoelectric conversion systems for nuclear-powered pacemakers, in: M. Schaldach, S. Furman (Eds.), Engineering in medicine, Springer, Berlin Heidelberg, 1975, p. 435.

[41] A.N. Mansour, W. Wong-Ng, Q. Huang, W. Tang, A. Thompson, J. Sharp, Structural characterization of Bi_2Te_3 and Sb_2Te_3 as a function of temperature using neutron powder diffraction and extended X-Ray absorption fine structure techniques, J. Appl. Phys. 116 (083513) (2014).
[42] I.T. Witting, T.C. Chasapis, F. Ricci, M. Peters, N.A. Heinz, G. Hautier, G.J. Snyder, The thermoelectric properties of bismuth telluride, Adv. Electron. Mater. 5 (2019) 1800904.
[43] I.T. Witting, F. Ricci, T.C. Chasapis, G. Hautier, G.J. Snyder, The thermoelectric properties of N-Type bismuth telluride: Bismuth selenide alloys $Bi_2Te_{3-x}Se_x$, Research 2020 (2020) 1.
[44] H. Goldsmid, B. Telluride, Its alloys as materials for thermoelectric generation, and, Materials 7 (2014) 2577.
[45] E.J. Wheelwright, P.A. Fuqua, Promethium and plutonium as fuels for pacemaker power sources, in: Engineering in medicine, Springer, Berlin Heidelberg, 1975, pp. 383–400.
[46] F. De Bruycker, K. Boboridis, D. Manara, P. Pöml, M. Rini, R.J.M. Konings, Reassessing the melting temperature of PuO_2, Mater. Today 13 (2010) 52.
[47] T. Murata, K. Shibata, Evaluation of the (α,n) reaction nuclear data for light nuclei, J. Nucl. Sci. Technol. 39 (2002) 76.
[48] M. Takano, M. Akabori, Y. Arai, K. Minato, Lattice thermal expansions of NpN, PuN and AmN, J. Nucl. Mater. 376 (2008) 114.
[49] J.D. Hixson, Comparison of radioisotope-fueled energy sources for a prosthetic device, in: Isotopes and radiation technology, first ed. 9, US Atomic Energy Commission, 1971.
[50] A.W. Penn, M.H. Brown, J. Myatt, M.J. Poole, The atomic energy research establishment miniature nuclear battery, Nucl. Technol. 14 (1972) 89.
[51] J.A.B. Gibson, J. Marshall, J. Docherty, Surface dosimetry using thermoluminescent discs IAEA-SM-143/6, in: Proceedings of advances in physical and biological radiation detectors, International Atomic Energy Agency, Vienna, 1970, p. 45.
[52] L.J. Mullins, G.M. Matlack, J. Bubernak, J.A. Leary, Characterization and Properties of Medical Grade 238Pu Fuels, No. LA-DC-72-473, Los Alamos National Laboratory, 1972.
[53] L.J. Mullins, G.M. Matlack, J. Bubernak, J.A. Leary, Characterization and properties of medical grade ^{238}Pu fuels, in: Power from radioisotopes, OECD, Madrid, 1972, p. 621, 887R.
[54] L.J. Mullins, Preparation and Evaluation of Medical-Grade Plutonium-238 Fuels, July 1, 1967-Jun 30, 1971, No. LA-4940, Los Alamos National Laboratory, 1972.
[55] T.H. Smith, J. Greenborg, W.E. Matheson, Benefit/risk analysis of cardiac pacemakers powered by betacel ® Promethium-147 batteries, Nucl. Technol. 26 (1975) 54.
[56] L.C. Olsen, Review of betavoltaic energy conversion, in: Proceedings of the 12th space photovoltaic research and technology conference, 1993, p. 256.
[57] M. Lewis, S.E. Seeman, Performance experience with prototype betacel nuclear batteries, Nucl. Technol. 17 (1973) 160.
[58] L.C. Olsen, P. Cabauy, B.J. Elkind, Betavoltaic power sources, Phys. Today Off. 65 (2012) 35.
[59] S.I. Maximenko, J.E. Moore, C.A. Affouda, P.P. Jenkins, Optimal semiconductors for ^3H and ^{63}Ni betavoltaics, Sci. Rep. 9 (2019) 10892.
[60] C.A. Klein, Bandgap, Dependence and related features of radiation ionization energies in semiconductors, J. Appl. Phys. 39 (1968) 2029.
[61] M.G. Spencer, T. Alam, High power direct energy conversion by nuclear batteries, Appl. Phys. Rev. 6 (031305) (2019).
[62] K. Kanaya, S. Okayama, Penetration and energy-loss theory of electrons in solid targets, J. Phys. Appl. Phys. 5 (1972) 43.

[63] A.V. Sachenko, A.I. Shkrebtii, R.M. Korkishko, V.P. Kostylyov, M.R. Kulish, I. O. Sokolovskyi, Efficiency analysis of betavoltaic elements, Solid State Electron. 111 (2015) 147.
[64] P. Rappaport, The electron-voltaic effect in p – n junctions induced by beta-particle bombardment, Phys. Rev. 93 (1954) 246.
[65] P. Rapaport, J.J. Loferski, E.G. Linder, The electron-voltaic effect in germanium and silicon Pn junctions, RCA Rev. 17 (1956) 100.
[66] L.C. Olsen, Betavoltaic energy conversion, Energy Convers. 13 (1973) 117.
[67] J.H. Jarrett, H.H. Van Tuyl, Promethium isotopic power data sheets, No. BNWL-1309, Battelle Northwest Laboratory, 1970. https://www.osti.gov/servlets/purl/4136914.
[68] P. Aldebert, J.P. Traverse, Etude par diffraction neutronique des structures de haute temperature de La_2O_3 et Nd_2O_3, Mater. Res. Bull. 14 (1979) 303.
[69] W.F. Sheely, Pacific Northwest laboratory, division of isotope development programs quarterly report, May 1969–July 1969., No. BNWL–1177, 1969, 4192674.
[70] V. Elkina, M. Kurushkin, Promethium: To strive, to seek, to find and not to yield, Front. Chem. 8 (2020) 588.
[71] H.H. Van Tuyl, Promethium isotopic power data sheets, No. BNWC-45, Battelle Northwest Laboratory, 1965. https://www.osti.gov/servlets/purl/4588014.
[72] C.R. Anderson, Hanford fission products contemplated facility plan, no. HW-71179, Hanford Atomic Products Operation, 1961.
[73] C. Chauvel, T. Lavergne, A. Cohen, P. Ducimètre, J. Yves le Heuzey, J. Valty, L. Guize, *Radioisotopic pacemaker: Long-term clinical results*, pacing clin, Electrophysiol 18 (1995) 286.
[74] V. Parsonnet, A lifetime pacemaker revisited, N. Engl. J. Med. 357 (2007) 2638.
[75] S. Boulé, C. Kouakam, This is the end, Int. J. Cardiol. 223 (2016) 805.

Th9. Novel ceramics processes for nuclear applications

8

Takaaki Koyanagi[1] *and Christian Deck*[2]
[1]Materials Science and Technology Division, Oak Ridge National Laboratory, Oak Ridge, TN, United States; [2]General Atomics, Electromagnetic Systems, San Diego, CA, United States

1. Introduction

Discovery of novel processing routes actively continues in the ceramics community, and that is also the case for nuclear applications. Drivers of new technology development in nuclear power include reduced cost, improved safety, social acceptability, and new opportunities [1]. The approaches include improved fuels for current-fleet reactors, advanced reactor concepts (that operate at higher temperatures), fusion energy development, and space propulsion. Ceramic materials play key roles in all of these [2]. Novel ceramic processing and innovative materials, components, and designs enabled by advanced processing technologies will contribute to the nuclear technology evolution. This chapter presents the current status of novel processing methods for nuclear ceramics, which have commonly been applied across various structural and functional ceramics. These novel techniques include electric current-assisted processing, fabrication using nano-sized precursor powders, and additive manufacturing.

A variety of materials and applications are covered: uranium-based fuels for light water reactors (UO_2) [3], advanced fission reactors (TRISO-based fuels [4], and uranium silicides [5]); reactor core structural materials for fission and fusion reactors (silicon carbide, or SiC, and its composites [6,7]); hydride-based moderators [8]; neutron absorbers (B_4C) [9]; reflectors for fast reactors [10], nuclear waste forms (pyrochlore) [11]; and ceramics for high-temperature use in nuclear reactors: transition metal diborides [12,13] and MAX phases [14].

The nuclear ceramics of interest in this chapter are mostly high-temperature ceramics. They are frequently fabricated by powder sintering. A general challenge in sintering high-temperature ceramic powders is the harsh sintering conditions required, including high temperature and high pressure, because of the inert nature of the

Ceramics for Nuclear Applications. https://doi.org/10.1016/B978-0-12-823865-3.00004-3

materials. Sintering at extremely high furnace temperatures could limit the selection of material systems when composite ceramics are fabricated (e.g., the maximum processing temperature could be limited by the reactivity of the composite constituents). High energy consumption and long processing times are also concerns for high-temperature processing. In cases where pressure-assisted sintering is not an option, the high pressures required during sintering could be an obstacle to complex component geometries and near-net-shape manufacturing. A traditional solution to reduce sintering temperature and pressure is the use of sintering additives [15] and ceramic precursor powders (e.g., reaction sintering) [16]. However, sintered ceramics contain secondary phases attributed to the additives and unreacted precursors, which are known to degrade the properties and performance of the material [17]. This chapter presents a recently developed approach to address the challenges of powder sintering by using localized heating; an electric current-assisted method enables rapid heating of the powder compact without furnace heating. Extremely localized heating without external pressure is accomplished by laser sintering and laser chemical vapor deposition methods, which are used for additive manufacturing of ceramic components. Regarding the different approach from localized heating, nanopowder sintering is employed to reduce the sintering temperature by taking advantage of the extremely high surface area within the volume of the green body. This chapter presents the successes of the recently developed sintering approach in moderating the sintering conditions.

The processing methods showcased in this chapter are intended for fabricating "nuclear-grade" materials with the potential to meet the performance requirements of neutronic properties and resistance to the nuclear reactor environment. For example, a reactor core-structure is intended to be neutron-transparent to the extent possible for good fuel economy (unless it is used as a neutron absorber), which means that the processing should not rely on neutron-absorbing elements (e.g., boron-based sintering additives, which undergo the $^{10}B[n, \alpha]^{7}Li$ transmutation reaction). Regarding resistance to the reactor environment, stability under neutron irradiation is a common primary requirement for nuclear reactor components. It is known that the presence of a substantial content of impurity phases is not ideal for neutron irradiation resistance because of potential radiation-induced cracking caused by different levels of volumetric swelling between the matrix and the impurity phases [18,19]. In addition, an amorphous structure may not be suitable because of the dimensional instability (significant shrinkage) associated with radiation-induced amorphous-to-crystal transformations [20,21]. In addition to irradiation, reactor core components need to withstand corrosive coolant environments, which vary depending on the reactor type and can include molten salts, liquid metals, water with specific chemistry control, and other heat transfer fluids. The presence of impurity and secondary phases largely affects the corrosion resistance, which could limit the use of processing additives [22]. These performance requirements are considered in the selection of processing methods presented in this chapter.

In Section 9.2, recent progress in nuclear ceramics powder sintering technologies is highlighted. Field-assisted sintering and sintering of nanosize powders are specific focuses. Section 9.3 discusses state-of-the-art additive manufacturing (AM) technologies applicable to the fabrication of nuclear reactor components. The topics include selection of AM methods, demonstration of nuclear ceramic components with complex shapes, and quality control methods unique to AM technology. A comparison of these processing techniques is provided in Table 8.1. As explained later, each processing method has its intrinsic advantages and limitations. Some conclusions are drawn in Section 9.4.

2. Powder sintering

2.1 Electric current–assisted sintering

2.1.1 Electric current assisted sintering of monolithic ceramics

Electric current–assisted sintering (ECAS), including spark plasma sintering (SPS) is becoming a common processing technique for ceramics and a replacement for conventional hot-pressing in the research field [23,24]. The first generation of the SPS apparatus with an open-air system was invented by a pioneer in the field, Dr Masao Tokita more than 50 years ago [24]. The pulsed electric current method of powder sintering was originally invented in Japan, and its use has spread worldwide [24]. During the past 3 decades, the use of SPS processing has expanded remarkably in the field of research and development of ceramic materials, in both scientific academia and industry. A typical current fifth-generation SPS system apparatus (Fig. 8.1) incorporates a

Table 8.1 Comparison of the different novel ceramics processing methods.

Processing technology	Advantages	Technology challenges
Field-assisted sintering	• Short processing time • Low energy consumption • Low nominal sintering temperature • Mitigation of grain growth	• Nonuniform temperature distribution • Difficulty in measuring sintering temperature
Sintering nanosize powders	• Fine-grain structure • Better slurry infiltration • Slightly lower sintering temperature	• Availability of nanosize powders • Control of powder surface chemistry
Additive manufacturing: **(1)** Binder jet–chemical vapor infiltration hybrid **(2)** Selective laser sintering **(3)** Selective laser chemical vapor deposition	 **(1)** High-purity component, no shrinkage during densification **(2)** Applicable to various ceramic systems **(3)** Potential for fabricating pure ceramics	 **(1)** Density is affected by its geometry **(2)** Low density, microstructure control **(3)** Dimensional restrictions, heterogeneous microstructure

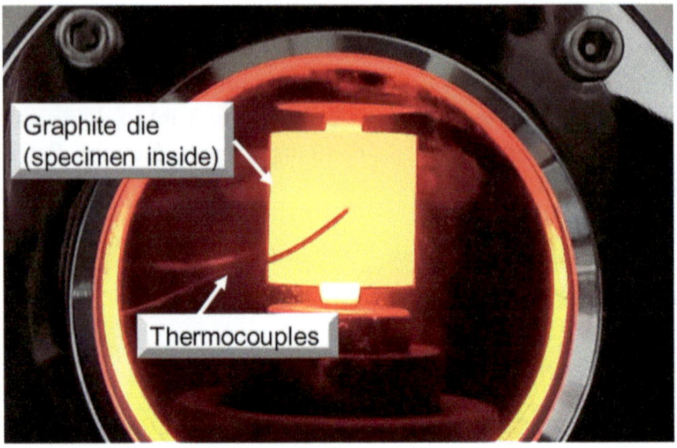

Figure 8.1 An example of an electric current–assisted sintering apparatus.

water-cooled vacuum chamber system and an advanced DC pulse generator, and temperature control systems as standard functions. It exploits the same punch-and-die system concept as the hot-pressing process. A powder or predensified powder compact is placed in the die and subsequently pressed by two punches with uniaxial stress. A high electric current is applied to the system, leading to efficient Joule heating of the die and powder, depending on the electrical conductivity of the materials.

As a result of the rapid heating and ECAS mechanisms, densification is completed in a short time and potentially at a lower furnace temperature compared with hot-press processing, which is a significant advantage in reducing energy consumption. For example, Jacobsen et al. [10] reported on the use of spark plasma sintering to densify zirconium silicide (Zr_3Si_2) compacts for fast reactor reflector material applications (Fig. 8.2). When using conventional hot pressing, the consolidation took well over 1 h, and attempts to consolidate Zr_3Si_2 powders using conventional hot pressing in less time (<15 min) were not successful. In contrast, sintered pellets of up to 92% relative density were obtained using ECAS with the consolidation occurring in approximately 8 min as the temperature ramped from 700°C up to a peak of 1100°C, roughly an order of magnitude faster than the conventional hot press processing.

A wide range of nuclear ceramics have been demonstrated by ECAS processing: UO_2 [25], U_3Si_2 [26], SiC [27], B_4C [28], pyrochlore-structure oxides [29,30], MAX phases [31], TiC [32], Zr_3Si_2 reflector materials [10], and ultrahigh-temperature diboride ceramics [33,34]. The references are only selected publications. In those studies, dense ceramic microstructures have successfully been demonstrated and fine-grain structures achieved as designed. The research highlights include very short processing times (a typical heating rate of 100–200°C/min and holding times of $\leq$10 min at a maximum processing temperature). In addition, the nominal processing temperature is often lower than that in conventional hot-pressing. For example, ECAS of SiC powder with a small amount of Al_2O_3–Y_2O_3 additives demonstrated a nearly fully dense sintered body at a sintering temperature of 1800°C, while

Figure 8.2 (Left) a 12.5 mm diameter Zr_3Si_2 pellet sintered via conventional hot pressing, (center) a 15 mm diameter Zr_3Si_2 pellet sintered using SPS, (right) Zr_3Si_2 pellet consolidation curve using SPS showing densification from ~20% to >90% between ~500 and ~950 s of process time [10].

hot-pressing of the same feedstock powders required a sintering temperature of 2000°C to achieve similar density [27]. Regarding B_4C, a dense specimen was fabricated by ECAS at a sintering temperature of 1700°C, which is 400–500°C lower than the temperature of hot-pressing [28]. The short processing times and reduced processing durations compared to hot pressing can produce a finer-grained ceramic material. These achievements in ceramics processing offer future research opportunities in testing of ECAS ceramics in nuclear reactor–related environments.

Regarding the microstructures of nuclear ceramics, a general conclusion has not been drawn as to whether a larger or a smaller grain structure is preferable for material performance in nuclear reactors.

There is a strong interest in achieving fine-grained structures in ECAS ceramics for nuclear applications because of the potential benefits for mechanical properties and irradiation resistance. The latter is often a limiting factor of material life [35]. Decreasing the grain size is normally expected to strengthen a material, according to the Hall–Petch relation. Thermal creep resistance is also dependent on grain size and grain boundary structure, and a fine-grain structure could be an advantage or disadvantage in that regard [36]. Regarding irradiation resistance, irradiation-induced defect formation is affected by grain size because grain boundaries act as defect sinks when the defects are effectively mobile [35]. Previous studies have demonstrated that the grain boundary sink effect mitigates cavity swelling of the irradiated materials [37,38] and radiation-induced amorphization [39] under certain irradiation conditions. The unprecedented radiation resistance of fine-grain materials resulting from their microstructural stability had a significant impact on the development of nuclear materials, and the effects of irradiation on nanostructures are actively studied for both metals and ceramics.

In contrast to the expected mechanical and irradiation-resistance benefits of a fine-grained ceramic, there are some potential downsides. The thermal conductivity of ceramics is expected to decrease with decreasing grain size because the number of interfaces that can serve as heat transport barriers increases significantly [40]. Some reactor core applications (e.g., fuel cladding and fusion blanket structure) require high thermal conductivity for efficient heat transport. In addition, grain boundaries could be preferentially attacked by corrosive reactor coolants. Grain boundary attack results in grain fallout, which accelerates consumption of the material [22,41]. For microstructural optimization, the ECAS process provides access to small-grain structures not conventionally obtained by traditional sintering methods.

A better understanding of ECAS gained from studies for both nonnuclear and nuclear applications has advanced the technology. For example, nonuniform temperature distribution in the specimen during sintering [42] is a common technological challenge in ECAS processing, and this has been addressed. The electrical resistivity of the specimen and tool greatly affects the temperature distribution. In addition, the electrical resistance of the tool assembly (which can be changed by modification of the tool dimensions and/or tool material) has an impact on the temperature distribution within the sample. In most of the cases presented in this chapter, graphite tools have lower electrical resistivity than the specimen, and so electric current travels mainly through the graphite tool. Therefore, the specimen is heated mainly by the Joule heat produced in the tool and transferred to the specimen by thermal conduction and radiation. As a result, the temperature at the edge of the specimen is higher than at the center. This uneven temperature distribution could be substantial in a large specimen. The temperature distribution could also be an issue for a specimen with a more complex shape than the typical disc geometry. To address this issue, finite element modeling is often used to simulate temperature distribution in the assembly and has shown promise for the simulation of sintering experiments [42–44]. This modeling approach could be used to optimize the geometries of the tools and the specimen to minimize temperature distributions within the specimen. Such a predictive modeling capability will also be useful for industrial implementation of ECAS techniques.

Another related issue addressed is the measurement of specimen temperature during sintering. In a conventional ECAS apparatus, the temperature of the graphite die and/or punch is measured by a thermocouple inserted into the graphite die (Fig. 8.1) or a pyrometer focused on a near-through hole in the graphite die and punch. Such a system cannot measure the temperature of the specimen accurately. Direct measurement of the specimen temperature was demonstrated by a sacrificial thermocouple inserted into the specimen [43]. This approach provided valuable experimental information and was used to understand ECAS mechanisms and improve the predictive modeling, although it is not practical for industrial production cycles. Currently, it is a common practice to use finite element modeling to understand the temperature and its distribution.

Although ECAS is an attractive ceramics manufacturing method, it generally requires high-pressure during sintering, which makes the fabrication of components with complex geometries challenging. Unique setups of graphite die have demonstrated pressure less ECAS of pellet specimens [45]. It is expected that the pressureless ECAS technology will be further advanced toward component fabrication.

2.1.2 Electric current–assisted sintering of composite ceramics

ECAS has been applied not only to monolithic ceramics but also to composite materials. The driving force for using ECAS is its short processing time, which minimizes reactions between the composite constituents and degradations of phases by annealing (in addition to the ECAS sintering mechanism benefits, including energy and time savings). For example, advanced uranium-based composite fuels have been explored using ECAS processing. The dopants or secondary phases have included Gd_2O_3 [46] and uranium borides [47] for control of fuel reactivity and Mo [48], BeO [49], SiC [50], and diamond [51] for enhancement of thermal conductivity. The volume fraction of the secondary phases was ~10 vol %, except for the uranium borides. In this section, we discuss two other examples of ECAS-processed ceramic composites with higher secondary phase fractions: (1) a Fully Ceramic Microencapsulated (FCM) fuel and (2) a hydride-based composite moderator.

FCM fuels consist of tristructural isotropic (TRISO) fuel particles embedded in a SiC matrix, and they have been designed for light water reactors and advanced fission

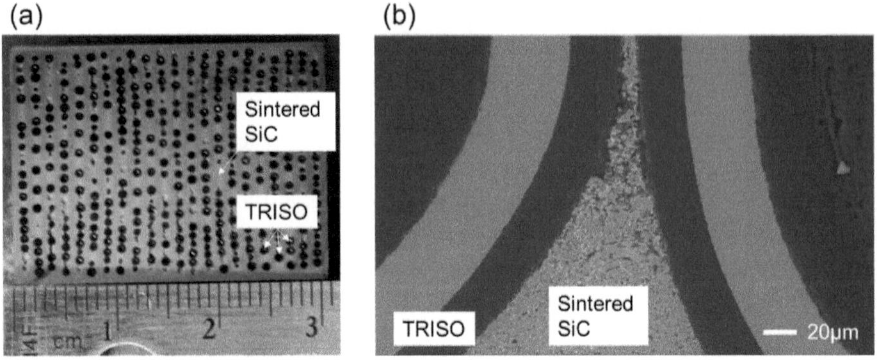

Figure 8.3 Cross-section of FCM fuel pellet (a) and scanning electron microscope micrograph of sintered fuel compact (b) [54]. SiC green preforms hosting TRISO particles were stacked to form a cylindrical pellet and sintered to achieve the systematic arrangement of the TRISO particles.

reactor applications [52,53] (Fig. 8.3). A TRISO particle consists of a spherical fuel kernel that is coated with successive layers consisting of porous carbon (the buffer layer), a dense inner pyrocarbon layer, SiC, and an outer pyrocarbon layer [4]. The SiC matrix acts as an effective barrier to fission product release, provides environmental stability under operating and transient conditions as well as in long-term storage, and provides proliferation resistance. FCM fuel was fabricated via sintering of a SiC nanopowder green body containing TRISO particles. An alumina-yttria system was used for the sintering additives. Minimization of the annealing of the TRISO particle is required to avoid fuel degradation. More specifically, lowering the SiC sintering temperature is intended to minimize grain growth of the SiC layer in the TRISO particle. This is beneficial because the characteristics of the SiC grain boundaries in the TRISO particle are controlled for better retention of specific fission products [55]. Therefore, ECAS was used in recent studies rather than hot-pressing because of the relatively shorter processing time at a similar sintering temperature [56]. The ECAS of the FCM fuels was successful, achieving a relative pellet density of ~95% and a TRISO loading of 34 vol % [54]. The sintering was reasonably complete even in the area near the TRISO particles (Fig. 8.3b). The porosity in this region was due to differential dimensional stabilities during ECAS (matrix shrinkage vs. nonshrinkage of TRISO particles) [54]. This local nonuniform shrinkage is a technical challenge in the sintering of composite ceramics in general. SiC is not the only material being considered as a matrix in FCM fuel. An FCM with an NbC matrix has been explored for nuclear thermal propulsion because of its high melting point and resistance to hot hydrogen [57]. Encapsulation of the fuel by an environmentally resistant ceramic matrix formed by ECAS has other potential applications because it offers the benefits of enhanced fission gas retention and improved environmental resistance of the fuel form.

The other example of an ECAS ceramic composite is zirconium hydride entrained in an inert ceramic matrix, which is an outstanding moderator for fast neutrons in terms of the moderating ratio [58]. Adding a ceramic matrix to zirconium hydride would benefit the structural robustness of the moderator system. Magnesium oxide was selected as a matrix candidate because of its neutron transparency and chemical compatibility with zirconium hydride. The composite moderator was fabricated via ECAS of magnesium oxide and a zirconium hydride powder mixture, as shown in Fig. 8.4. The fraction of zirconium hydride was ~35 wt%. Because the hydride decomposes at elevated temperatures, the sintering was conducted by using an ECAS system for a short period of time: a sintering time of 10 min and a maximum temperature of 900°C for 5 min [59]. In addition, a partial hydrogen atmosphere was used to stabilize the hydride. The current challenge is maximizing the hydride content without losing density in the composite. Even with the rapid sintering, ~20% of the hydrogen was lost. The loss of hydrogen is expected to be greater in the case of hot-pressing with a slower heating rate. Other metal hydrides with better thermal stability (e.g., yttrium hydride [60]) could be an option to mitigate the challenge of hydrogen losses. In addition, as the hydride prevented densification of the matrix, the trade-off between hydride content and composite density must be balanced when selecting sintering conditions, and processing parameters chosen to preserve the hydride phase will produce a lower density composite and vice versa. Although the development of the composite moderator is still in progress, the hydride-encapsulated composite is a good example of how ECAS processing enables new ceramic composites that cannot be made by hot-pressing.

The recent progress in the development of the FCM fuel and the composite moderator shows that rapid ECAS processing has great potential for composite fabrication with minimal chemical interactions between the constituents and reduced degradation of the encapsulated phase during densification. These demonstrations encourage further discovery of new composite ceramics for existing and new nuclear applications.

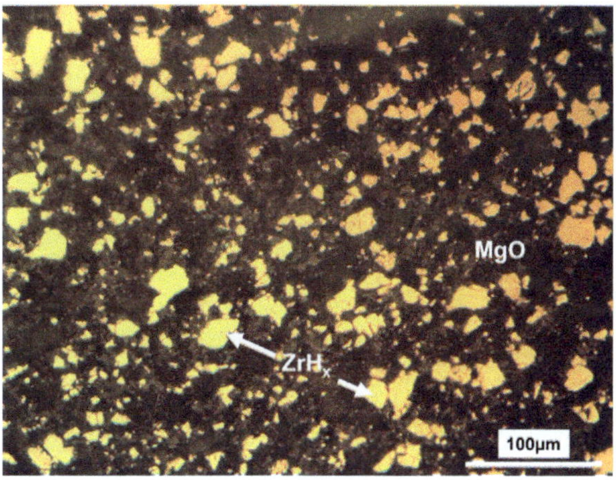

Figure 8.4 Magnesium oxide–zirconium hydride composite moderator fabricated by ECAS [59].

2.1.3 Electric current–assisted joining

ECAS processing has also been employed for joining ceramic components. Among studies of various nuclear ceramics, the joining of SiC in an ECAS apparatus has been actively studied for potential fusion and fission structural applications [61]. These joining studies used an exotic joining agent as an aid to reduce the joining temperature compared with the conditions seen in direct joining. The popular interlayers are titanium [62–64] and titanium-based materials such as Ti_3SiC_2 [65]. The joining experiments used a simple plate geometry for the substrate. The motivation for using an ECAS apparatus, rather than conventional hot-pressing, is its ability to tailor the microstructure of the bonding layer; the rapid heating and cooling during ECAS allows exploration of the joint microstructure, on which the temperature history during heating and cooling has a limited effect. In addition, the short processing time is beneficial for minimizing the effects of annealing on the properties of the substrate.

Fig. 8.5 highlights the difference in the microstructures of titanium diffusion-bonded SiC by ECAS and by hot-pressing. The joint phases were identified by electron backscatter diffraction, and in this system, a dominant MAX phase Ti_3SiC_2 joint material is formed. The ECAS joining of a SiC/titanium foil/SiC sandwich occurred in a cylindrical graphite mold, and the specimen was heated under pulsed current at a rate of 50 °C/min with a dwell time of 5 min at a furnace temperature of 1700°C in vacuum under a uniaxial pressure of 50 MPa [61,62]. The hot-pressing joining was conducted at 1500°C with a heating rate of 10°C/min for 1 h in vacuum, under an uniaxial pressure of 17 MPa [61]. Because of the different heating mechanisms and temperature monitoring, the processing parameters to achieve the joining of SiC were different between ECAS and hot-pressing. The ECAS joint had a dominant Ti_3SiC_2 phase (~90%) in the bonding layer, in addition to a minor TiC phase near the middle of the bonding layer. The hot-pressed joint had a Ti_3SiC_2 phase near the joint interface and a mixed structure of Ti_3SiC_2 and $TiSi_2$ at the center of the joint layer. Since the

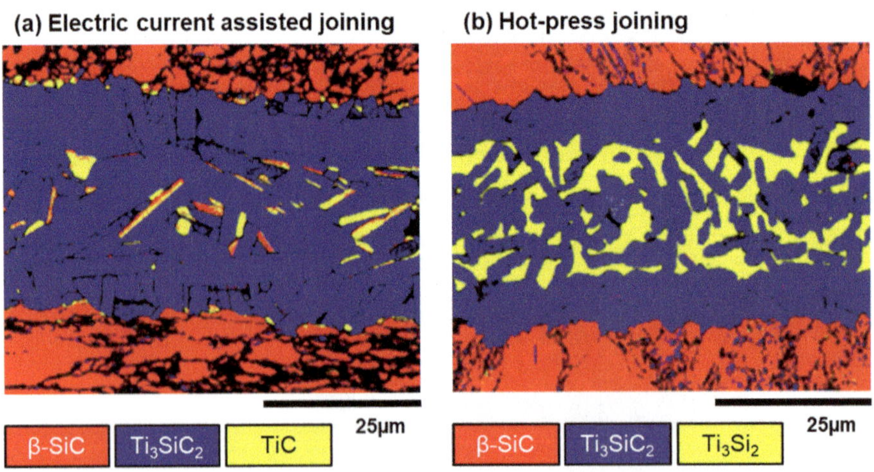

Figure 8.5 Electron backscatter diffraction phase maps of SiC joined with titanium interlayers in (a) an ECAS apparatus and (b) a hot-press [61].

joint phases were formed by interdiffusion, the differences in heating profile and holding time largely affected the resultant joint phases. More specifically, the rapid ECAS-based joining allowed the formation of a main phase of Ti_3SiC_2, which was metastable at 1700°C [63]. The comparison of the joint microstructures shows a unique joint microstructure formed by a rapid joining process. Although the strengths of both joints shown in Fig. 8.5 were high even after neutron irradiation (torsion shear strength of >70 MPa [61,66]), the tailorable joint microstructure is important to identifying promising joint phases that meet other performance requirements, including high-temperature stability, irradiation resistance, and chemical compatibility in nuclear reactors.

Previous studies of ECAS-based joining of ceramics demonstrated well-bonded microstructures and robustness of the plate joints. Note that this is a pressure-assisted joining technique, which may limit the practical application of this process at the current maturity level to assemblies where pressure application can be performed without damage to the individual components. Pressureless joining is promising for the integration of large, complex-shaped components. An applied stress provides better joint quality because it allows better contact between the substrate and the joint phase during joining [67], which is difficult in pressureless joining. Further technology development of ECAS-based joining is warranted to advance the integration of ceramic components.

2.1.4 Flash sintering

Flash sintering is a class of ECAS. In flash sintering, an electric field is directly applied to the green specimen and the current is forced to flow within the ceramic body. At a specific onset combination of electric field/furnace temperature, the material densifies in an extremely short time, typically from several seconds to a few minutes [68]. Flash sintering can be done with a dieless ECAS setup (Fig. 8.6) or with an

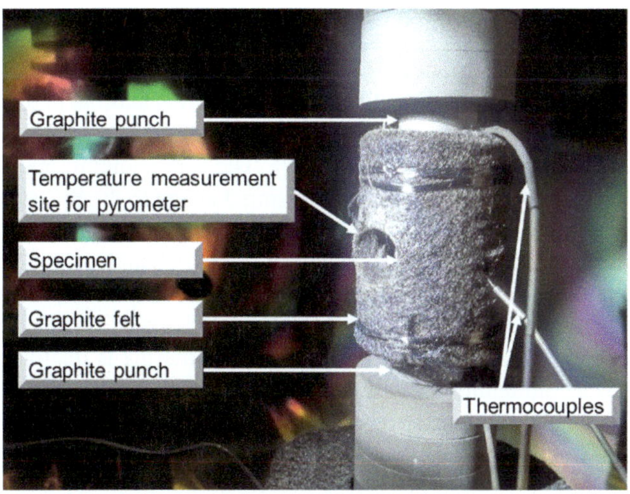

Figure 8.6 Experimental setup for flash sintering [69].

electrically insulating die; an adapted dilatometer/mechanical loading frame with insulator components and electrodes attached to the sample; or a vertical tube furnace with the sample suspended horizontally from the platinum wires acting as electrodes to force or enhance current through the electrically conductive green body [70]. With the dieless ECAS configuration shown in Fig. 8.6, the specimen surface temperature can be measured by a pyrometer and thermocouples (the temperature at the center of the sample could be considerably different from that at the surface [70]). Owing to the rapid sintering process and the presence of an electric current in the green compact, flash sintering mechanisms can be different from those of conventional ECAS [68]. ECAS studies presented in the previous section typically assembled a green body in an electrically conductive graphite die. Thus, most of the current flowed through the graphite die for an electrically insulating specimen, but part of the current was able to pass through the electrically conductive specimen.

Flash sintering of UO_2 fuel has been reported by researchers at Los Alamos National Laboratory [71–73]. Flash sintering of stoichiometric and hyper-stoichiometric pellets was demonstrated at a furnace temperature of 600°C within 2–3 min [73]. This is in contrast to conventional sintering at 1000°C for ~60 min in the case of hyper-stoichiometric UO_2 [73]. An in situ high-energy X-ray diffraction study suggested that a mechanism based on thermal runaway Joule heating can be generalized to explain the onset of flash sintering [72]. Although the flash sintering of a high-density fuel proved to be feasible, the nonuniform microstructure of the pellet across the radial direction [73] presented a challenge. It is expected that the temperature will be higher at the core than at the outer surfaces because of heat losses from the surfaces.

Flash sintering was also applied to the joining of ceramics. Tatarko et al. designed a novel assembly approach using a graphite felt to enhance the current passing through the specimen; with it, they implemented flash joining (joining time <10 s) at a relatively low temperature of ~1250°C for SiC-based ceramic matrix composites, using a titanium foil interlayer [74]. Li et al. joined electrically conductive, nitrogen-doped SiC at 2150°C in 1 min by enhancing the passage of a current through the material with a dieless ECAS configuration [69]. Additionally, a comparison of flash joining and conventional ECAS with a graphite die found that it is feasible to reduce the nominal joining temperature of direct-bonded SiC ceramics by ~150°C by enhancing the passage of current through the material [69]. The rapid joining and low nominal joining temperature are important for minimizing damage to the SiC substrate during the process.

Other examples of flash sintering of nuclear-related ceramics have been reported: pure TiB_2 sintered within 20 or 40 s with up to 98% density in the central area [75], pure ZrB_2 densified up to 95% in 35 s [76], and boron carbide densified up to 99% in 1 min with Ti and C sintering additives [77]. Flash sintering is relatively new to the nuclear ceramics community, and research opportunities exist in the flash sintering of electrically conductive nuclear ceramics for the fabrication of unique materials and energy and time savings.

2.2 Sintering of nanosize powders

As is briefly explained in Section 9.2.1, a fine-grain structure in nuclear ceramics is of interest for the unique mechanical properties and radiation stability it can provide. Sintering nano-size powders is one way to fabricate such materials [78,79]. Because of the high surface-to-volume ratio of a nanopowder compact, low-temperature mass transport by surface diffusion could promote sintering, which would consequently prevent grain growth [79]. The short processing time for ECAS nano-size ceramic powders is also helpful in maintaining a fine-grain structure [79,80].

Nanopowder sintering has been actively studied for SiC-based materials [17,81–85], and this work has been aided by the availability of commercial nanopowders supplied by different vendors [86]. ECAS SiC nanopowders have also been demonstrated [81–84]. Other examples of nanopowder sintering for nuclear applications can be seen for pyrochlore oxides [80,87] and boron carbide [88]. In the case of $Gd_2Zr_2O_7$ pyrochlore ceramic, ECAS of nanocrystalline powders demonstrated the submicron grain structure, which has not been realized by hot-pressing [80]. In SiC nanopowders, the powder surface is terminated by free carbon, siliconoxycarbide, or both [86]. Because of the larger surface area, the amounts of impurities may be significant and may affect sintering behavior. This issue is an important aspect of ceramic nanopowder sintering in general: the surface condition greatly affects dispersion behavior in solvents [86] and can impact impurity content in the resulting ceramics.

In some cases, the use of nano-sized powders provides significant benefits over the use of micron-sized powders for the fabrication of composite ceramics. An example can be seen in the microstructure of the FCM fuel in Fig. 8.3b. In this case, the nominal diameter of the SiC nanopowder was roughly 40 nm. Because they were so small, the SiC particles infiltrated into the narrow spaces between the TRISO particles. The effective infiltration of nano-size SiC particles has also been used in processing a high-density SiC fiber-reinforced, SiC matrix (SiC/SiC) composite [89,90]. The SiC matrix was formed via hot-pressing of a carbon-coated SiC fiber fabric, embedded in a SiC nanopowder slurry with ~10 wt% alumina-yttria sintering additives, at 1800°C for 2 h in an argon atmosphere under a pressure of 20 MPa. Scanning electron microscopy (SEM) micrographs of the cross-section of a SiC/SiC composite in Fig. 8.7 show infiltration of the SiC matrix into the fiber bundle region. Even the narrow intratow area surrounded by thin SiC fibers 10 μm in diameter has been filled by the SiC matrix. Additionally, SiC nanopowder was used to lower the sintering temperature of the SiC/SiC composite. Dong et al. fabricated a similar type of hot-pressed SiC/SiC composites at ~1750°C [91], which is lower than the hot-pressing temperature of 1880°C for the similar sintering additive system [92]. This is the reason why processing hot-pressed SiC/SiC composites uses nanopowder or submicron SiC powder feedstock [93,94]. Interbundle infiltration is a technological obstacle for the chemical vapor infiltration (CVI) process, which is the most commonly used method of fabricating nuclear-grade SiC/SiC composites [95–97]. The near full density of sintered SiC composites has demonstrated superior performance in thermal conductivity [89], gas tightness [98], and matrix cracking resistance [99].

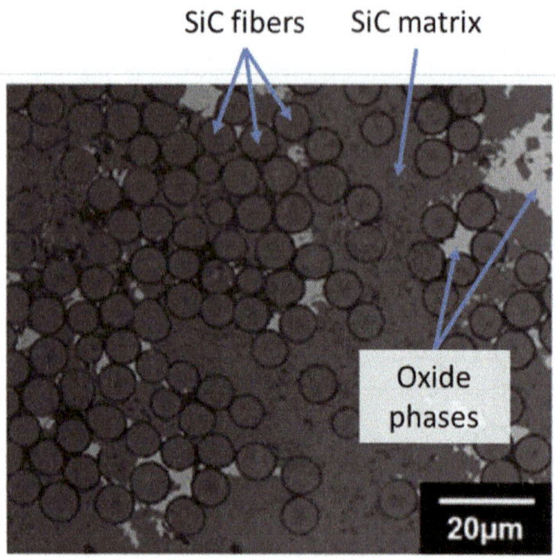

Figure 8.7 SEM micrographs of sintered SiC/SiC composite. The processing details can be found in Ref. [90].

Liquid phase sintering (LPS) of SiC nanopowder results in a unique microstructure, unlike that produced by LPS of micron-size SiC powders. The small grains and high grain boundary density result in a fine distribution of the secondary phases attributed to the sintering additives; Fig. 8.8 shows the distribution of secondary phases in LPS SiC fabricated with a nanopowder feedstock (SiC powder and ZrO_2–CeO_2–AlO_2 sintering additives) [22]. Scanning transmission electron

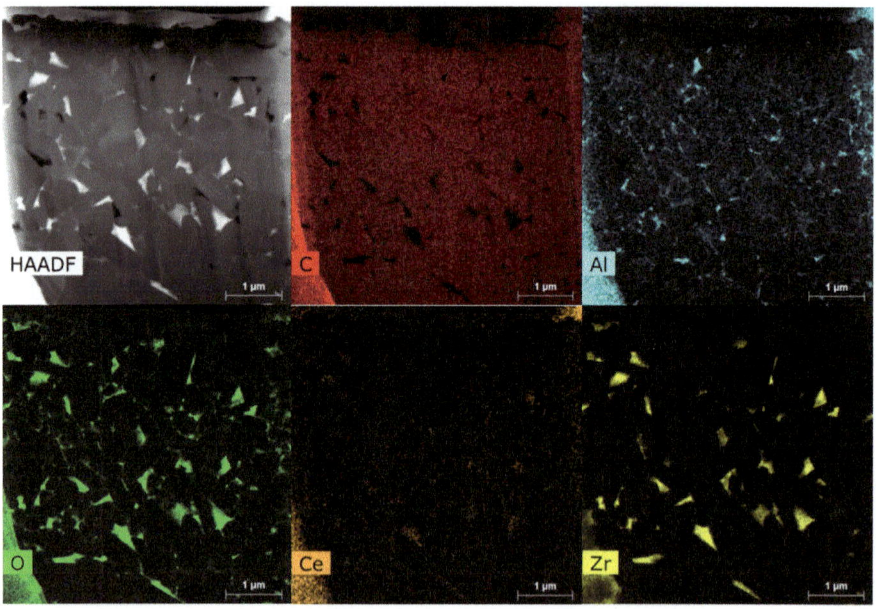

Figure 8.8 Fine distribution of secondary phases in liquid phase–sintered SiC using a nanopowder feedstock.

microscopy/energy-dispersive X-ray spectroscopy analysis shows the oxide phases in the grain pockets. Secondary phase films with a thickness of a few nanometers were also observed along the grain boundaries at higher magnification [22]. Such secondary phase distributions have also been reported in the literature [22,100,101]. The microstructures of conventional LPS SiC ceramics, in contrast, show SiC grain islands in a network of secondary phases [92,102,103].

One of the benefits of the fine distribution of secondary phases is improved radiation resistance. Fine-grain LPS SiC is less susceptible to mechanical degradation by neutron irradiation at elevated temperatures [90,104] than is conventional LPS SiC [105]. One mechanism known to be involved in irradiation-induced strength deterioration is microcracking caused by misfit stresses due to the differential swelling between the secondary phase and the SiC [18]. The smaller grain size is thought to aid resistance to differential swelling–induced strain, based on a fracture mechanics analysis by Davidge et al. [106]. In addition, the isolated secondary phase is preferred for corrosion resistance, as the secondary phases are preferentially attacked in SiC in general. Preferential corrosion was indicated in hydrothermal corrosion environments related to normal light water reactor operating conditions [22,107].

Nanopowder sintered SiC ceramics can offer fine-grain structure and distribution of the secondary phases; this can offer potential advantages to both the irradiation and corrosion resistance, and to the fabrication of dense composite ceramics due to improved particle infiltration of nanopowders. Although the application of nanopowders for SiC ceramics synthesis has been highlighted in this section, the processing and material design approaches discussed can be applied to other relevant nuclear ceramic materials. A key to nanopowder sintering is the availability of commercial nanopowders. Research in this area is ongoing for future production of other carbide materials (e.g., TiC [108], ZrC [109], and boron carbide [110]).

3. Additive manufacturing of nuclear ceramics

3.1 Additive manufacturing processing options

The use of Additive Manufacturing (AM) is growing exponentially in a wide variety of applications [111]. The same is true for the use of nuclear ceramics for fuels and structural applications [112–114]. AM offers remarkable advantages for the rapid prototyping and fabrication of objects with complex geometries. The latter is especially important for ceramic materials because near-net shaping, machining, and integration are common technology challenges to their deployment [115,116]. The design of new reactor components without the constraints on geometry imposed by traditional fabrication methods could bring a technological revolution [117,118]. Although ceramics processing has a long history in near-net-shape fabrication [119–121], AM technology has the potential to further extend the design space. For example, a new design for a nuclear demonstration reactor aided by AM has been pursued under the US Department of Energy's Transformational Challenge Reactor (TCR) Program [122]; the fuel element consisted of TRISO fuel particles embedded inside a 3D printed SiC matrix (Fig. 8.9).

Processes for AM of general ceramics involve preforming a green body based on computer-aided design (CAD), followed by densification. One-step, layer-by-layer consolidation processes during AM have also been demonstrated for specific cases. Fig. 8.10 shows feedstocks, AM methods, and densification processes that are popular for ceramics [124]. The feedstock materials can be preceramic polymer-based, powder-based, or reactive chemical–based inputs. The choice of feedstock type depends on the AM method. Methods that require a densification process include stereolithography, direct ink writing, gel casting, binder jet printing, and laminated object manufacturing. AM methods capable of layer-by-layer consolidation include selective laser sintering (SLS) and selective laser chemical vapor deposition (CVD). SLS may require an additional densification process (e.g., annealing in a furnace and hot isostatic pressing) if the sintering is incomplete or if the density is low. It is worth mentioning that selective laser melting—a popular AM technique for metals [111,117]—is not common for high-temperature ceramics because of the porous microstructures and rough surface finishes, and low dimensional accuracy [124].

The appropriate selection of an AM method depends on the component size, feature size, speed of the additive process, whether structural support is required during AM, surface quality, cost, and dimensional restrictions. An extensive summary of the details for the AM of general ceramics can be found elsewhere [124]. The processing options specific to SiC for nuclear applications have been detailed by Koyanagi et al. [123]. Among various material combinations, AM methods, and densification processes, the following sections showcase AM of ceramics potentially useful for nuclear applications. The densification methods are intended to produce a crystalline microstructure via powder sintering or CVD/CVI to achieve improved mechanical and environmental resistance compared to ceramics with an amorphous microstructure. The final composition should also be high purity to avoid irradiation-induced performance degradation, which can be caused by the differential response of secondary and impurity phases. AM involving polymer pyrolysis, which often forms

Figure 8.9 TCR fuel component: additively manufactured 3D printed SiC shell containing TRISO fuel particles [122].

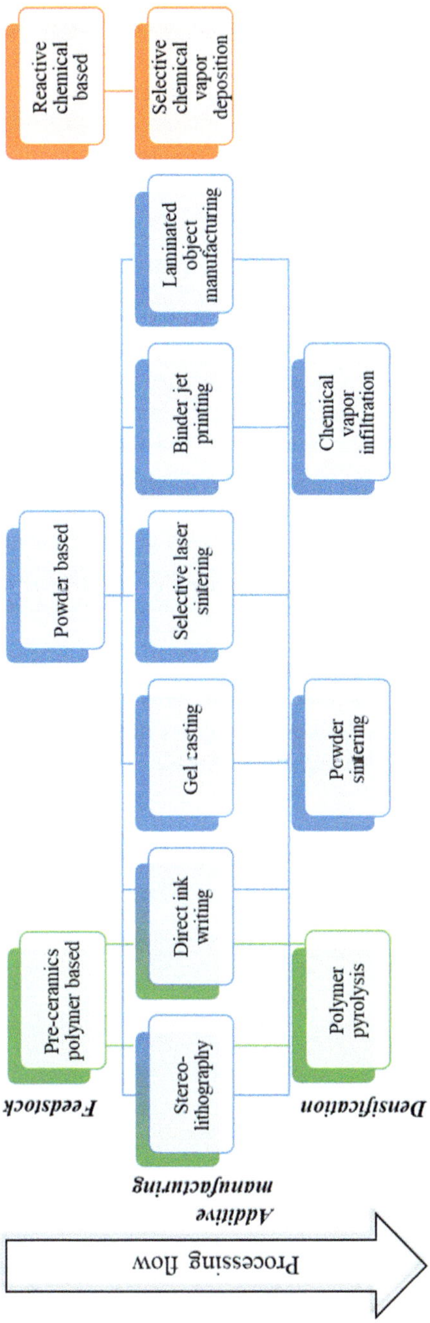

Figure 8.10 Additive manufacturing options for ceramic materials [123].

amorphous-like structures, which is generally undesirable for nuclear applications, would require additional processing steps (heat treatment, etc.) to create a more crystalline structure. There is also a general challenge of powder-based AM techniques to incorporate continuous reinforcing fiber into the component to form an AM-based fibre-reinforced composite. Absent continuous fiber reinforcement, AM-based ceramics would be expected to have lower fracture toughness compared to a ceramic matrix composite, and specific application requirements must be considered when selecting a component fabrication route.

3.2 Binder jet–chemical vapor infiltration hybrid SiC 3D printing

Fig. 8.11 shows a SiC component fabricated by a combination of binder jet AM and CVI (BJ-CVI 3D-printed SiC). The fine features of the logo and characters on the side walls of the cubic object (~0.5 mm width at minimum) represent the ability of AM to fabricate a 3D object with complex features. This technology has been used to fabricate a core component of a demonstration nuclear reactor [122], a significant advancement in processing. Previously, a similar processing approach without AM was used to fabricate SiC components; a green body with a simple geometry was prepared by conventional powder compaction by applying stress instead of by AM, although CVI densification was also used [125]. The BJ-CVI process is superior to this process in its ability to form complex shapes with finer resolution features.

In the BJ-CVI 3D printing process, a ceramic green body was first prepared by binder jet printing using a mixture of a SiC powder and an organic binder, based on CAD. Then the green body went through a binder-curing process followed by CVI SiC densification. The component density ranged from 70% to 90% of the

Figure 8.11 Pure SiC component fabricated by a combination of binder jet AM and CVI [122,123].

theoretical density for SiC, depending on the CVI condition and component geometry. There are limits to the maximum achievable density because the CVI infiltration is likely to be incomplete in areas far from the surfaces if a component is too thick. Therefore, component size and thickness considerations will impact the characteristics of the completed part. The details of the fabrication conditions can be found elsewhere [112,126].

A highlight of this demonstration of AM was the production of a high-purity, crystalline SiC component that retained the dimensions specified in the CAD, which was not common in previous demonstrations of AM SiC processing [112,123]. Shrinkage during densification and the presence of undesirable phases, such as an amorphous SiC phase and secondary phases, have been reported [123]. A high degree of purity and a highly crystalline microstructure are both important to the radiation resistance of SiC at elevated temperatures (>~150°C) [105].

The BJ-CVI hybrid approach could also be applicable to other nuclear ceramic systems so long as feedstock powder and CVD processing are available. CVD is a well-established processing method, and its use to produce some nuclear-related high-temperature ceramics has been demonstrated: UO_2 [127], ZrC [128], TiC [129], boron carbide [130], TiB_2 [131], and pyrochlore oxide [132]. Therefore, a wide range of applications could be proposed using this AM method.

3.3 Selective laser sintering

SLS-based AM has potential applications in the nuclear area because powder sintering is a popular densification method, as explained in Section 9.2. Densification can be achieved by solid-state diffusion, chemical reaction of precursor powders (reaction sintering), LPS, and melting powders. Unless the powders melt during the layer-by-layer deposition process, as was demonstrated for zirconium diboride AM [133], fabrication of a fully dense component is expected to be challenging because of the low green density. Preparation of a green body requires isostatic pressing at a pressure of >100 MPa for dense pressure-sintered ceramics [134,135]. Therefore, postAM heat treatment by conventional furnace or hot isostatic pressing has been used for further densification to achieve >90% relative density [136]. Generally, postAM densification is avoided because the shrinkage makes dimensional control of the object difficult.

Novel SLS of SiC powders without postAM densification has been carried out at PolarOnyx. A powder bed fusion system for direct melting of ceramic powders was modified from PolarOnyx's commercial system. The feedstock was only SiC powder. Sintering SiC without sintering additives is challenging because crystalline SiC does not melt (it sublimates) and also because of the inert nature of SiC (this necessitates temperatures of ~2000°C to achieve sintering [137]). This method has been used in a demonstration to fabricate a cube (50 × 50 × 50 mm) with solid outer walls on four sides and an internal structure of parallel channels with walls 1 mm thick (Fig. 8.12). The wall structure had excellent uniformity based on the CAD. The details of the processing conditions were reported by Koyanagi et al. [123]. The unique aspect of SLS is the short laser pulse width

Figure 8.12 Selective laser-sintered AM SiC component [123].

of 1 ns, which is key to controlling residual stress. Residual stress is one of the most critical parameters that must be optimized; it can cause disconnection of the part from its supporting structure or peeling after printing of individual layers. A short pulse was needed for successful layer-by-layer consolidation. The fabrication of a 3D object was the starting point for deployment. Optimization of the processing parameters to produce the required physical and mechanical properties for nuclear reactor applications warrants future research and development.

Short-pulse SLS was also applied at PolarOnyx to fabricate a boron carbide–aluminum alloy composite for use as a neutron absorber [138]. The unique geometry was intended for more efficient neutron absorption compared with the same component with a simpler geometry. Encapsulation of a ceramic neutron absorber in an aluminum alloy was also demonstrated by Hehr et al. [139]. In that case, ceramic AM was not used; rather, ultrasonic AM of aluminum alloy powders was conducted to incorporate Eu_2O_3 neutron absorber discs. These demonstrations indicate that AM is a game-changing technology for the fabrication of more efficient neutron absorbers.

Although SLS could be used to produce a variety of nuclear ceramics, its applications are limited by the difficulty of pressureless sintering. Potential applications are expected to include both nuclear fuel and structural components. Preliminary results for the laser sintering of uranium silicide fuel, an accident-tolerant nuclear fuel candidate, have been reported. The direct SLS synthesis from UF_4 to U_3Si_2 provides significant savings in both production time and manufacturing cost by eliminating the conversion process from UF_4 to U metal in the traditional process and by eliminating postsintering machining process. The current challenges are forming silicide compounds with an exact stoichiometry and without an impurity phase [114]. A better understanding of laser–ceramics interactions [140] will also be required for process optimization.

3.4 Selective laser CVD

Laser CVD is an AM process that uses a gas-phase precursor chemical feedstock. Maintaining the focus of the laser during CVD allows the local deposition of ceramics [141]. As CVD can produce a dense ceramic microstructure, postAM densification is not required. This is an advantage for precise dimensional control of AM products. A laser focused to a micron-scale spot makes laser CVD-based AM ideal for fabricating small objects with dense, fine features. The presence of a laser is known to accelerate the speed of CVD. Zhang et al. reported 40 μm/h deposition of CVD SiC, 10 times faster than conventional CVD speeds [141]. Nonetheless, the deposition speed limits the application of CVD for producing large-scale components because it is still relatively slow compared with other AM methods listed in Fig. 8.10. But laser CVD could find a wide range of applications because of the availability of CVD for use in producing various ceramics.

Laser CVD AM of SiC has been developed at Free Form Fibers [142]. The continuous deposition enabled the production of long SiC fibers with diameters in the 30–50 μm range (Fig. 8.13). The fibers are being considered for use in a fiber-reinforced ceramic matrix composite for nuclear applications [143]. Long, continuous fibers are formed by the continuous laser focus at the same location when the substrate is pulled away from the laser. The CVD SiC is deposited from the gas-phase reaction on the fiber tip and thus additively forms long fibers. Because of the transverse Gaussian intensity profile of the laser, the laser CVD SiC fiber possesses a heterogeneous microstructure (crystalline to amorphous transition from the fiber core to the edge) that is not seen in traditional CVD SiC. If the amorphous structure is not suitable for the application, postAM heat treatment will be required. Note that laser CVD fiber

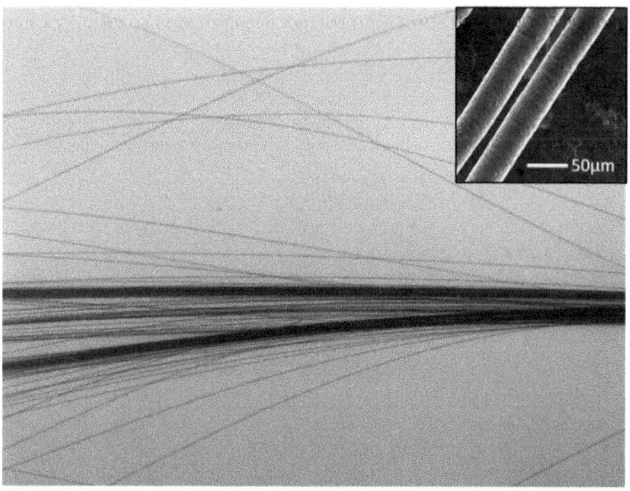

Figure 8.13 Continuous laser CVD SiC fiber fabricated by the laser CVD process at free form fibers.

has been demonstrated for other high-temperature ceramics at Free Form Fibers, which include boron carbide, tungsten carbide, and boron nitride.

Selective laser CVD SiC technology has another potential nuclear application: the encapsulation of uranium-based fuel for light water reactors, with enhanced containment of fission gas products within the fuel component. The preliminary fabrication of the products and numerical analysis of the reactor performance have been reported [144]. In addition, the use of selective laser CVD has been demonstrated in the joining of ceramic components. Harrison et al. joined SiC tubes by laser CVD SiC [145]. A SiC joint phase was formed in a manner similar to the formation of a SiC phase in laser CVD SiC fiber. A different laser beam setting was explored to improve the hermeticity and hardness of the joint section.

For joining purposes, General Atomics also demonstrated selective area CVD joining to seal the end-plug on a SiC composite fuel cladding, using a process based on a global CVD joining technique [146]. When induction heating was used as a local heating method, a larger area could be deposited than was obtained with laser CVD joining (Fig. 8.14). The joint showed promising mechanical properties and hermeticity for the cladding applications [116,146].

3.5 Future research opportunities

AM for nuclear ceramics is still in an embryonic stage. Many of the examples presented in this chapter were demonstrated in recent years. In addition, the type of materials that can be used is limited, although AM has a wide range of potential applications. Most current activities focus mainly on demonstrating the processing methodology. In addition to the study of irradiation effects, there are research opportunities to study the microstructures unique to AM ceramics and to show how they affect components' thermomechanical properties. Moreover, the impacts of the design

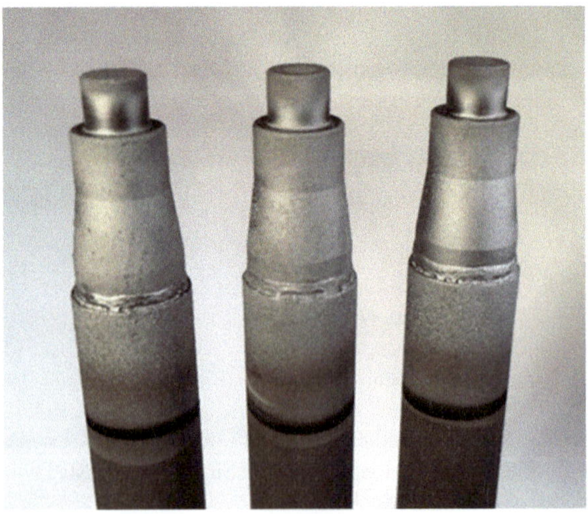

Figure 8.14 Example of cladding-endplug joints formed using a hybrid preceramic polymer/CVI process using localized induction heating at general atomics.

freedom enabled by AM on the performance, efficiency, and safety of nuclear ceramic components remain unexplored.

Additionally, a new opportunity exists to collect rich 3D information by recording the existence of defects throughout a material during production. Such information, which is being used to ascertain the quality of additively manufactured metal components [147–149], is expected to provide opportunities to accelerate nuclear-grade component qualification [150], which is not possible with traditional ceramics processing.

4. Concluding remarks

Processing technologies continue to advance in the ceramics community. Nuclear ceramics technology has benefitted from these advancements and has adapted them to explore new materials and designs for better reactor performance and to tailor the microstructure to improve material properties. This chapter presents recent progress in ECAS, nanopowder sintering, and AM of nuclear fuel and of ceramic materials for reactor structural and functional components. These relatively new processing methods have demonstrated complex geometries and/or material microstructures that traditional processing could not realize. Those methods are applied not only to monolithic ceramics but also to ceramic composites. The processing advancements will save processing energy and time, improve component performance, and support potential new designs and applications of ceramics in nuclear reactors. Future research opportunities exist particularly in comprehensive material characterizations under reactor-relevant applications, industrial-scale production, and qualification.

Acknowledgments

Preparation of this chapter was supported by the US Department of Energy, Office of Fusion Energy Sciences, under contract DE-C05-00OR22725 with UT-Battelle LLC.

References

[1] S.J. Zinkle, G.S. Was, Materials challenges in nuclear energy, Acta Mater. 61 (2013) 735–758, https://doi.org/10.1016/j.actamat.2012.11.004.

[2] G.S. Was, D. Petti, S. Ukai, S. Zinkle, Materials for future nuclear energy systems, J. Nucl. Mater. 527 (2019) 151837, https://doi.org/10.1016/j.jnucmat.2019.151837.

[3] J. Rest, M.W.D. Cooper, J. Spino, J.A. Turnbull, P. Van Uffelen, C.T. Walker, Fission gas release from UO2 nuclear fuel: A review, J. Nucl. Mater. 513 (2019) 310–345, https://doi.org/10.1016/j.jnucmat.2018.08.019.

[4] P.A. Demkowicz, B. Liu, J.D. Hunn, Coated particle fuel: historical perspectives and current progress, J. Nucl. Mater. 515 (2019) 434–450, https://doi.org/10.1016/j.jnucmat.2018.09.044.

[5] M.R. Finlay, G.L. Hofman, J.L. Snelgrove, Irradiation behaviour of uranium silicide compounds, J. Nucl. Mater. 325 (2004) 118–128, https://doi.org/10.1016/j.jnucmat.2003.11.009.

[6] Y. Katoh, L.L. Snead, Silicon carbide and its composites for nuclear applications – historical overview, J. Nucl. Mater. 526 (2019) 151849, https://doi.org/10.1016/j.jnucmat.2019.151849.

[7] T. Koyanagi, Y. Katoh, T. Nozawa, L.L. Snead, S. Kondo, C.H. Henager, M. Ferraris, T. Hinoki, Q. Huang, Recent progress in the development of SiC composites for nuclear fusion applications, J. Nucl. Mater. 511 (2018) 544–555, https://doi.org/10.1016/j.jnucmat.2018.06.017.

[8] M.T. Simnad, The U-ZrHx alloy: its properties and use in TRIGA fuel, Nucl. Eng. Des. 64 (1981) 403–422, https://doi.org/10.1016/0029-5493(81)90135-7.

[9] M. Mashhadi, E. Taheri-Nassaj, V.M. Sglavo, H. Sarpoolaky, N. Ehsani, Effect of Al addition on pressureless sintering of B4C, Ceram. Int. 35 (2009) 831–837, https://doi.org/10.1016/j.ceramint.2008.03.003.

[10] G.M. Jacobsen, H. Choi, J.A. Turso, A.M. Johnsen, A.J. Bascom, X. Wei, E.A. Olevsky, Fabrication and characterization of zirconium silicide for application to gas-cooled fast reactors, Nucl. Technol. (2021) 1–10, https://doi.org/10.1080/00295450.2021.1877504.

[11] R.C. Ewing, W.J. Weber, J. Lian, Nuclear waste disposal—pyrochlore (A2B2O7): nuclear waste form for the immobilization of plutonium and "minor" actinides, J. Appl. Phys. 95 (2004) 5949–5971, https://doi.org/10.1063/1.1707213.

[12] T. Koyanagi, Y. Katoh, C. Ang, D. King, G.E. Hilmas, W.G. Fahrenholtz, Response of isotopically tailored titanium diboride to neutron irradiation, J. Am. Ceram. Soc. 102 (2019) 85–89, https://doi.org/10.1111/jace.16036.

[13] A. Bhattacharya, C.M. Parish, T. Koyanagi, C.M. Petrie, D. King, G. Hilmas, W. G. Fahrenholtz, S.J. Zinkle, Y. Katoh, Nano-scale microstructure damage by neutron irradiations in a novel Boron-11 enriched TiB2 ultra-high temperature ceramic, Acta Mater. 165 (2019) 26–39, https://doi.org/10.1016/j.actamat.2018.11.030.

[14] Z.M. Sun, Progress in research and development on MAX phases: a family of layered ternary compounds, Int. Mater. Rev. 56 (2011) 143–166, https://doi.org/10.1179/1743280410Y.0000000001.

[15] R.M. German, P. Suri, S.J. Park, Review: liquid phase sintering, J. Mater. Sci. 44 (2009) 1–39, https://doi.org/10.1007/s10853-008-3008-0.

[16] T. Hase, H. Suzuki, T. Iseki, Formation process of β-SiC during reaction-sintering, J. Nucl. Mater. (1976), https://doi.org/10.1016/0022-3115(76)90006-4.

[17] K. Shimoda, N. Eiza, J.-S. Park, T. Hinoki, A. Kohyama, S. Kondo, High-temperature mechanical property improvements of SiC ceramics by NITE process, Mater. Trans. 47 (2006) 1204–1208, https://doi.org/10.2320/matertrans.47.1204.

[18] R.B. Matthews, Irradiation damage in reaction-bonded silicon carbide, J. Nucl. Mater. 51 (1974) 203–208, https://doi.org/10.1016/0022-3115(74)90003-8.

[19] T. Koyanagi, Y. Katoh, J.O. Kiggans, T. Hinoki, H.E. Khalifa, C.P. Deck, C.A. Back, Irradiation resistance of silicon carbide joint at light water reactor–relevant temperature, J. Nucl. Mater. 488 (2017) 150–159, https://doi.org/10.1016/j.jnucmat.2017.03.017.

[20] R.G. Elliman, J.S. Williams, W.L. Brown, A. Leiberich, D.M. Maher, R.V. Knoell, Ion-beam-induced crystallization and amorphization of silicon, Nucl. Instrum. Methods Phys. Res. Sect. B Beam Interact. Mater. Atoms 19–20 (1987) 435–442, https://doi.org/10.1016/S0168-583X(87)80086-1.

[21] G. Youngblood, R. Jones, A. Kohyama, L. Snead, Radiation response of SiC-based fibers, J. Nucl. Mater. 258–263 (1998) 1551–1556, https://doi.org/10.1016/S0022-3115(98)00217-7.

[22] C.M. Parish, K.A. Terrani, Y.-J. Kim, T. Koyanagi, Y. Katoh, Microstructure and hydrothermal corrosion behavior of NITE-SiC with various sintering additives in LWR coolant environments, J. Eur. Ceram. Soc. 37 (2017), https://doi.org/10.1016/j.jeurceramsoc.2016.11.033.

[23] R. Raj, M. Cologna, J.S.C. Francis, Influence of externally imposed and internally generated electrical fields on grain growth, diffusional creep, sintering and related phenomena in ceramics, J. Am. Ceram. Soc. 94 (2011) 1941–1965, https://doi.org/10.1111/j.1551-2916.2011.04652.x.

[24] M. Tokita, Progress of Spark Plasma Sintering (SPS) method, systems, ceramics applications and industrialization, Ceramics 4 (2021) 160–198, https://doi.org/10.3390/ceramics4020014.

[25] L. Ge, G. Subhash, R.H. Baney, J.S. Tulenko, E. McKenna, Densification of uranium dioxide fuel pellets prepared by spark plasma sintering (SPS), J. Nucl. Mater. 435 (2013) 1–9, https://doi.org/10.1016/j.jnucmat.2012.12.010.

[26] B. Gong, T. Yao, P. Lei, J. Harp, A.T. Nelson, J. Lian, Spark plasma sintering (SPS) densified U3Si2 pellets: Microstructure control and enhanced mechanical and oxidation properties, J. Alloys Compd. 825 (2020) 154022, https://doi.org/10.1016/j.jallcom.2020.154022.

[27] N. Tamari, T. Tanaka, K. Tanaka, I. Kondoh, M. Kawahara, M. Tokita, Effect of spark plasma sintering on densification and mechanical properties of silicon carbide, J. Ceram. Soc. Japan. 103 (1995) 740–742, https://doi.org/10.2109/jcersj.103.740.

[28] B.M. Moshtaghioun, F.L. Cumbrera-Hernández, D. Gómez-García, S. de Bernardi-Martín, A. Domínguez-Rodríguez, A. Monshi, M.H. Abbasi, Effect of spark plasma sintering parameters on microstructure and room-temperature hardness and toughness of fine-grained boron carbide (B4C), J. Eur. Ceram. Soc. 33 (2013) 361–369, https://doi.org/10.1016/j.jeurceramsoc.2012.08.028.

[29] R. Kumar, K. Singh, D. Chakravarty, A. Chowdhury, Attaining near-theoretical densification in nanograined pyrochlore La2Zr2O7 (LZ) ceramic at 1150°C by spark plasma sintering, Scr. Mater. 117 (2016) 37–40, https://doi.org/10.1016/j.scriptamat.2016.02.014.

[30] L. Wang, X. Shu, X. Lu, Y. Wu, Y. Ding, S. Zhang, Rapid synthesis of high densified single phase ceramic Gd2Zr2O7 by spark plasma sintering, Mater. Lett. 196 (2017) 403–405, https://doi.org/10.1016/j.matlet.2017.03.061.

[31] C. Magnus, W.M. Rainforth, Spark plasma sintering (SPS) synthesis and tribological behaviour of MAX phase composite of the family Tin+1SiCn (n = 2), Wear 438–439 (2019) 203062, https://doi.org/10.1016/j.wear.2019.203062.

[32] A. Babapoor, M.S. Asl, Z. Ahmadi, A.S. Namini, Effects of spark plasma sintering temperature on densification, hardness and thermal conductivity of titanium carbide, Ceram. Int. 44 (2018) 14541–14546, https://doi.org/10.1016/j.ceramint.2018.05.071.

[33] D. Demirskyi, H. Borodianska, Y. Sakka, O. Vasylkiv, Ultra-high elevated temperature strength of TiB2-based ceramics consolidated by spark plasma sintering, J. Eur. Ceram. Soc. 37 (2017) 393–397, https://doi.org/10.1016/j.jeurceramsoc.2016.08.009.

[34] S.-Q. Guo, T. Nishimura, Y. Kagawa, J.-M. Yang, Spark plasma sintering of zirconium diborides, J. Am. Ceram. Soc. 91 (2008) 2848–2855, https://doi.org/10.1111/j.1551-2916.2008.02587.x.

[35] G.S. Was, Fundamentals of Radiation Materials Science: Metals and Alloys, second ed., 2016, https://doi.org/10.1007/978-1-4939-3438-6.

[36] J. Pelleg, Creep in ceramics, Solid Mech. Its Appl. (2017) 41–61, https://doi.org/10.1007/978-3-319-50826-9_4.

[37] O. El-Atwani, J.E. Nathaniel, A.C. Leff, B.R. Muntifering, J.K. Baldwin, K. Hattar, M. L. Taheri, The role of grain size in He bubble formation: Implications for swelling resistance, J. Nucl. Mater. 484 (2017) 236–244, https://doi.org/10.1016/j.jnucmat.2016.12.003.

[38] O. El-Atwani, E. Esquivel, E. Aydogan, E. Martinez, J.K. Baldwin, M. Li, B. P. Uberuaga, S.A. Maloy, Unprecedented irradiation resistance of nanocrystalline tungsten with equiaxed nanocrystalline grains to dislocation loop accumulation, Acta Mater. 165 (2019) 118–128, https://doi.org/10.1016/j.actamat.2018.11.024.

[39] J. Zhang, J. Lian, A.F. Fuentes, F. Zhang, M. Lang, F. Lu, R.C. Ewing, Enhanced radiation resistance of nanocrystalline pyrochlore Gd2(Ti0.65Zr0.35)2O7, Appl. Phys. Lett. 94 (2009) 243110, https://doi.org/10.1063/1.3155855.

[40] Y. Wang, K. Fujinami, R. Zhang, C. Wan, N. Wang, Y. Ba, K. Koumoto, Interfacial thermal resistance and thermal conductivity in nanograined SrTiO3, Appl. Phys. Express. 3 (2010) 031101, https://doi.org/10.1143/APEX.3.031101.

[41] P.J. Doyle, S. Zinkle, S.S. Raiman, Hydrothermal corrosion behavior of CVD SiC in high temperature water, J. Nucl. Mater. 539 (2020) 152241, https://doi.org/10.1016/j.jnucmat.2020.152241.

[42] O. Guillon, J. Gonzalez-Julian, B. Dargatz, T. Kessel, G. Schierning, J. Räthel, M. Herrmann, Field-assisted sintering technology/spark plasma sintering: mechanisms, materials, and technology developments, Adv. Eng. Mater. 16 (2014) 830–849, https://doi.org/10.1002/adem.201300409.

[43] G. Lee, E.A. Olevsky, C. Manière, A. Maximenko, O. Izhvanov, C. Back, J. McKittrick, Effect of electric current on densification behavior of conductive ceramic powders consolidated by spark plasma sintering, Acta Mater. 144 (2018) 524–533, https://doi.org/10.1016/j.actamat.2017.11.010.

[44] E.A. Olevsky, C. Garcia-Cardona, W.L. Bradbury, C.D. Haines, D.G. Martin, D. Kapoor, Fundamental aspects of spark plasma sintering: II. Finite element analysis of scalability, J. Am. Ceram. Soc. 95 (2012) 2414–2422, https://doi.org/10.1111/j.1551-2916.2012.05096.x.

[45] R. Yamanoglu, Pressureless spark plasma sintering: a perspective from conventional sintering to accelerated sintering without pressure, Powder Metall. Met. Ceram. 57 (2019) 513–525, https://doi.org/10.1007/s11106-019-00010-1.

[46] E.K. Papynov, O.O. Shichalin, A.A. Belov, A.S. Portnyagin, I.Y. Buravlev, V. Y. Mayorov, A.E. Sukhorada, E.A. Gridasova, A.D. Nomerovskiy, V.O. Glavinskaya, I. G. Tananaev, V.I. Sergienko, Spark plasma sintering of UO2 fuel composite with Gd2O3 integral fuel burnable absorber, Nucl. Eng. Technol. 52 (2020) 1756–1763, https://doi.org/10.1016/j.net.2020.01.032.

[47] E. Kardoulaki, D.M. Frazer, J.T. White, U. Carvajal, A.T. Nelson, D.D. Byler, T. A. Saleh, B. Gong, T. Yao, J. Lian, K.J. McClellan, Fabrication and thermophysical properties of UO2-UB2 and UO2-UB4 composites sintered via spark plasma sintering, J. Nucl. Mater. 544 (2021) 152690, https://doi.org/10.1016/j.jnucmat.2020.152690.

[48] J. Buckley, J.D. Turner, T.J. Abram, Uranium dioxide - Molybdenum composite fuel pellets with enhanced thermal conductivity manufactured via spark plasma sintering, J. Nucl. Mater. 523 (2019) 360–368, https://doi.org/10.1016/j.jnucmat.2019.05.059.

[49] B. Li, Z. Yang, J. Jia, Y. Zhong, X. Liu, P. Zhang, R. Gao, T. Liu, R. Li, H. Huang, M. Sun, D. Mazhao, High temperature thermal physical performance of BeO/UO2

composites prepared by spark plasma sintering (SPS), Scr. Mater. 142 (2018) 70–73, https://doi.org/10.1016/j.scriptamat.2017.08.031.
[50] S. Yeo, E. Mckenna, R. Baney, G. Subhash, J. Tulenko, Enhanced thermal conductivity of uranium dioxide–silicon carbide composite fuel pellets prepared by Spark Plasma Sintering (SPS), J. Nucl. Mater. 433 (2013) 66–73, https://doi.org/10.1016/j.jnucmat.2012.09.015.
[51] Z. Chen, G. Subhash, J.S. Tulenko, Spark plasma sintering of diamond-reinforced uranium dioxide composite fuel pellets, Nucl. Eng. Des. 294 (2015) 52–59, https://doi.org/10.1016/j.nucengdes.2015.08.021.
[52] K.A. Terrani, J.O. Kiggans, Y. Katoh, K. Shimoda, F.C. Montgomery, B.L. Armstrong, C.M. Parish, T. Hinoki, J.D. Hunn, L.L. Snead, Fabrication and characterization of fully ceramic microencapsulated fuels, J. Nucl. Mater. 426 (2012) 268–276, https://doi.org/10.1016/j.jnucmat.2012.03.049.
[53] H.-M. Kim, Y.-W. Kim, K.-Y. Lim, Pressureless sintered silicon carbide matrix with a new quaternary additive for fully ceramic microencapsulated fuels, J. Eur. Ceram. Soc. 39 (2019) 3971–3980, https://doi.org/10.1016/j.jeurceramsoc.2019.06.008.
[54] C. Ang, L. Snead, Y. Kato, A logical approach for zero-rupture Fully Ceramic Micro-encapsulated (FCM) fuels via pressure-assisted sintering route, J. Nucl. Mater. 531 (2020) 151987, https://doi.org/10.1016/j.jnucmat.2020.151987.
[55] E. López-Honorato, H. Zhang, D. Yang, P. Xiao, Silver diffusion in silicon carbide coatings, J. Am. Ceram. Soc. 94 (2011) 3064–3071, https://doi.org/10.1111/j.1551-2916.2011.04544.x.
[56] C. Ang, G. Singh, L. Snead, Y. Katoh, Preliminary study of sintering zero-rupture Fully Ceramic Microencapsulated (FCM) fuel, Int. J. Appl. Ceram. Technol. 16 (2019) 1699–1707, https://doi.org/10.1111/ijac.13294.
[57] C. Ang, L. Snead, K. Benensky, Niobium carbide as a technology demonstrator of ultra-high temperature ceramics for fully ceramic microencapsulated fuels, Int. J. Ceram. Eng. Sci. 1 (2019) 92–102, https://doi.org/10.1002/ces2.10014.
[58] X. Hu, K.A. Terrani, B.D. Wirth, Hydrogen desorption kinetics from zirconium hydride and zirconium metal in vacuum, J. Nucl. Mater. 448 (2014) 87–95, https://doi.org/10.1016/j.jnucmat.2014.01.028.
[59] C. Ang, M. Koehler, E. Miller, M. Scott, L. Snead, B. Cheng, D. Sprouster, N.R. Brown, Composite moderators: zirconium hydrides aentrained and encapsulated by MgO, in: 45th Int. Conf. Expo Adv. Ceram. Compos. (ICACC 2021 Virtual), 2021.
[60] X. Hu, K.A. Terrani, Thermomechanical properties and microstructures of yttrium hydride, J. Alloys Compd. 867 (2021) 158992, https://doi.org/10.1016/j.jallcom.2021.158992.
[61] T. Koyanagi, Y. Katoh, T. Hinoki, C. Henager, M. Ferraris, S. Grasso, Progress in development of SiC-based joints resistant to neutron irradiation, J. Eur. Ceram. Soc. 40 (2020) 1023–1034, https://doi.org/10.1016/j.jeurceramsoc.2019.10.055.
[62] S. Grasso, P. Tatarko, S. Rizzo, H. Porwal, C. Hu, Y. Katoh, M. Salvo, M.J. Reece, M. Ferraris, Joining of β-SiC by spark plasma sintering, J. Eur. Ceram. Soc. 34 (2014) 1681–1686, https://doi.org/10.1016/j.jeurceramsoc.2013.12.023.
[63] S. Rizzo, S. Grasso, M. Salvo, V. Casalegno, M.J. Reece, M. Ferraris, Joining of C/SiC composites by spark plasma sintering technique, J. Eur. Ceram. Soc. 34 (2014) 903–913, https://doi.org/10.1016/j.jeurceramsoc.2013.10.028.
[64] L.-X. Wu, L.-L. Zhu, W.-M. Guo, S.-K. Sun, W.-B. Niu, J.-X. Xue, J.-H. Zhai, H.-B. Ma, R.-L. Lin, H.-T. Lin, K. Plucknett, Y.-H. Liao, T. Liu, Q.-S. Ren, Effects of the joining process on the microstructure and properties of liquid-phase-sintered SiC-SiC

joints formed with Ti foil, J. Eur. Ceram. Soc. 41 (2021) 225–232, https://doi.org/10.1016/j.jeurceramsoc.2020.07.072.

[65] P. Tatarko, V. Casalegno, C. Hu, M. Salvo, M. Ferraris, M.J. Reece, Joining of CVD-SiC coated and uncoated fibre reinforced ceramic matrix composites with pre-sintered Ti3SiC2 MAX phase using Spark Plasma Sintering, J. Eur. Ceram. Soc. 36 (2016) 3957–3967, https://doi.org/10.1016/j.jeurceramsoc.2016.06.025.

[66] Y. Katoh, L.L. Snead, T. Cheng, C. Shih, W.D. Lewis, T. Koyanagi, T. Hinoki, C. H. Henager, M. Ferraris, Radiation-tolerant joining technologies for silicon carbide ceramics and composites, J. Nucl. Mater. 448 (2014) 497–511, https://doi.org/10.1016/j.jnucmat.2013.10.002.

[67] M. Hambuch, F. Gora, K. Beart, F. Wittmann, A. Roosen, Joining of sintered alumina substrates and LTCC green tapes via cold low-pressure lamination, Int. J. Appl. Ceram. Technol. 11 (2014) 443–450, https://doi.org/10.1111/ijac.12200.

[68] M. Biesuz, V.M. Sglavo, Flash sintering of ceramics, J. Eur. Ceram. Soc. 39 (2019) 115–143, https://doi.org/10.1016/j.jeurceramsoc.2018.08.048.

[69] H. Li, T. Koyanagi, C. Ang, Y. Katoh, Electric current–assisted direct joining of silicon carbide, J. Eur. Ceram. Soc. 41 (2021) 3072–3081, https://doi.org/10.1016/j.jeurceramsoc.2020.05.072.

[70] C.E.J. Dancer, Flash sintering of ceramic materials, Mater. Res. Express 3 (2016) 102001, https://doi.org/10.1088/2053-1591/3/10/102001.

[71] A.M. Raftery, J.G. Pereira da Silva, D.D. Byler, D.A. Andersson, B.P. Uberuaga, C. R. Stanek, K.J. McClellan, Onset conditions for flash sintering of UO2, J. Nucl. Mater. 493 (2017) 264–270, https://doi.org/10.1016/j.jnucmat.2017.06.022.

[72] D.J. Sprouster, E. Kardoulaki, R. Weidner, A.M. Raftery, M. Elbakhshwan, R. Pokharel, H.M. Reiche, D.D. Byler, S.K. Ghose, E. Dooryhee, K.J. McClellan, L.E. Ecker, In situ X-ray characterization of uranium dioxide during flash sintering, Materialia 2 (2018) 176–182, https://doi.org/10.1016/j.mtla.2018.07.006.

[73] J.A. Valdez, D.D. Byler, E. Kardoulaki, J.S.C. Francis, K.J. McClellan, Flash sintering of stoichiometric and hyper-stoichiometric urania, J. Nucl. Mater. 505 (2018) 85–93, https://doi.org/10.1016/j.jnucmat.2018.03.049.

[74] P. Tatarko, S. Grasso, T.G. Saunders, V. Casalegno, M. Ferraris, M.J. Reece, Flash joining of CVD-SiC coated Cf/SiC composites with a Ti interlayer, J. Eur. Ceram. Soc. 37 (2017) 3841–3848, https://doi.org/10.1016/j.jeurceramsoc.2017.05.016.

[75] S. Failla, S. Fu, D. Sciti, S. Grasso, Flash spark plasma sintering of pure TiB2, Open Ceram 5 (2021) 100075, https://doi.org/10.1016/j.oceram.2021.100075.

[76] S. Grasso, T. Saunders, H. Porwal, O. Cedillos-Barraza, D.D. Jayaseelan, W.E. Lee, M. J. Reece, Flash Spark Plasma Sintering (FSPS) of pure ZrB_2, J. Am. Ceram. Soc. 97 (2014) 2405–2408, https://doi.org/10.1111/jace.13109.

[77] B. Niu, F. Zhang, J. Zhang, W. Ji, W. Wang, Z. Fu, Ultra-fast densification of boron carbide by flash spark plasma sintering, Scr. Mater. 116 (2016) 127–130, https://doi.org/10.1016/j.scriptamat.2016.02.012.

[78] M.J. Mayo, Processing of nanocrystalline ceramics from ultrafine particles, Int. Mater. Rev. 41 (1996) 85–115, https://doi.org/10.1179/imr.1996.41.3.85.

[79] R. Chaim, M. Levin, A. Shlayer, C. Estournes, Sintering and densification of nanocrystalline ceramic oxide powders: a review, Adv. Appl. Ceram. 107 (2008) 159–169, https://doi.org/10.1179/174367508X297812.

[80] J. Qi, Z. Huang, M. Zhou, S. Peng, N. Ma, T. Lu, Transparent sub-mircon Gd2Zr2O7 ceramic prepared by spark plasma sintering using nanocrystalline powders, J. Eur. Ceram. Soc. 38 (2018) 2256–2258, https://doi.org/10.1016/j.jeurceramsoc.2017.11.014.

[81] C. Lorrette, A. Réau, L. Briottet, Mechanical properties of nanostructured silicon carbide consolidated by spark plasma sintering, J. Eur. Ceram. Soc. 33 (2013) 147–156, https://doi.org/10.1016/j.jeurceramsoc.2012.07.030.
[82] D.O. Moskovskikh, Y.-C. Lin, A.S. Rogachev, P.J. McGinn, A.S. Mukasyan, Spark plasma sintering of SiC powders produced by different combustion synthesis routes, J. Eur. Ceram. Soc. 35 (2015) 477–486, https://doi.org/10.1016/j.jeurceramsoc.2014.09.014.
[83] T. Yamamoto, H. Kitaura, Y. Kodera, T. Ishii, M. Ohyanagi, Z.A. Munir, Consolidation of nanostructured β-SiC by spark plasma sintering, J. Am. Ceram. Soc. 87 (2004) 1436–1441, https://doi.org/10.1111/j.1551-2916.2004.01436.x.
[84] S. Grasso, T. Saunders, H. Porwal, M. Reece, Ultra-high temperature spark plasma sintering of α-SiC, Ceram. Int. 41 (2015) 225–230, https://doi.org/10.1016/j.ceramint.2014.08.062.
[85] K. Kijima, H. Noguchi, M. Konishi, Sintering of ultrafine SiC powders prepared by plasma CVD, J. Mater. Sci. 24 (1989) 2929–2933, https://doi.org/10.1007/BF02385649.
[86] K. Shimoda, T. Koyanagi, Surface properties and dispersion behaviors of SiC nanopowders, Colloids Surfaces A Physicochem. Eng. Asp. 463 (2014), https://doi.org/10.1016/j.colsurfa.2014.09.013.
[87] L. An, A. Ito, T. Goto, Highly transparent lutetium titanium oxide produced by spark plasma sintering, J. Eur. Ceram. Soc. 31 (2011) 237–240, https://doi.org/10.1016/j.jeurceramsoc.2010.09.010.
[88] J. Dai, W. Lee, E.D. Gomez, N. Yamamoto, Demonstrating low-temperature sintering of boron carbide powders, Int. J. Ceram. Eng. Sci. 1 (2019) 178–184, https://doi.org/10.1002/ces2.10028.
[89] Y. Katoh, S.M. Dong, A. Kohyama, Thermo-mechanical properties and microstructure of silicon carbide composites fabricated by nano-infiltrated transient eutectoid process, Fusion Eng. Des. (2002) 723–731, https://doi.org/10.1016/S0920-3796(02)00180-1.
[90] T. Koyanagi, K. Ozawa, T. Hinoki, K. Shimoda, Y. Katoh, Effects of neutron irradiation on mechanical properties of silicon carbide composites fabricated by nano-infiltration and transient eutectic-phase process, J. Nucl. Mater. 448 (2014) 478–486, https://doi.org/10.1016/j.jnucmat.2013.10.005.
[91] S. Dong, Y. Katoh, A. Kohyama, Preparation of SiC/SiC composites by hot pressing, using Tyranno-SA fiber as reinforcement, J. Am. Ceram. Soc. 86 (2003) 26–32, https://doi.org/10.1111/j.1151-2916.2003.tb03272.x.
[92] D. Sciti, A. Bellosi, Effects of additives on densification, microstructure and properties of liquid-phase sintered silicon carbide, J. Mater. Sci. (2000), https://doi.org/10.1023/A:1004881430804.
[93] K. Yoshida, T. Yano, Room and high-temperature mechanical and thermal properties of SiC fiber-reinforced SiC composite sintered under pressure, J. Nucl. Mater. 283–287 (2000) 560–564, https://doi.org/10.1016/S0022-3115(00)00212-9.
[94] Y.-W. Kim, Y.-S. Chun, J.Y. Park, W.S. Ryu, D.-S. Park, Process for dense 2D SiC fiber-SiC matrix composites, Met. Mater. Int. 13 (2007) 197–200, https://doi.org/10.1007/BF03027804.
[95] Y. Katoh, K. Ozawa, C. Shih, T. Nozawa, R.J. Shinavski, A. Hasegawa, L.L. Snead, Continuous SiC fiber, CVI SiC matrix composites for nuclear applications: properties and irradiation effects, J. Nucl. Mater. 448 (2014) 448–476, https://doi.org/10.1016/j.jnucmat.2013.06.040.
[96] C.P. Deck, H.E. Khalifa, B. Sammuli, T. Hilsabeck, C.A. Back, Fabrication of SiC-SiC composites for fuel cladding in advanced reactor designs, Prog. Nucl. Energy (2012) 38–45, https://doi.org/10.1016/j.pnucene.2011.10.002.

[97] C. Sauder, Ceramic matrix composites: nuclear applications, in: Ceram. Matrix Compos. Mater. Model. Technol, 2014, pp. 609–646, https://doi.org/10.1002/9781118832998.ch22.
[98] T. Hino, E. Hayashishita, Y. Yamauchi, M. Hashiba, Y. Hirohata, A. Kohyama, Helium gas permeability of SiC/SiC composite used for in-vessel components of nuclear fusion reactor, Fusion Eng. Des. 73 (2005) 51–56, https://doi.org/10.1016/j.fusengdes.2005.01.002.
[99] K. Shimoda, A. Kohyama, T. Hinoki, High mechanical performance SiC/SiC composites by NITE process with tailoring of appropriate fabrication temperature to fiber volume fraction, Compos. Sci. Technol. 69 (2009) 1623–1628, https://doi.org/10.1016/j.compscitech.2009.03.011.
[100] K.A. Terrani, J.O. Kiggans, C.M. Silva, C. Shih, Y. Katoh, L.L. Snead, Progress on matrix SiC processing and properties for fully ceramic microencapsulated fuel form, J. Nucl. Mater. 457 (2015) 9–17, https://doi.org/10.1016/j.jnucmat.2014.10.034.
[101] T. Koyanagi, S. Kondo, T. Hinoki, The influence of sintering additives on the irradiation resistance of NITE SiC, J. Nucl. Mater. 417 (2011), https://doi.org/10.1016/j.jnucmat.2010.12.181.
[102] F. Rodríguez-Rojas, A.L. Ortiz, F. Guiberteau, M. Nygren, Oxidation behaviour of pressureless liquid-phase-sintered α-SiC with additions of 5Al2O3+3RE2O3 (RE=La, Nd, Y, Er, Tm, or Yb), J. Eur. Ceram. Soc. 30 (2010) 3209–3217, https://doi.org/10.1016/j.jeurceramsoc.2010.07.013.
[103] O. Borrero-López, A.L. Ortiz, F. Guiberteau, N.P. Padture, Effect of liquid-phase content on the contact-mechanical properties of liquid-phase-sintered α-SiC, J. Eur. Ceram. Soc. 27 (2007) 2521–2527, https://doi.org/10.1016/j.jeurceramsoc.2006.09.012.
[104] K. Terrani, C. Ang, L.L. Snead, Y. Katoh, Irradiation stability and thermo-mechanical properties of NITE-SiC irradiated to 10 dpa, J. Nucl. Mater. 499 (2018) 242–247, https://doi.org/10.1016/j.jnucmat.2017.11.040.
[105] L.L. Snead, T. Nozawa, Y. Katoh, T.S. Byun, S. Kondo, D.A. Petti, Handbook of SiC properties for fuel performance modeling, J. Nucl. Mater. 371 (2007) 329–377, https://doi.org/10.1016/j.jnucmat.2007.05.016.
[106] R.W. Davidge, G. Tappin, Internal strain energy and the strength of brittle materials, J. Mater. Sci. 3 (1968) 297–301, https://doi.org/10.1007/BF00741965.
[107] T. Koyanagi, Y. Katoh, K.A. Terrani, Y.-J. Kim, J.O. Kiggans, T. Hinoki, Hydrothermal corrosion of silicon carbide joints without radiation, J. Nucl. Mater. 481 (2016), https://doi.org/10.1016/j.jnucmat.2016.09.027.
[108] D. Sarkar, M. Chu, S.-J. Cho, Y. Il Kim, B. Basu, Synthesis and morphological analysis of titanium carbide nanopowder, J. Am. Ceram. Soc. 92 (2009) 2877–2882, https://doi.org/10.1111/j.1551-2916.2009.03316.x.
[109] J.P. Yasnó, R.F.K. Gunnewiek, R.H.G.A. Kiminami, Microwave synthesis of ultra-high temperature ceramic ZrC nanopowders, Adv. Powder Technol. 30 (2019) 1348–1355, https://doi.org/10.1016/j.apt.2019.04.010.
[110] Y. Gu, L. Chen, Y. Qian, W. Zhang, J. Ma, Synthesis of nanocrystalline boron carbide via a solvothermal reduction of CCl4 in the presence of amorphous boron powder, J. Am. Ceram. Soc. 88 (2004) 225–227, https://doi.org/10.1111/j.1551-2916.2004.00023.x.
[111] M. Attaran, The rise of 3-D printing: the advantages of additive manufacturing over traditional manufacturing, Bus. Horiz. 60 (2017) 677–688, https://doi.org/10.1016/j.bushor.2017.05.011.
[112] K. Terrani, B. Jolly, M. Trammell, 3D printing of high-purity silicon carbide, J. Am. Ceram. Soc. 103 (2020) 1575–1581, https://doi.org/10.1111/jace.16888.

[113] A. Bergeron, J.B. Crigger, Early progress on additive manufacturing of nuclear fuel materials, J. Nucl. Mater. 508 (2018) 344–347, https://doi.org/10.1016/j.jnucmat.2018.05.058.

[114] J. Rosales, I.J. van Rooyen, C.J. Parga, Characterizing surrogates to develop an additive manufacturing process for U3Si2 nuclear fuel, J. Nucl. Mater. 518 (2019) 117–128, https://doi.org/10.1016/j.jnucmat.2019.02.026.

[115] A. Sommers, Q. Wang, X. Han, C. T'Joen, Y. Park, A. Jacobi, Ceramics and ceramic matrix composites for heat exchangers in advanced thermal systems—a review, Appl. Therm. Eng. 30 (2010) 1277–1291, https://doi.org/10.1016/j.applthermaleng.2010.02.018.

[116] C.P. Deck, G.M. Jacobsen, J. Sheeder, O. Gutierrez, J. Zhang, J. Stone, H.E. Khalifa, C. A. Back, Characterization of SiC-SiC composites for accident tolerant fuel cladding, J. Nucl. Mater. 466 (2015) 1–15, https://doi.org/10.1016/j.jnucmat.2015.08.020.

[117] M.K. Thompson, G. Moroni, T. Vaneker, G. Fadel, R.I. Campbell, I. Gibson, A. Bernard, J. Schulz, P. Graf, B. Ahuja, F. Martina, Design for additive manufacturing: trends, opportunities, considerations, and constraints, CIRP Ann. 65 (2016) 737–760, https://doi.org/10.1016/j.cirp.2016.05.004.

[118] B.H. Jared, M.A. Aguilo, L.L. Beghini, B.L. Boyce, B.W. Clark, A. Cook, B.J. Kaehr, J. Robbins, Additive manufacturing: toward holistic design, Scr. Mater. 135 (2017) 141–147, https://doi.org/10.1016/j.scriptamat.2017.02.029.

[119] H.T. Larker, R. Lundberg, Near net shape production of monolithic and composite high temperature ceramics by hot isostatic pressing (HIP), J. Eur. Ceram. Soc. 19 (1999) 2367–2373, https://doi.org/10.1016/S0955-2219(99)00109-0.

[120] M.J. Edirisinghe, J.R.G. Evans, Review: Fabrication of engineering ceramics by injection moulding. II. Techniques, Int. J. High Technol. Ceram. 2 (1986) 249–278, https://doi.org/10.1016/0267-3762(86)90020-2.

[121] P. Greil, Near net shape manufacturing of polymer derived ceramics, J. Eur. Ceram. Soc. 18 (1998) 1905–1914, https://doi.org/10.1016/S0955-2219(98)00129-0.

[122] K.A. Terrani, B.C. Jolly, M.P. Trammell, G. Vasudevamurthy, D. Schappel, B. Ade, G. W. Helmreich, H. Wang, A.M. Rossy, B.R. Betzler, A.T. Nelson, Architecture and properties of TCR fuel form, J. Nucl. Mater. 547 (2021) 152781, https://doi.org/10.1016/j.jnucmat.2021.152781.

[123] T. Koyanagi, K. Terrani, S. Harrison, J. Liu, Y. Katoh, Additive manufacturing of silicon carbide for nuclear applications, J. Nucl. Mater. 543 (2021), https://doi.org/10.1016/j.jnucmat.2020.152577.

[124] Z. Chen, Z. Li, J. Li, C. Liu, C. Lao, Y. Fu, C. Liu, Y. Li, P. Wang, Y. He, 3D printing of ceramics: a review, J. Eur. Ceram. Soc. 39 (2019) 661–687, https://doi.org/10.1016/j.jeurceramsoc.2018.11.013.

[125] Method of making chemical vapor composites, US 20060057287A1, 2006. https://patents.justia.com/patent/20060057287.

[126] T.S. Byun, B.E. Garrison, H. Wang, A. Trofimov, T.G. Lach, C.M. Parich, M. D. Richardson, B.D. Jolly, M. Trammell, A. Schumacher, G. Vasudevamurthy, T. Koyanagi, K.A. Terrani, Mechanical and Thermophysical Properties of 3D-Printed SiC-FY20, Oak Ridge, TN, 2020.

[127] R.N. Singh, R.L. Coble, Growth of uranium dioxide single crystals by chemical vapor deposition, J. Cryst. Growth 21 (1974) 261–266, https://doi.org/10.1016/0022-0248(74)90013-X.

[128] Y.S. Won, V.G. Varanasi, O. Kryliouk, T.J. Anderson, L. McElwee-White, R.J. Perez, Equilibrium analysis of zirconium carbide CVD growth, J. Cryst. Growth 307 (2007) 302–308, https://doi.org/10.1016/j.jcrysgro.2007.05.039.

[129] W. Schintlmeister, O. Pacher, K. Pfaffinger, T. Raine, Structure and strength effects in CVD titanium carbide and titanium nitride coatings, J. Electrochem. Soc. 123 (1976) 924–929, https://doi.org/10.1149/1.2132969.

[130] A.O. Sezer, J.I. Brand, Chemical vapor deposition of boron carbide, Mater. Sci. Eng. B. 79 (2001) 191–202, https://doi.org/10.1016/S0921-5107(00)00538-9.

[131] T. Takahashi, H. Kamiya, Chemical vapor deposition of titanium diboride, J. Cryst. Growth 26 (1974) 203–209, https://doi.org/10.1016/0022-0248(74)90247-4.

[132] Y. Zhang, J. Yu, R. Tu, T. Goto, Preparation of Ca2Ti2O6 pyrochlore films by MOCVD, J. Phys. Conf. Ser. 152 (2009) 012032, https://doi.org/10.1088/1742-6596/152/1/012032.

[133] C.-N. Sun, M.C. Gupta, K.M.B. Taminger, Electron beam sintering of zirconium diboride, J. Am. Ceram. Soc. 93 (2010) 2484–2486, https://doi.org/10.1111/j.1551-2916.2010.03832.x.

[134] Y. Zhou, H. Tanaka, S. Otani, Y. Bando, Low-temperature pressureless sintering of alpha-SiC with Al4C3-B4C-C additions, J. Am. Ceram. Soc. 82 (2004) 1959–1964, https://doi.org/10.1111/j.1151-2916.1999.tb02026.x.

[135] A. Gubernat, L. Stobierski, P. Łabaj, Microstructure and mechanical properties of silicon carbide pressureless sintered with oxide additives, J. Eur. Ceram. Soc. 27 (2007) 781–789, https://doi.org/10.1016/j.jeurceramsoc.2006.04.009.

[136] K. Liu, T. Wu, D.L. Bourell, Y. Tan, J. Wang, M. He, H. Sun, Y. Shi, J. Chen, Laser additive manufacturing and homogeneous densification of complicated shape SiC ceramic parts, Ceram. Int. 44 (2018) 21067–21075, https://doi.org/10.1016/j.ceramint.2018.08.143.

[137] T.S. Suzuki, T. Uchikoshi, Y. Sakka, Densification of SiC by colloidal processing and SPS without sintering additives, Adv. Appl. Ceram. 113 (2014) 85–88, https://doi.org/10.1179/1743676113Y.0000000133.

[138] S. Bai, H.J. Lee, J. Liu, 3D printing with mixed powders of boron carbide and Al alloy, Appl. Sci. 10 (2020) 3055, https://doi.org/10.3390/app10093055.

[139] A. Hehr, J. Wenning, K. Terrani, S.S. Babu, M. Norfolk, Five-axis ultrasonic additive manufacturing for nuclear component manufacture, JOM 69 (2017) 485–490, https://doi.org/10.1007/s11837-016-2205-6.

[140] I. Choi, H.Y. Jeong, H. Shin, G. Kang, M. Byun, H. Kim, A.M. Chitu, J.S. Im, R. S. Ruoff, S.-Y. Choi, K.J. Lee, Laser-induced phase separation of silicon carbide, Nat. Commun. 7 (2016) 13562, https://doi.org/10.1038/ncomms13562.

[141] S. Zhang, R. Tu, T. Goto, High-speed epitaxial growth of β-SiC film on Si(111) single crystal by laser chemical vapor deposition, J. Am. Ceram. Soc. 95 (2012) 2782–2784, https://doi.org/10.1111/j.1551-2916.2012.05354.x.

[142] J. Pegna, J. Schneiter, K. Williams, R. Goduguchinta, High Strength Ceramic Fibers and Methods of Fabrication, US Patent 10,047,015B2, 2018.

[143] D.R. Patel, T. Koyanagi, High-temperature creep properties of SiC fibers with different compositions, Fusion Sci. Technol. 75 (2019) 636–641, https://doi.org/10.1080/15361055.2019.1647029.

[144] W. Li, K. Shirvan, S. Harrison, J. Pegna, Innovative accident tolerant fuel concept enabled through direct manufacturing technology, Appl. Energy 264 (2020) 114742, https://doi.org/10.1016/j.apenergy.2020.114742.

[145] S. Harrison, H. Marcus, Gas-phase Selective Area Laser Deposition (SALD) joining of SiC, Mater. Des. 20 (1999) 147–152, https://doi.org/10.1016/S0261-3069(99)00021-7.

[146] H.E. Khalifa, C.P. Deck, O. Gutierrez, G.M. Jacobsen, C.A. Back, Fabrication and characterization of joined silicon carbide cylindrical components for nuclear

applications, J. Nucl. Mater. 457 (2015) 227–240, https://doi.org/10.1016/j.jnucmat.2014.11.071.

[147] S.K. Everton, M. Hirsch, P. Stravroulakis, R.K. Leach, A.T. Clare, Review of in-situ process monitoring and in-situ metrology for metal additive manufacturing, Mater. Des. 95 (2016) 431–445, https://doi.org/10.1016/j.matdes.2016.01.099.

[148] M. Seifi, A. Salem, J. Beuth, O. Harrysson, J.J. Lewandowski, Overview of materials qualification needs for metal additive manufacturing, JOM 68 (2016) 747–764, https://doi.org/10.1007/s11837-015-1810-0.

[149] M. Grasso, B.M. Colosimo, Process defects and in situ monitoring methods in metal powder bed fusion: a review, Meas. Sci. Technol. 28 (2017) 044005, https://doi.org/10.1088/1361-6501/aa5c4f.

[150] K. Terrani, Accelerating the deployment of advanced nuclear energy systems, Nucl. News 63 (2020) 34–37.

[151] K. Terrani, C. Ang, L.L. Snead, Y. Katoh, Irradiation stability and thermo-mechanical properties of NITE-SiC irradiated to 10 dpa, J. Nucl. Mater. 499 (2018) 242–247, https://doi.org/10.1016/j.jnucmat.2017.11.040.

Index

'*Note*: Page numbers followed by "f" indicate figures and "t" indicate tables.'

R

S